Nanotoxicology in *Caenorhabditis elegans*

Dayong Wang

Nanotoxicology in
Caenorhabditis elegans

 Springer

Dayong Wang
Medical School
Southeast University
Nanjing, Jiangsu, China

ISBN 978-981-13-0232-9   ISBN 978-981-13-0233-6 (eBook)
https://doi.org/10.1007/978-981-13-0233-6

Library of Congress Control Number: 2018939186

Printed on acid-free paper

This Springer imprint is published by the registered company Springer Nature Singapore Pte Ltd.
The registered company address is: 152 Beach Road, #21-01/04 Gateway East, Singapore 189721, Singapore

Preface

Engineered nanomaterials (ENMs) are defined as materials with one or more dimensions of less than 100 nm. ENMs have been promised to bring advances in numerous areas. Meanwhile, the potential toxicity of ENMs on human health and environmental organisms receives the great attention. Besides the typical properties of model animals, *Caenorhabditis elegans* is very sensitive to environmental toxicants, which makes it a wonderful *in vivo* assay system for toxicological study. Now, *C. elegans* has been widely accepted and used in both the toxicity assessment and the toxicological study. The data on different aspects of nanotoxicology has been obtained in *C. elegans*.

This book will focus on the knowledge system on the toxicity and the translocation of different ENMs and the underlying mechanisms at different aspects in nematodes. In Chap. 1, we first introduced the value and limitations of *C. elegans* in toxicological study. In Chaps. 2 and 3, we introduced the endpoints for toxicity assessment of ENMs and exposure routes of ENMs, which are important for suitable design for the exposure to certain ENMs in nematodes. In Chap. 4, we introduced the toxic effects of certain ENMs at different aspects in nematodes. In Chaps. 5, 6, and 7, we introduced the physicochemical basis, the underlying cellular and physiological mechanisms, and the underlying molecular mechanisms of nanotoxicity formation in nematodes. In Chap. 8, we introduced the distribution and translocation of ENMs in nematodes. Finally, in Chaps. 9, 10, and 11, we introduced the related information on the confirmation of ENMs with low-toxicity or nontoxicity property, as well as the surface chemical modification or pharmacological prevention to reduce or prevent the toxicity of ENMs in nematodes.

Therefore, this book aims at providing a comprehensive summary of the knowledge on the nanotoxicology obtained in the *in vivo* assay system of *C. elegans*. Meanwhile, this book reflects the establishment of the knowledge system of model animal toxicology in *C. elegans* to a great degree. Such a knowledge system has the important prediction potential for our understanding the possible biological effects and behavior and the underlying mechanisms of environmental toxicants or stresses in mammals and human beings.

Nanjing, Jiangsu, China   Dayong Wang

Contents

1 Values of *C. elegans* in Toxicological Study 1
    1.1   Introduction 1
    1.2   Raise of a Series of Useful Sublethal Endpoints for Toxicity
              Assessment of Environmental Toxicants 2
    1.3   High-Throughput Screen and Identification of Chemicals 3
    1.4   Toxicity Assessment of Environmental Toxicants Under
              Susceptible Genetic Backgrounds 3
    1.5   Toxicity Assessment of Environmental Toxicants
              at Environmentally Relevant Concentrations................. 4
    1.6   Understanding the In Vivo Physicochemical,
              Cellular, and Physiological Mechanisms of Toxicity Induced
              by Environmental Toxicants.............................. 4
    1.7   Elucidation of Toxicological Mechanisms of Environmental
              Toxicants in Certain Targeted Organs 5
    1.8   Elucidation of Underlying Molecular Mechanisms of Toxicity
              Induced by Environmental Toxicants....................... 5
    1.9   Distribution and Translocation of Environmental Toxicants 5
    1.10 Confirmation of Chemical with Low-Toxicity or Non-toxicity
              Property 6
    1.11 Limitations of *C. elegans* in the Toxicological Study 6
    References... 6

2 Endpoints for Toxicity Assessment of Nanomaterials 11
    2.1   Introduction 11
    2.2   Lethality... 12
    2.3   Morphology and Development 12
    2.4   Reproduction 12
    2.5   Neuronal Development and Function 15
    2.6   Intestinal Development and Function 18
    2.7   Epidermal Development 21
    2.8   Innate Immune Response................................. 22

2.9   Lifespan   23
2.10   Metabolism   24
2.11   Oxidative Stress   26
2.12   Transgenic Strains Reflecting Stress Response or Oxidative
       Stress   27
2.13   Perspectives   29
References...   29

3   **Exposure Routes of Nanomaterials**...........................   33
3.1   Introduction   33
3.2   Exposure Routes of ENMs................................   34
3.3   Toxicity Assessment of ENMs at Environmentally Relevant
       Concentrations   40
3.4   Perspectives   42
References...   43

4   **Toxic Effects of Certain Nanomaterials**   45
4.1   Introduction   45
4.2   Metal Nanoparticles.....................................   46
4.3   Rare Earth Fluoride Nanocrystals (NCs)...................   55
4.4   Quantum Dots (QDs)....................................   55
4.5   Carbon Nanotubes   57
4.6   Fullerenol..   58
4.7   Graphene and Its Derivatives............................   59
4.8   Nanotoxicity Formation Under the Oxidative Stress Condition ...   61
4.9   Nanotoxicity Formation Under the Susceptible
       Mutation Backgrounds...................................   61
4.10   Perspectives   63
References...   63

5   **Physicochemical Basis for Nanotoxicity Formation**..............   67
5.1   Introduction   67
5.2   Size..   68
5.3   Surface Charge...   72
5.4   Shape   72
5.5   Surface Groups...   72
5.6   Impurity   74
5.7   The Underlying Chemical Mechanism for the Oxidative
       Stress Induced by ENMs   75
5.8   Perspectives   76
References...   77

6   **Cellular and Physiological Mechanisms of Nanotoxicity Formation** .   79
6.1   Introduction   79
6.2   Cellular Mechanisms for Nanotoxicity Formation   80
6.3   Physiological Mechanisms for Nanotoxicity Formation   100
6.4   Perspectives   105
References...   105

7   Molecular Mechanisms of Nanotoxicity Formation 109
   7.1   Introduction 109
   7.2   Molecular Basis for Induction of Oxidative Stress
          in ENM-Exposed Nematodes 110
   7.3   Important Signaling Pathways Involved in the Regulation
          of Nanotoxicity Formation in Nematodes 113
   7.4   Molecular Basis for Nanotoxicity Formation Based
          on Omics Study in Nematodes............................ 145
   7.5   Alteration in Molecular Machinery for Important
          Biochemical Events in ENM-exposed Nematodes 162
   7.6   Functional Analysis of Human Genes in Nematodes 164
   7.7   Perspectives 164
   References.. 166

8   Distribution and Translocation of Nanomaterials 169
   8.1   Introduction 169
   8.2   Methods for the Analysis on Distribution and Translocation
          of ENMs.. 170
   8.3   Primary Targeted Organ: Intestine......................... 180
   8.4   Secondary Targeted Organ: Reproductive Organs............. 185
   8.5   Secondary Targeted Organ: Neurons 187
   8.6   Cellular Metabolism of ENMs in Nematodes 189
   8.7   Transgenerational Translocation of Nanomaterials 190
   8.8   Crucial Role of Intestinal Barrier Against the Toxicity of ENMs
          in Nematodes 192
   8.9   Chemical Control of Translocation of ENMs in Nematodes...... 195
   8.10   Molecular Signals Regulating the Distribution
          and Translocation of ENMs.............................. 197
   8.11   Chemical Metabolism and Degradation of ENMs in the Body
          of Animals 199
   8.12   Perspectives 201
   References.. 202

**9   Confirmation of Nanomaterials with Low-Toxicity or Non-toxicity
      Property** 205
   9.1   Introduction 205
   9.2   Confirmation of ENMs with Low-Toxicity or Non-toxicity
          Property 206
   9.3   Confirmation of the Relative Safe Property of Titanium Dioxide
          Nanoparticles (TiO$_2$-NPs) at Realistic Concentrations 216
   9.4   Recovery Response of Toxicity of TiO$_2$-NPs After Transfer
          to the Normal Conditions in Nematodes 218
   9.5   Perspectives 220
   References.. 221

**10 Surface Chemical Modification to Reduce the Toxicity
  of Nanomaterials** 227
  10.1 Introduction 227
  10.2 PEG Modification 228
  10.3 ZnS Surface Coating 233
  10.4 FBS Surface Coating 238
  10.5 COOH Modification. 239
  10.6 Beneficial Effects of Other Surface Modifications 242
  10.7 Perspectives 242
  References. 243

**11 Pharmacological Prevention of the Toxicity Induced
  by Environmental Nanomaterials** 247
  11.1 Introduction 247
  11.2 Value of Antioxidants in Being Against the Toxicity of ENMs. ... 249
  11.3 Value of Natural Compounds in Being Against the Toxicity
    of ENMs. 257
  11.4 Beneficial Function of Lactic Acid (LAB) Against the
    GO Toxicity by Maintaining Normal Intestinal Permeability 265
  11.5 Perspectives 269
  References. 269

Chapter 1
Values of *C. elegans* in Toxicological Study

Abstract The model animal of nematode *Caenorhabditis elegans* has become an important in vivo alternative assay system for toxicological study of different environmental toxicants or stresses. We here introduced the several important values of *C. elegans* in the toxicological study for environmental toxicants or stresses. Meanwhile, we also discussed the limitations of nematodes for the toxicological study of environmental toxicants or stresses.

Keywords Alternative assay system · Value · Limitation · *Caenorhabditis elegans*

1.1 Introduction

So far, the model animal of nematode *Caenorhabditis elegans* has gradually become an important in vivo alternative assay system for both the toxicity assessment and the toxicological study of different environmental toxicants or stresses [1–3]. *C. elegans* is a free-living nematode mainly found in the liquid phase of soils. *C. elegans* is one of the most thoroughly studied model animals and has the typical properties of model organisms, such as well-defined anatomy, short life cycle, short lifespan, small size, perfect reproductive capacity, availability of many useful genetic sources, and ease in handling [4]. Moreover, the nematodes can be easily cultivated in a laboratory and reproduced in thousands of individuals, which allow the offer of an advantage assay system suitable for asking the in vivo underlying mechanisms for the observed toxicity of environmental toxicants or stresses. *C. elegans* has the ecological significance due to its important roles in the nutrient cycling in the soil. Especially, *C. elegans* has been proven to be very sensitive to the adverse effects at different aspects induced by different environmental toxicants or stresses [1–3].

In this chapter, we discussed the several important values of *C. elegans* in the toxicological study for environmental toxicants or stresses. The mainly introduced values are:

1. Raise of a series of useful sublethal endpoints for toxicity assessment of environmental toxicants
2. High-throughput screen and identification of chemicals

© Springer Nature Singapore Pte Ltd. 2018
D. Wang, *Nanotoxicology in* Caenorhabditis elegans,
https://doi.org/10.1007/978-981-13-0233-6_1

3. Toxicity assessment of environmental toxicants under susceptible genetic backgrounds
4. Toxicity assessment of environmental toxicants at environmentally relevant concentrations
5. Understanding the in vivo physicochemical, cellular, and physiological mechanisms of toxicity induced by environmental toxicants
6. Elucidation of toxicological mechanisms of environmental toxicants in certain targeted organs
7. Elucidation of underlying molecular mechanisms of toxicity induced by environmental toxicants
8. Distribution and translocation of environmental toxicants
9. Confirmation of chemical with low-toxicity or non-toxicity property

Moreover, we further discussed the limitations of nematodes for the toxicological study of environmental toxicants or stresses.

1.2   Raise of a Series of Useful Sublethal Endpoints for Toxicity Assessment of Environmental Toxicants

In nematodes, besides the endpoint of lethality, many important sublethal endpoints associated with development, reproduction, neuronal development and function, intestinal development and function, epidermal development, innate immune response, lifespan, metabolism, oxidative stress, and transgenic strains reflecting stress response or oxidative stress have been further employed and raised [5–9]. Among these raised sublethal endpoints, the endpoints associated with the development and the reproduction are widely used ones involved in the evaluation of possible toxicity of environmental toxicants or stresses on the growth and the development of nematodes [6–9]. Additionally, the endpoints associated with the oxidative stress and the transgenic strains reflecting stress response or oxidative stress are also widely used endpoints to reflect the possible toxicity of environmental toxicants or stresses on nematodes [10–15]. Especially, the raised some useful transgenic strains reflecting stress response or oxidative stress can help us directly detect the potential toxicity induction of environmental toxicants or stresses on nematodes based on the noticeable induction of fluorescent signals [13–15]. The endpoints associated with the lifespan and innate immune response are important to detect the potential long-term effects of certain environmental toxicants or stresses on nematodes [16–18]. More importantly, some useful endpoints associated with the possible damage on the functions of primary targeted organs, such as the intestine and the epidermal, of certain environmental toxicants have been raised in nematodes [19–21]. The useful endpoints associated with the possible damage on the functions of secondary targeted organs, such as the reproductive organs and the neurons, have also been raised in nematodes [8, 9, 22]. These useful sublethal endpoints will largely open some new windows for us in understanding the underlying toxicological mechanisms of environmental toxicants and environmental stresses in organisms.

1.3 High-Throughput Screen and Identification of Chemicals

Due to the important properties of small size and easy cultivation in the laboratory of nematodes, this model animal is very valuable for high-throughput screen and identification of chemicals. Using *C. elegans* as an in vivo assay system, one of the values in the high-throughput screen is the high-throughput toxicity assessment of environmental toxicants or chemicals. For example, *C. elegans* was used in the high-throughput evaluation of possible toxicity of 20 engineered nanomaterials (ENMs) at 4 concentrations using body length, locomotion speed, and lifespan as the toxicity assessment endpoints [23]. Using *C. elegans* as an in vivo assay system, another important value in the high-throughput screen is to identify the susceptible or resistant genetic loci affecting the toxicity formation of certain environmental toxicants. For example, *C. elegans* was used in the high-throughput identification of genetic loci affecting the toxicity and the translocation of graphene oxide (GO), an important carbon-based ENMs, based on the screen of 20 strains with mutations of genes required for stress response or oxidative stress [24]. Seven genes were identified, and their mutations altered both the translocation and toxicity of GO in nematodes [24]. Mutations of the *hsp-16.48*, *gas-1*, *sod-2*, *sod-3*, or *aak-2* resulted in greater GO translocation into the body and toxic effects on both primary and secondary targeted organs; however, mutations of the *isp-1* or *clk-1* caused significantly decreased GO translocation into the body and toxicity on both primary and secondary targeted organs [24].

1.4 Toxicity Assessment of Environmental Toxicants Under Susceptible Genetic Backgrounds

Due to the role of classic model animal, so far, there are many useful genetic mutants that are available for researchers in the related fields. Meanwhile, it is easy to perform RNAi knockdown of any interested gene in the nematodes. These research backgrounds provide a solid foundation to systematically perform the toxicity assessment of environmental toxicants under susceptible genetic backgrounds. With the TiO_2-nanoparticles (TiO_2-NPs) as an example, *sod-2*, *sod-3*, *mtl-2*, and *hsp-16.48* mutants were susceptible for TiO_2-NP toxicity on reproduction and locomotion behavior; *sod-2*, *sod-3*, and *mtl-2* mutants were susceptible for TiO_2-NP toxicity on survival and intestinal development; and *mtl-2* mutant was susceptible for TiO_2-NP toxicity on development [25]. Mutations of these genes, together with sensitive endpoints, will have the potential in assessing the possible TiO_2-NP toxicity at the concentration of 0.0001 μg/L [25].

1.5   Toxicity Assessment of Environmental Toxicants at Environmentally Relevant Concentrations

Considering the sensitivity property of *C. elegans* to environmental toxicants or stresses, *C. elegans* has been gradually used in the toxicity assessment of environmental toxicants at environmentally relevant concentrations [26–30]. In nematodes, at least acute exposure, prolonged exposure, chronic exposure, one-generation exposure, and transgenerational exposure have been raised as the useful exposure routes for toxicity assessment of environmental toxicants. Among these exposure routes, at least the prolonged exposure and the chronic exposure have the potential in assessing the possible toxicity of certain environmental toxicants at environmentally relevant concentrations. Further with TiO_2-NPs as an example, it has been shown that TiO_2-NPs at the concentration of 0.01 µg/L could cause the significant reduction in brood size, decrease in locomotion behavior, and induction of intestinal autofluorescence in nematodes after prolonged exposure from L1-larvae to adult day-1 [25]. Moreover, after chronic exposure from adult day-1 to adult day-8, TiO_2-NPs (4 and 10 nm) at concentrations more than 0.01 µg/L could significantly decrease the locomotion behavior in nematodes [30].

1.6   Understanding the In Vivo Physicochemical, Cellular, and Physiological Mechanisms of Toxicity Induced by Environmental Toxicants

As a model animal, *C. elegans* provide an important in vivo assay system to systematically examine the potential roles of different physicochemical properties of certain environmental toxicants, such as the ENMs, in the toxicity formation in organisms. In nematodes, the important contribution of physicochemical properties, such as size, surface charge, shape, surface groups, and impurity, in the toxicity formation of ENMs have been examined [5, 12, 17, 27–29, 31–35]. Moreover, the underlying chemical mechanism for the oxidative stress induced by ENMs has also been elucidated in nematodes [36].

Besides this, several aspects of cellular mechanisms of toxicity formation of certain environmental toxicants, such as ENMs, have been examined in nematodes. These raised cellular mechanisms of toxicity formation of environmental toxicants include release of metal ion, oxidative stress, intestinal permeability, defecation behavior, bioavailability to targeted organs, acceleration in aging process, innate immune response, mitochondrial damage and DNA damage, developmental fate, and deficit in cellular endocytosis in intestinal cells [5, 10, 16–19, 37–46]. Moreover, several aspects of physiological mechanisms of toxicity formation of certain environmental toxicants, such as ENMs, have also been determined. These raised physiological mechanisms of toxicity formation of environmental toxicant include environmental factors, exposure, physiological state of nematodes, developmental stages, and hormesis of nematodes [5, 26, 31, 34–36, 47, 48].

1.7 Elucidation of Toxicological Mechanisms of Environmental Toxicants in Certain Targeted Organs

With ENMs as the example, the ENMs such as the GO could be distributed and translocated into both the primary targeted organs, such as the intestine, and the secondary targeted organs, such as the reproductive organs of gonad and spermatheca and the neurons, in nematodes [18, 42]. Using series of tissue-specific RNAi knockdown tools, we can perform the RNAi knockdown of certain genes in certain tissues in nematodes. Meanwhile, there are different tissue-specific promoters available in nematodes, which can help us to express certain genes in certain tissues. With the aid of these techniques, we can systematically elucidate the possible toxicological mechanisms of environmental toxicants in certain targeted organs in nematodes.

1.8 Elucidation of Underlying Molecular Mechanisms of Toxicity Induced by Environmental Toxicants

In nematodes, their many basic physiological processes, stress responses, signal transduction pathways, and epigenetic marks are conserved compared with those in mammals and humans. Additionally, the completion of *C. elegans* genome has approximately 45% of the genes with the human homologues, including numerous disease-related genes. So far, some important signaling pathways, such as insulin, p38 MAPK, Wnt, ERK, and oxidative stress-related signaling, have been identified to be involved in the regulation of response of nematodes to environmental toxicants, such as carbon-based ENMs, in nematodes [13, 15, 24, 49–51]. Moreover, some important microRNAs and long noncoding RNAs have also been identified to be involved in the regulation of response of nematodes to carbon-based ENMs in nematodes [16, 52–56].

1.9 Distribution and Translocation of Environmental Toxicants

Distribution and translocation of environmental toxicants is one of the crucial cellular contributors for the toxicity induction of certain environmental toxicants. The property of transparent body of nematodes allows us directly visualize the distribution and the translocation of certain environmental toxicants, such as some ENMs. Some powerful techniques have already been employed to determine the distribution and the translocation of ENMs, and the distribution and the translocation patterns of some important ENMs have been well described in nematodes with the aid of these powerful techniques [57–64]. Moreover, using these techniques, the behavior and the regulation of distribution and translocation of different ENMs in the primary or the secondary targeted organs, as well as the patterns of transgenerational translocation of ENMs, have been systematically investigated in nematodes [7, 13, 18–20, 24, 28, 29, 35, 42, 43, 60]. *C. elegans* is also helpful for the elucidation of dynamic cellular, molecular, and chemical metabolisms of environmental toxicants, such as the ENMs, in the body of nematodes [18, 21, 24, 29, 60, 65, 66].

1.10   Confirmation of Chemical with Low-Toxicity or Non-toxicity Property

Due to the sensitivity of *C. elegans* to environmental toxicants, *C. elegans* not only acts as a wonderful in vivo assay model for assessing ecotoxicological effects of certain environmental toxicants but also serves as a useful assay model for the confirmation of low-toxicity or relative non-toxicity property of environmental chemicals. With ENMs as the example, the relative non-toxicity property of some important ENMs, such as graphite, graphene quantum dots (GQDs), carboxyl-functionalized graphene (G-COOH), $Gd@C_{82}(OH)_{22}$, and fluorescent nanodiamond (FND), has been confirmed in nematodes [7, 21, 64, 67].

1.11   Limitations of *C. elegans* in the Toxicological Study

Although *C. elegans* has many important values in both the toxicity assessment and the toxicological study of environmental toxicants or stresses, the limitations of nematodes in the toxicological study still exist. One of the important limitations is that the nematodes do not have some important organs, such as the heart, liver, lung, and kidney, which exist in the mammals, since the nematodes do not have the related developmental process of mesoderm during the development. Another important limitation is that the genome for human or mammals may be more complex information and structure than that of nematodes. Therefore, some important molecular signaling pathways may be not able to be detected in the in vivo assay system of *C. elegans*.

References

1. Leung MC, Williams PL, Benedetto A, Au C, Helmcke KJ, Aschner M, Meyer JN (2008) *Caenorhabditis elegans*: an emerging model in biomedical and environmental toxicology. Toxicol Sci 106:5–28
2. Zhao Y-L, Wu Q-L, Li Y-P, Wang D-Y (2013) Translocation, transfer, and *in vivo* safety evaluation of engineered nanomaterials in the non-mammalian alternative toxicity assay model of nematode *Caenorhabditis elegans*. RSC Adv 3:5741–5757
3. Charão MF, Souto C, Brucker N, Barth A, Sjornada D, Fagundez D, Ávila DS, Eifler-Lima VL, Sguterres S, Rpohlmann A, Garcia SC (2015) *Caenorhabditis elegans* as an alternative

in vivo model to determine oral uptake, nanotoxicity, and efficacy of melatonin-loaded lipid-core nanocapsules on paraquat damage. Int J Nanomedicine 10:5093–5106

4. Brenner S (1974) The genetics of *Caenorhabditis elegans*. Genetics 77:71–94
5. Ma H, Bertsch PM, Glenn TC, Kabengi NJ, Williams PL (2009) Toxicity of manufactured zinc oxide nanoparticles in the nematode *Caenorhabditis elegans*. Environ Toxicol Chem 28(6):1324–1330
6. Wang D-Y, Wang Y (2008) Phenotypic and behavioral defects caused by barium exposure in nematode *Caenorhabditis elegans*. Arch Environ Contam Toxicol 54(3):447–453
7. Yang J-N, Zhao Y-L, Wang Y-W, Wang H-F, Wang D-Y (2015) Toxicity evaluation and translocation of carboxyl functionalized graphene in *Caenorhabditis elegans*. Toxicol Res 4:1498–1510
8. Zhao Y-L, Wu Q-L, Wang D-Y (2016) An epigenetic signal encoded protection mechanism is activated by graphene oxide to inhibit its induced reproductive toxicity in *Caenorhabditis elegans*. Biomaterials 79:15–24
9. Liu P-D, He K-W, Li Y-X, Wu Q-L, Yang P, Wang D-Y (2012) Exposure to mercury causes formation of male-specific structural deficits by inducing oxidative damage in nematodes. Ecotoxicol Environ Saf 79:90–100
10. Li Y-X, Yu S-H, Wu Q-L, Tang M, Pu Y-P, Wang D-Y (2012) Chronic Al$_2$O$_3$-nanoparticle exposure causes neurotoxic effects on locomotion behaviors by inducing severe ROS production and disruption of ROS defense mechanisms in nematode *Caenorhabditis elegans*. J Hazard Mater 219-220:221–230
11. Wu Q-L, Liu P-D, Li Y-X, Du M, Xing X-J, Wang D-Y (2012) Inhibition of ROS elevation and damage on mitochondrial function prevents lead-induced neurotoxic effects on structures and functions of AFD neurons in *Caenorhabditis elegans*. J Environ Sci 24(4):733–742
12. Wu Q-L, Nouara A, Li Y-P, Zhang M, Wang W, Tang M, Ye B-P, Ding J-D, Wang D-Y (2013) Comparison of toxicities from three metal oxide nanoparticles at environmental relevant concentrations in nematode *Caenorhabditis elegans*. Chemosphere 90:1123–1131
13. Zhao Y-L, Zhi L-T, Wu Q-L, Yu Y-L, Sun Q-Q, Wang D-Y (2016) p38 MAPK-SKN-1/ Nrf signaling cascade is required for intestinal barrier against graphene oxide toxicity in *Caenorhabditis elegans*. Nanotoxicology 10(10):1469–1479
14. Zhao L, Rui Q, Wang D-Y (2017) Molecular basis for oxidative stress induced by simulated microgravity in nematode *Caenorhabditis elegans*. Sci Total Environ 607-608:1381–1390
15. Zhao Y-L, Yang R-L, Rui Q, Wang D-Y (2016) Intestinal insulin signaling encodes two different molecular mechanisms for the shortened longevity induced by graphene oxide in *Caenorhabditis elegans*. Sci Rep 6:24024
16. Wu Q-L, Zhao Y-L, Zhao G, Wang D-Y (2014) microRNAs control of *in vivo* toxicity from graphene oxide in *Caenorhabditis elegans*. Nanomed Nanotechnol Biol Med 10:1401–1410
17. Yu S-H, Rui Q, Cai T, Wu Q-L, Li Y-X, Wang D-Y (2011) Close association of intestinal autofluorescence with the formation of severe oxidative damage in intestine of nematodes chronically exposed to Al$_2$O$_3$-nanoparticle. Environ Toxicol Pharmacol 32:233–241
18. Wu Q-L, Zhao Y-L, Fang J-P, Wang D-Y (2014) Immune response is required for the control of *in vivo* translocation and chronic toxicity of graphene oxide. Nanoscale 6:5894–5906
19. Ding X-C, Wang J, Rui Q, Wang D-Y (2018) Long-term exposure to thiolated graphene oxide in the range of μg/L induces toxicity in nematode *Caenorhabditis elegans*. Sci Total Environ 616–617:29–37
20. Kim SW, Nam S, An Y (2012) Interaction of silver nanoparticles with biological surfaces of *Caenorhabditis elegans*. Ecotoxicol Environ Saf 77:64–70
21. Zhao Y-L, Liu Q, Shakoor S, Gong JR, Wang D-Y (2015) Transgenerational safe property of nitrogen-doped graphene quantum dots and the underlying cellular mechanism in *Caenorhabditis elegans*. Toxicol Res 4:270–280
22. Yu X-M, Guan X-M, Wu Q-L, Zhao Y-L, Wang D-Y (2015) Vitamin E ameliorates the neurodegeneration related phenotypes caused by neurotoxicity of Al$_2$O$_3$-nanoparticles in *C. elegans*. Toxicol Res 4:1269–1281

23. Jung S, Qu X, Aleman-Meza B, Wang T, Riepe C, Liu Z, Li Q, Zhong W (2015) Multi-endpoint, high-throughput study of nanomaterial toxicity in *Caenorhabditis elegans*. Environ Sci Technol 49:2477–2485

24. Wu Q-L, Zhao Y-L, Li Y-P, Wang D-Y (2014) Molecular signals regulating translocation and toxicity of graphene oxide in nematode *Caenorhabditis elegans*. Nanoscale 6:11204–11212

25. Wu Q-L, Zhao Y-L, Li Y-P, Wang D-Y (2014) Susceptible genes regulate the adverse effects of TiO_2-NPs at predicted environmental relevant concentrations on nematode *Caenorhabditis elegans*. Nanomed: Nanotechnol Biol Med 10:1263–1271

26. Zhang H, He X, Zhang Z, Zhang P, Li Y, Ma Y, Kuang Y, Zhao Y, Chai Z (2011) Nano-CeO_2 exhibits adverse effects at environmental relevant concentrations. Environ Sci Technol 45(8):3725–3730

27. Li Y-X, Wang W, Wu Q-L, Li Y-P, Tang M, Ye B-P, Wang D-Y (2012) Molecular control of TiO_2-NPs toxicity formation at predicted environmental relevant concentrations by Mn-SODs proteins. PLoS One 7(9):e44688

28. Nouara A, Wu Q-L, Li Y-X, Tang M, Wang H-F, Zhao Y-L, Wang D-Y (2013) Carboxylic acid functionalization prevents the translocation of multi-walled carbon nanotubes at predicted environmental relevant concentrations into targeted organs of nematode *Caenorhabditis elegans*. Nanoscale 5:6088–6096

29. Wu Q-L, Li Y-X, Li Y-P, Zhao Y-L, Ge L, Wang H-F, Wang D-Y (2013) Crucial role of biological barrier at the primary targeted organs in controlling translocation and toxicity of multi-walled carbon nanotubes in nematode *Caenorhabditis elegans*. Nanoscale 5:11166–11178

30. Wu Q-L, Wang W, Li Y-X, Li Y-P, Ye B-P, Tang M, Wang D-Y (2012) Small sizes of TiO_2-NPs exhibit adverse effects at predicted environmental relevant concentrations on nematodes in a modified chronic toxicity assay system. J Hazard Mater 243:161–168

31. Wu S, Lu J-H, Rui Q, Yu S-H, Cai T, Wang D-Y (2011) Aluminum nanoparticle exposure in L1 larvae results in more severe lethality toxicity than in L4 larvae or young adults by strengthening the formation of stress response and intestinal lipofuscin accumulation in nematodes. Environ Toxicol Pharmacol 31:179–188

32. Collin B, Oostveen E, Tsyusko OV, Unrine JM (2014) Influence of natural organic matter and surface charge on the toxicity and bioaccumulation of functionalized ceria nanoparticles in *Caenorhabditis elegans*. Environ Sci Technol 48:1280–1289

33. Gorka DE, Osterberg JS, Gwin CA, Colman BP, Meyer JN, Bernhardt ES, Gunsch CK, DiGulio RT, Liu J (2015) Reducing environmental toxicity of silver nanoparticles through shape control. Environ Sci Technol 49:10093–10098

34. Ellegaard-Jensen L, Jensen KA, Johansen A (2012) Nano-silver induces dose-response effects on the nematode *Caenorhabditis elegans*. Ecotoxicol Environ Saf 80:216–223

35. Meyer JN, Lord CA, Yang XY, Turner EA, Badireddy AR, Marinakos SM, Chilkoti A, Wiesner MR, Auffan M (2010) Intracellular uptake and associated toxicity of silver nanoparticles in *Caenorhabditis elegans*. Aquat Toxicol 100(2):140–150

36. Zhang W, Wang C, Li Z, Lu Z, Li Y, Yin J, Zhou Y, Gao X, Fang Y, Nie G, Zhao Y (2012) Unraveling stress-induced toxicity properties of graphene oxide and the underlying mechanism. Adv Mater 24:5391–5397

37. Franklin NM, Rogers NJ, Apte SC, Batley GE, Gadd GE, Casey PS (2007) Comparative toxicity of nanoparticulate ZnO, bulk ZnO, and $ZnCl_2$ to a freshwater microalga (*Pseudokirchneriella subcapitata*): the importance of particle solubility. Environ Sci Technol 41:8484–8490

38. Roh J, Sim SJ, Yi J, Park K, Chung KH, Ryu D, Choi J (2009) Ecotoxicity of silver nanoparticles on the soil nematode *Caenorhabditis elegans* using functional ecotoxicogenomics. Environ Sci Technol 43:3933–3940

39. Yang X, Gondikas AP, Marinakos SM, Auffan M, Liu J, Hsu-Kim H, Meyer JN (2012) Mechanism of silver nanoparticle toxicity is dependent on dissolved silver and surface coating in *Caenorhabditis elegans*. Environ Sci Technol 46:1119–1127

40. Hsu PL, O'Callaghan M, Al-Salim N, Hurst MRH (2012) Quantum dot nanoparticles affect the reproductive system of *Caenorhabditis elegans*. Environ Toxicol Chem 31:2366–2374

41. Lim D, Roh J, Eom H, Choi J, Hyun J, Choi J (2012) Oxidative stress-related PMK-1 p38 MAPK activation as a mechanism for toxicity of silver nanoparticles to reproduction in the nematode *Caenorhabditis elegans*. Environ Toxicol Chem 31:585–592

42. Wu Q-L, Yin L, Li X, Tang M, Zhang T, Wang D-Y (2013) Contributions of altered permeability of intestinal barrier and defecation behavior to toxicity formation from graphene oxide in nematode *Caenorhabditis elegans*. Nanoscale 5(20):9934–9943

43. Zhao L, Qu M, Wong G, Wang D-Y (2017) Transgenerational toxicity of nanopolystyrene particles in the range of μg/L in nematode *Caenorhabditis elegans*. Environ Sci Nano 4:2356–2366

44. Pluskota A, Horzowski E, Bossinger O, von Mikecz A (2009) In *Caenorhabditis elegans* nanoparticle-bio-interactions become transparent: silica-nanoparticles induce reproductive senescence. PLoS One 4(8):e6622

45. Scharf A, Piechulek A, von Mikecz A (2013) Effect of nanoparticles on the biochemical and behavioral aging phenotype of the nematode *Caenorhabditis elegans*. ACS Nano 7(12):10695–10703

46. Wang Q, Zhou Y, Song B, Zhong Y, Wu S, Cui R, Cong H, Su Y, Zhang H, He Y (2016) Linking subcellular disturbance to physiological behavior and toxicity induced by quantum dots in *Caenorhabditis elegans*. Small 23:3143–3154

47. Ma H, Kabengi NJ, Bertsch PM, Unrine JM, Glenn TC, Williams PL (2011) Comparative phototoxicity of nanoparticulate and bulk ZnO to a free-living nematode *Caenorhabditis elegans*: the importance of illumination mode and primary particle size. Environ Pollut 159:1473–1480

48. Tyne W, Little S, Spurgeon DJ, Svendsen C (2015) Hormesis depends upon the life-stage and duration of exposure: examples for a pesticide and a nanomaterial. Ecotoxicol Environ Saf 120:117–123

49. Zhi L-T, Qu M, Ren M-X, Zhao L, Li Y-H, Wang D-Y (2017) Graphene oxide induces canonical Wnt/β-catenin signaling-dependent toxicity in *Caenorhabditis elegans*. Carbon 113:122–131

50. Qu M, Li Y-H, Wu Q-L, Xia Y-K, Wang D-Y (2017) Neuronal ERK signaling in response to graphene oxide in nematode *Caenorhabditis elegans*. Nanotoxicology 11(4):520–533

51. Ren M-X, Zhao L, Lv X, Wang D-Y (2017) Antimicrobial proteins in the response to graphene oxide in *Caenorhabditis elegans*. Nanotoxicology 11(4):578–590

52. Zhao L, Wan H-X, Liu Q-Z, Wang D-Y (2017) Multi-walled carbon nanotubes-induced alterations in microRNA *let-7* and its targets activate a protection mechanism by conferring a developmental timing control. Part Fibre Toxicol 14:27

53. Zhi L-T, Yu Y-L, Jiang Z-X, Wang D-Y (2017) *mir-355* functions as an important link between p38 MAPK signaling and insulin signaling in the regulation of innate immunity. Sci Rep 7:14560

54. Zhuang Z-H, Li M, Liu H, Luo L-B, Gu W-D, Wu Q-L, Wang D-Y (2016) Function of RSKS-1-AAK-2-DAF-16 signaling cascade in enhancing toxicity of multi-walled carbon nanotubes can be suppressed by *mir-259* activation in *Caenorhabditis elegans*. Sci Rep 6:32409

55. Zhao Y-L, Wu Q-L, Li Y-P, Nouara A, Jia R-H, Wang D-Y (2014) *In vivo* translocation and toxicity of multi-walled carbon nanotubes are regulated by microRNAs. Nanoscale 6(8):4275–4284

56. Wu Q-L, Zhou X-F, Han X-X, Zhuo Y-Z, Zhu S-T, Zhao Y-L, Wang D-Y (2016) Genome-wide identification and functional analysis of long noncoding RNAs involved in the response to graphene oxide. Biomaterials 102:277–291

57. Lim SF, Riehn R, Ryu WS, Khanarian N, Tung C, Tank D, Austin RH (2006) *In vivo* and scanning electron microscopy imaging of upconverting nanophosphors in *Caenorhabditis elegans*. Nano Lett 6(2):169–174

58. Chen J, Guo C, Wang M, Huang L, Wang L, Mi C, Li J, Fang X, Mao C, Xu S (2011) Controllable synthesis of NaYF4 : Yb,Er upconversion nanophosphors and their application to *in vivo* imaging of *Caenorhabditis elegans*. J Mater Chem 21(8):2632

59. Zhou J, Yang Z, Dong W, Tang R, Sun L, Chun-Hua Yan C (2011) Bioimaging and toxicity assessments of near-infrared upconversion luminescent NaYF4:Yb,Tm nanocrystals. Biomaterials 32:9059–9067
60. Qu Y, Li W, Zhou Y, Liu X, Zhang L, Wang L, Li Y, Iida A, Tang Z, Zhao Y, Chai Z, Chen C (2011) Full assessment of fate and physiological behavior of quantum dots utilizing *Caenorhabditis elegans* as a model organism. Nano Lett 11:3174–3183
61. Zhang L, Xia J, Zhao Q, Liu L, Zhang Z (2010) Functional graphene oxide as a nanocarrier for controlled loading and targeted delivery of mixed anticancer drugs. Small 6:537–544
62. Cranfield CG, Dawe A, Karloukovski V, Dunin-Borkowski RE, de Pomerai D, Dobson J (2004) Biogenic magnetite in the nematode *Caenorhabditis elegans*. Proc R Soc Lond B 271(Suppl):S436–S439
63. Gao Y, Liu N, Chen C, Luo Y, Li Y, Zhang Z, Zhao Y, Zhao B, Iida A, Chai Z (2008) Mapping technique for biodistribution of elements in a model organism, *Caenorhabditis elegans*, after exposure to copper nanoparticles with microbeam synchrotron radiation X-ray fluorescence. J Anal At Spectrom 23:1121–1124
64. Zanni E, De Bellis G, Bracciale MP, Broggi A, Santarelli ML, Sarto MS, Palleschi C, Uccelletti D (2012) Graphite nanoplatelets and *Caenorhabditis elegans*: insights from an *in vivo* model. Nano Lett 12:2740–2744
65. Zhao Y-L, Wang X, Wu Q-L, Li Y-P, Wang D-Y (2015) Translocation and neurotoxicity of CdTe quantum dots in RMEs motor neurons in nematode *Caenorhabditis elegans*. J Hazard Mater 283:480–489
66. Zhao Y-L, Wang X, Wu Q-L, Li Y-P, Tang M, Wang D-Y (2015) Quantum dots exposure alters both development and function of D-type GABAergic motor neurons in nematode *Caenorhabditis elegans*. Toxicol Res 4:399–408
67. Zhang W, Sun B, Zhang L, Zhao B, Nie G, Zhao Y (2011) Biosafety assessment of Gd@ C82(OH)22 nanoparticles on *Caenorhabditis elegans*. Nanoscale 3:2636–2641

Chapter 2
Endpoints for Toxicity Assessment of Nanomaterials

Abstract Selecting the suitable endpoints is very important for both toxicity assessment and toxicological study of environmental toxicants, including engineered nanomaterials (ENMs). The well-described cellular, developmental, and genetic backgrounds of model animal of *Caenorhabditis elegans* provide many useful sublethal toxicity assessment endpoints. We here mainly introduced the sublethal endpoints associated with development, reproduction, neuronal development and function, intestinal development and function, epidermal development, innate immune response, lifespan, metabolism, oxidative stress, and transgenic strains reflecting stress response or oxidative stress in nematodes.

Keywords Endpoint · Sublethal · Nanotoxicity · *Caenorhabditis elegans*

2.1 Introduction

Both the toxicity assessment and the exposure route are very important for the design and the selection of suitable exposure to certain environmental toxicants, including engineered nanomaterials (ENMs), with certain research aims. *Caenorhabditis elegans* is a classic model animal with the property of well-described cellular, developmental, and genetic backgrounds [1]. Meanwhile, due to the wonderful sensitivity to different environmental toxicants or environmental stresses [2–4], *C. elegans* has been widely used in both the toxicity assessment and the toxicological study using lethality and sublethal endpoints. Especially, so far, many useful sublethal endpoints have been raised and employed in the toxicity assessment and the toxicological study of different environmental toxicants in nematodes [2–4].

Besides the endpoint of lethality, in this chapter, we systematically introduced the value and the related detailed information for the sublethal endpoints associated with development, reproduction, neuronal development and function, intestinal development and function, epidermal development, innate immune response, lifespan, metabolism, oxidative stress, and transgenic strains reflecting stress response or oxidative stress. These sublethal endpoints will enlarge the windows for us in understanding the underlying toxicological mechanisms of environmental toxicants and environmental stresses in organisms.

© Springer Nature Singapore Pte Ltd. 2018
D. Wang, *Nanotoxicology in* Caenorhabditis elegans,
https://doi.org/10.1007/978-981-13-0233-6_2

2.2 Lethality

Lethality is a normally used endpoint in the field of environmental science or toxicology. For lethality assay in nematodes, a 1.0 mL aliquot of test solution is added to each of the wells of testing plate, which will be subsequently loaded with at least 50 nematodes for each of the examined concentrations [5]. Following the exposure to certain toxicants, the wells are observed under a dissecting microscopy, where the inactive ones can be scored. The nematodes will be judged to be dead if they would not respond to the stimulus using a small, metal wire. The lethality is evaluated by the percentage of survival animals. Five replicates are suggested to be performed.

In the field of environmental science or toxicology, the median lethal concentrations (LC50s or LC20s) for certain environmental toxicants can be further calculated and raised based on the obtained data on lethality [6].

During the analysis of lethality, the lethality and the paralysis are suggested to be carefully distinguished. The nematodes are judged to be paralyzed if they cannot move after being touched with a small metal wire, although they still have the normal pharynx pumping behavior [7]. Fifty nematodes are suggested to be examined for each treatment. Five replicates are suggested to be performed.

2.3 Morphology and Development

So far, only limited endpoints have been raised to assess the effect of different environmental toxicants or environmental stresses on the development. One of the important endpoint about the development is to evaluate the morphology of the body of nematodes [8]. For example, exposure to high concentrations of barium (75 mM or 200 mM) could cause the appearance of vulva abnormality with ratios of approximately 5% and 9%, respectively, whereas no vulva abnormality was observed in nematodes exposed to barium at the concentration of 2.5 μM [8]. The vulva of control nematodes has the slightly smooth surface; however, abnormal vulva observed in barium exposed nematodes formed a large protuberance [8].

Another normally used endpoint to reflect the growth or development is the body length [9]. The length is measured by calculating the flat surface length along the body of nematodes using the Image-Pro® Express software. Ten replicates are suggested to be performed.

2.4 Reproduction

Reproductive organs, such as the gonad and the spermatheca, are important targeted organs of ENMs, such as carbon-based ENMs of graphene oxide (GO) and multi-walled carbon nanotubes (MWCNTs) in nematodes [10–13].

2.4.1 Endpoints Assessing the Development of Reproductive Organs

2.4.1.1 Germline Apoptosis

The germline apoptosis assay can be performed with a modified acridine orange (AO) staining method [14]. The examined nematodes are transferred into a Costar 24-well plate containing the aliquots of 25 mg/mL of AO. To facilitate the uptake of dye, E. coli OP50 can be added prior to the staining. The nematodes are incubated at 20 °C for 60 min and allowed to recover for 60 min on the bacterial lawns in order to repel the excessive dye in intestinal lumen. The nematodes are mounted onto agar pads on slides and examined under an epifluorescence microscopy (Olympus BX41, Olympus Corporation, Japan). The gonad arm in the posterior part of the body is suggested to be scored. Twenty nematodes are suggested to be examined per treatment, and three replicates are suggested to be performed.

2.4.1.2 Number of Apoptotic Cells per Gonad Arm Using CED-1::GFP Transgenic Strain

Apoptotic cells can be measured using MD701 transgenic nematodes, which carry the CED-1::GFP fusion protein, and can help us directly visualize apoptotic cells [14]. In transgenic strain of MD701, the CED-1::GFP is expressed in the nuclei of gonadal sheath cells. After the exposure, the CED-1::GFP transgenic nematodes are mounted onto agar pads on slides in 5 M NaN_3 and examined under a fluorescence microscope (Olympus BX41, Olympus Corporation, Japan). The apoptotic cells would be labeled with a bright GFP-positive circle, while intact cells would be uniformly green in color. The number of apoptotic cells in the germline meiotic region of one gonad arm is scored. Twenty nematodes are suggested to be examined per treatment, and three replicates are suggested to be performed.

2.4.1.3 Assay of DNA Damage Using HUS-1::GFP Transgenic Strain

The DNA damage can be analyzed using WS1433 transgenic strain carrying the HUS-1::GFP fusion protein [14]. The HUS-1::GFP signals are localized in the nuclei of proliferating and meiotic germ cells. The HUS-1::GFP foci can be quantified by counting the number of bright foci present in the middle/late pachytene germ cells under a fluorescence microscope (Olympus BX41, Olympus Corporation, Japan). Twenty nematodes are suggested to be examined per treatment, and three replicates are suggested to be performed.

2.4.1.4 Assay of 40,6-Diamidino-2-Phenylindole (DAPI) Staining

The dye of DAPI can be used to assess nuclear morphology and/or to quantify the number of mitotic cells in germline [14]. After transferring the examined nematodes into the NaN_3 drop on a microscope slide, the pharynx was cut in order to outflow the gonad using of syringe needle. After removing NaN_3 using filter paper, add 10 mL of Carnoy's fluid (ethanol/trichloromethane/acetic acid = 6:3:1) for the fixation. Carnoy's fluid after 15 min is removed, and DAPI at a final concentration of 2 ng/mL in M9 buffer containing 0.5% Tween 20 is added. A coverslip is placed, and both sides of it are fixed with nail polish. The slides can be observed under a fluorescence microscope (Olympus BX41, Olympus Corporation, Japan) after approximately 5 min staining. Six replicates are suggested to be performed.

2.4.2 Endpoints Assessing the Function of Reproductive Organs

2.4.2.1 Brood Size

To assay the brood size, the number of offspring at each stage (beyond the egg stage) would be counted [15]. Twenty nematodes are suggested to be examined per treatment.

2.4.2.2 The Number of Oocytes

The number of oocytes can be counted as the oocyte number in single gonad arm, since in most conditions the intestine would shade the gonad arm [14]. Twenty nematodes are suggested to be examined per treatment.

2.4.2.3 Egg Ejection

The egg ejection can be determined by counting the eggs on the plate every hour after removing the single laying nematode to a new NGM plate [14]. Twenty nematodes are suggested to be examined per treatment.

2.4.2.4 Embryonic Lethality

The embryonic lethality can be analyzed by dividing the number of eggs remaining by the total number plated [14]. Twenty nematodes are suggested to be examined per treatment.

2.4.3 Reproduction of Male Nematodes

2.4.3.1 Male Formation Assay

After the exposure, the male nematodes can be identified with the specific features of radically distinct male gonad full of sperm joined to a posterior cloaca rather than a mid-body vulva, a smaller, thinner body, and a differentiated tail with a cellular fan, rays, and a hook [16]. To evaluate the ratio of male formation, the number of male nematodes in a population with an approximately 5000 nematodes is suggested to be counted. Experiments are suggested to be repeated three times.

2.4.3.2 Abnormal Male-Specific Structures

Abnormal males can be identified to have abnormal male-specific structures, including cellular fan and rays, in the differentiated tail of nematodes [16]. To evaluate the percentage of abnormal males, approximately 50 males are suggested to be picked into a NGM plate, and the number of abnormal male nematodes is counted. To ensure the consistent positions for taking pictures, the nematodes are sticked onto the dried agar pad to ensure the consistent developmental stage. Experiments are suggested to be repeated three times.

2.5 Neuronal Development and Function

The neurons are another important secondary targeted organ of ENMs, such as quantum dots (QDs), in nematodes [7, 17].

2.5.1 Neuronal Development

2.5.1.1 Analysis of Axonal Degeneration and Neuronal Loss of D-Type GABAergic Motor Neurons

The D-type GABAergic motor neurons can be visualized using the transgenic strain of *oxIs12* [18]. The numbers of ventral and dorsal cord gaps are quantified to reflect the axonal degeneration. The neuronal loss is examined by counting the number of cell bodies in the D-type GABAergic nervous system. Images are suggested to be photographed and examined on the same day to avoid the effects of light source variance on fluorescent intensity. Thirty nematodes are suggested to be examined per treatment, and three replicates are suggested to be performed.

2.5.1.2 Fluorescent Images of Certain Neurons

The AVL and the DVB neurons are selected as the examples [18]. In nematodes, the AVL and the DVB interneurons establish a neuronal basis for defecation behavior. The fluorescent images of AVL and DVB neurons can be captured with a Zeiss Axiocam MRm camera on a Zeiss Axioplan 2 Imaging System using a SlideBook software (Intelligent Imaging Innovations). The images are acquired with a Quantix cooled CCD camera, and illumination was provided by a 175 W xenon arc lamp and GFP filter sets. Relative sizes of fluorescent puncta for cell bodies of neurons can be measured as the maximum radius for assayed fluorescent puncta. Relative fluorescent intensities of fluorescent puncta for cell bodies of neurons can also be examined. The aim for examining the fluorescent imaging of neurons is to assess the damage on examined neurons. Thirty nematodes are suggested to be examined per treatment, and three replicates are suggested to be performed.

2.5.2   Neuronal Function

2.5.2.1   Movement Speed

Using a video camera interfaced with a computer containing the National Institutes of Health tracking software, the movements of individual nematodes can be tracked and recorded [6]. The average movement per worm per second can be further calculated.

2.5.2.2   Locomotion Behavior

The locomotion behavior is under the control of motor neurons in nematodes. The locomotion behavior of nematodes can be at least evaluated by the endpoints of head thrash and body bend. To assay head thrash, the examined nematodes were washed with K medium and transferred into a microtiter well containing 60 μL of K medium on the top of agar. After a 1-min recovery period, head thrashes were counted for 1 min. A head thrash is defined as a change in the direction of bending at the mid-body. To assay body bend, the examined nematodes were picked onto a second plate and scored for the number of body bends in an interval of 20 s. A body bend is defined as a change in the direction of the part of the nematodes corresponding to the posterior bulb of the pharynx along the y-axis, assuming that nematode was traveling along the x-axis. At least 30 nematodes are suggested to be examined per treatment.

2.5.2.3   Thermotaxis Perception and Thermotaxis Learning Assays

The thermotaxis perception and thermotaxis learning are under the control of at least of AFD sensory neurons and AIY interneurons in nematodes. The thermotaxis behavior can be determined using a radial temperature gradient was performed [18]. A radial thermal gradient can be generated on an agar surface in a 9 cm Petri dish, in which a steeper gradient, ranging from approximately 17 °C in the central area to approximately 25 °C at the periphery, will be formed. The radial thermal gradient can be further generated by placing a 45 ml glass vial (catalog no. 224833, Wheaton) containing frozen acetic acid on the bottom of 9 cm agar plate without OP50 and incubating the plate at 26 °C for 90 min in the presence of a constant humidity of 60%. The surface temperature of the plate containing a glass vial can be analyzed by a thermograph (TVS-610, Nippon Avionics). The examined nematodes would be transferred onto a fresh plate devoid of bacteria for 2 min at 20 °C (the training temperature). Individual nematodes are deposited on a 9 cm Petri dish with a thermal gradient and allowed to move freely for 1.5–2 h. Upon removal of the nematodes from the plates, the tracks left on the agar surface are photographed, and the nematodes in each temperature area are counted. Based on the tracks left by the examined nematodes, movement to thermal gradient of 25 °C is scored as thermophilic (T, the fraction of nematodes moving to a higher temperature than the training temperature); movement to thermal gradient of 17 °C is scored as cryophilic (C, the faction of nematodes moving to a lower temperature than the training temperature); movement across the thermal gradients (17–25 °C) is scored as athermotactic (A, the faction of nematodes moving between high temperature and low temperature); and movement at a thermal gradient of 20 °C is scored as isothermal tracking behavior (IT, the fraction of nematodes staying in the radial thermal gradient of 20 °C). Each data point represents 5 independent assays using 30 nematodes per treatment.

To examine the thermotaxis learning, approximately 50 examined nematodes were grown at 25 °C for 12 h in the presence of food on a 9 cm Petri dish and then shifted to a selected plate at 20 °C for different time intervals (0, 0.5, 1, 3, 12, and 18 h) [18]. The aim is to investigate the ability of examined nematodes to learn a new cultivation temperature (20 °C), which would be different from their original cultivation temperature (25 °C). The examined nematodes will be further analyzed for their IT behavior as described above. Fifty nematodes are suggested to be assayed per treatment, and five replicates are suggested to be performed.

2.5.2.4   Foraging Behavior Assay

The foraging behavior is under the control of RMEs motor neurons in nematodes [17]. On the NGM plates with food, the tip of the nose of nematodes with normal foraging behavior will move from side to side within a narrow arc of movement. In contrast, the nematodes with damaged RMEs motor neurons or abnormal foraging behavior will have a "loopy" foraging in which the flexures of the nose will become grossly exaggerated. Thirty nematodes are suggested to be examined per treatment, and three replicates are suggested to be performed.

2.5.2.5 Shrinking Behavior Assay

The shrinking behavior is under the control of D-type motor neurons in nematodes [7]. A nematode will normally move backward by propagating a sinusoidal wave from the tail to the head, when it tapes on the nose. However, once the function of D-type motor neurons is disrupted, nematodes will show shrinking behavior. The shrinking behavior is defined as the phenotype that nematode pulls in its head and shortens its body due to hypercontraction of the body wall muscles on both the sides of the body. Thirty nematodes are suggested to be examined per treatment, and three replicates are suggested to be performed.

2.5.2.6 Neurotransmission

Because treatment with aldicarb, an acetylcholinesterase (AChE) inhibitor, or levamisole, a nicotinic acetylcholine receptor (AChR) agonist, can cause hyperactive cholinergic synapses, muscle hypercontraction, and paralysis, the synaptic transmission can be detected using the aldicarb or levamisole in nematodes [18]. The nematodes defective in presynaptic Ca^{2+}-dependent vesicle release would be resistant to aldicarb, and the nematodes only lacking a functional AChR would be also resistant to levamisole. The examined nematodes are placed on freshly seeded NGM plates containing 1 mM aldicarb or 100 µM levamisole. After an 8 h treatment, the nematodes would be scored as motile if they can still exhibit locomotion and pharyngeal pumping when prodded three times. Five replicates are suggested to be performed.

2.6 Intestinal Development and Function

The intestine is normally considered as the crucial primary targeted organs of ENMs, such as GO and TiO_2-nanoparticles in nematodes [19, 20].

2.6.1 Intestinal Development

2.6.1.1 Analysis of Intestinal Development Based on Transmission Electron Microscopy (TEM) Assay

The damage on intestinal development by ENMs can be directly determined by TEM assay in nematodes. For example, GO could be translocated into the intestinal cells and distributed to be adjacent to or surrounding mitochondria [19]. Meanwhile, the disrupted ultrastructure of microvilli and the loss of many microvilli in intestinal cells of GO exposed nematodes could be detected [19].

To perform the TEM assay, the nematodes are fixed with 0.5% glutaraldehyde and 2% osmium tetroxide. The fixed nematodes are further embedded in araldite resin following the infiltration series (30% araldite/acetone for 4 h, 70% araldite/acetone for 5 h, 90% araldite/acetone overnight, and pure araldite for 8 h). Serial sections are collected using an Ultracut E microtome. The images are obtained on a Hitachi H-7100 electron microscope using a Gatan slow-scan digital camera.

2.6.1.2 Analysis of Expression Patterns of Genes Required for the Control of Intestinal Development

The effect of certain ENMs on the intestinal development can be further evaluated by the possible dysregulation of expression patterns of genes required for the control of intestinal development. For example, prolonged exposure to GO caused significantly decrease in expression levels of *par-6* and *pkc-3* and increase in expression level of *nhx-2* [19]. In *C. elegans*, *par-6* and *pkc-3* are required for the development of microvilli on intestinal cells, and *nhx-2* is required for development of basolateral domain of the intestine.

The method of reverse transcription and quantitative real-time polymerase chain reaction (qRT-PCR) can be used to determine this aspect. For this aim, the total RNA can be extracted using RNeasy Mini Kit (Qiagen). Approximately 6000 nematodes are used for each exposure. Total nematode RNA ($\sim$1 μg) is reverse transcribed using a cDNA synthesis kit (Bio-Rad Laboratories). Quantitative reverse-transcription PCR is run at the optimized annealing temperature. The relative quantification of targeted genes in comparison to the reference genes, such as the *tba-1* encoding a tubulin protein, was determined. At least two references genes are suggested to be selected. The final results can be expressed as the ratio of the relative expressions of the targeted and reference genes. Three replicates are suggested to be performed.

2.6.2 *Intestinal Function*

2.6.2.1 Intestinal Reactive Oxygen Species (ROS) Production

Induction of intestinal ROS production can reflect the activation of oxidative stress and the damage formed in intestinal cells after exposure to certain ENMs [18]. To examine the intestinal ROS production, the nematodes are transferred to 1 μM of 5′,6′-chloromethyl-2′,7′dichlorodihydro-fluorescein diacetate (CM-H$_2$DCFDA; Molecular Probes) to incubate for 3 h at 20 °C. The CM-H$_2$DCFDA can in particular detect the various intracellularly produced ROS species in organisms. The nematodes are mounted on 2% agar pads for the examination under a laser scanning confocal microscope (Leica, TCS SP2, Bensheim, Germany) at an excitation wavelength of 488 nm and an emission filter of 510 nm. The relative fluorescence intensity of the ROS signals in the intestine can be semi-quantified in comparison to the intestinal autofluorescence. The positive control of paraquat treatment can be selected in the assay. Thirty nematodes are suggested to be examined each treatment, and three replicates are suggested to be performed.

2.6.2.2 Intestinal Permeability

One of the methods to assess the intestinal permeability is the Nile red staining [21]. The Nile red (Molecular Probes, Eugene, OR) is dissolved in acetone to produce a 0.5 mg/mL stock solution and stored at 4 °C. The fresh stock solution was diluted in 1 × phosphate buffer saline (PBS) to 1 μg/mL, and 150 μL of the diluted solution can be used for Nile red staining. Both the fluorescence intestine of fluorescent signals in the intestine and the distribution of fluorescent signals in different tissues are needed to be analyzed. Considering the fact that the Nile red can also be used to label fat storage, the triglyceride content is also normally analyzed using an enzymatic kit (Wako Pure Chemical Ltd., Osaka, Japan) to exclude the possibility of fat storage induced by ENMs in nematodes.

Besides this, recently, the intestinal permeability can also be determined by staining with a blue food dye (5.0% wt/vol in water) in the presence of OP50 for 3 h [22]. After that, the nematodes are transferred onto normal NGM plates seeded with OP50 to analyze the blue food dye in the body cavity using a microscope. Twenty nematodes are suggested to be examined per treatment.

2.6.3 Defecation Behavior

2.6.3.1 Defecation Behavior

To measure the mean defecation cycle length, individual nematodes are examined for a fixed number of cycles, and a cycle period was defined as the interval between initiations of two successive posterior body wall muscle contraction steps [7]. Twenty nematodes are suggested to be examined per treatment.

2.6.3.2 Analysis and Fluorescent Images of Neurons Controlling the Defecation Behavior

This aspect of analysis will provide the underlying cellular basis for the observed damage of ENMs in defecation behavior in nematodes [7]. Pictures of the fluorescence of AVL and DVB neurons controlling the defecation behavior can be taken with a Zeiss Axiocam MRm camera on a Zeiss Axioplan 2 imaging system using SlideBook software (Intelligent Imaging Innovations). The images are acquired with a Quantix cooled charge-coupled device (CCD) camera, and the illumination is provided by a 175 W xenon arc lamp and GFP filter sets. The relative sizes of fluorescent puncta at the position of cell body of AVL or DVB neuron are measured as the maximum radius of the assayed fluorescent puncta.

2.6.3.3 Analysis of Expression Patterns of Genes Required for the Control of Defecation Behavior

This aspect of analysis using qRT-PCR methods will provide the underlying molecular basis for the observed damage of ENMs in defecation behavior in nematodes [19]. For example, prolonged exposure to GO (100 mg/L) could significantly decrease the expression levels of unc-101, itr-1, smp-1, iri-1, and cab-1 and increase the expression levels of fat-3, isp-1, unc-44, fat-2, clk-1, hlh-8, gat-1, lim-6, unc-93, mlg-2, ced-10, and egl-30 [19].

2.7 Epidermal Development

Not like the intestinal barrier, it is normally considered that the epidermal barrier is usually strong enough to help the nematodes against the damage from environmental toxicants. Nevertheless, a report was published in 2012 to imply that exposure to

Ag-NPs (10 mg/L) could cause the significant epidermal fissuring in nematodes using the scanning electron microscopy [23]. Additionally, the serious epidermal burst effects were further detected in Ag-NPs (100 mg/L) exposed nematodes [23].

2.8   Innate Immune Response

Under the normal conditions, the damage on innate immune response provides a crucial cellular basis for the effects of chronic exposure to certain ENMs on nematodes [24]. More importantly, the damage on innate immune response is the key cellular basis for the effects of ENMs exposure on nematodes under the condition of pathogen infection [25].

2.8.1   Analysis of Expression Patterns of Genes Encoding Antimicrobial Peptides

Using the qRT-PCR method, the expression patterns of genes encoding antimicrobial peptides can be determined in ENMs exposed nematodes [24]. For example, chronic exposure to GO (1 mg/L) could induce a significant decrease in the expression levels of *F08G5.6*, *pqm-1*, *K11D12.5*, *prx-11*, *spp-1*, *lys-7*, *lys-2*, *abf-2*, *acdh-1*, and *lys-8* in nematodes [24].

In nematodes, *nlp-29* encodes another antimicrobial peptide in nematodes. With the concern of this research aim, the transgenic strain of P*nlp-29::GFP* can also be employed. For example, chronic exposure to GO (1 mg/L) could also significantly decrease the expression of NLP-29::GFP expression in nematodes [24].

2.8.2   Analysis of Expression Patterns of Genes Encoding Important Signaling Pathway Regulating Innate Immune Response in Nematodes

In nematodes, p38 MAP kinase (MAPK) pathway is one of the key signaling pathways regulating the innate immune response to pathogen infection. Using qRT-PCR method, the effect of ENMs exposure on the expression patterns of genes encoding p38 MAPK signaling pathway can be further examined in nematodes. For example, chronic exposure to GO (1 mg/L) significantly decreased the expression levels of *nsy-1*, *sek-1*, and *pmk-1*, which encodes the core signaling cascade in the p38 MAPK signaling pathway in nematodes [24].

2.8.3   OP50 Accumulation

In nematodes after chronic to certain ENMs, the dysregulated innate immune response may be largely due to the severe accumulation of OP50 in the intestinal lumen in nematodes [24]. With the aid of strain of OP50::GFP, the accumulation of OP50 in the intestinal lumen can be analyzed in nematodes. For example, with an increase of GO exposure duration, a severe accumulation of OP50::GFP could be detected in the intestinal lumen of nematodes at least after adult day-4 [24]. In nematodes, animals can normally survive with OP50 as their food source.

2.8.4   Colony-Forming Unit (CFU) Assay

With *C. albicans* as an example, its CFU in the body of nematodes is determined after infection with *C. albicans* lawns for 24 h [25]. The nematodes washed three times with sterile M9 buffer to remove surface *C. albicans*. Each group of 50 nematodes is disrupted using a homogenizer and plated on YPD agar containing kanamycin (45 µg/mL), ampicillin (100 µg/mL), and streptomycin (100 µg/mL). The plates are incubated for 48 h at 37 °C. Colonies would be counted to determine the CFU per nematode. Ten replicates are suggested to be analyzed for each experiment.

2.9   Lifespan

2.9.1   Lifespan Curve Assay

To measure the lifespan curve, the hermaphrodites are normally transferred daily for the first 7 days of adulthood [26–28]. The nematodes are checked every 2 days and would be scored as dead when they did not move even after repeated taps with a pick. At least forty nematodes are suggested to be examined per treatment. For lifespan, graphs are representative of at least three trials. The lifespan data were analyzed using a two-tailed two sample *t*-test (Minitab Ltd., Coventry, UK).

Mean lifespan means the time when 50% of animals will die. Besides the lifespan curves, the mean lifespan can also be used as an important endpoint for longevity analysis.

2.9.2   Intestinal Autofluorescence

Intestinal autofluorescence can be caused by lysosomal deposits of lipofuscin, accumulated over time in aging nematodes. The images can be collected for endogenous intestine fluorescence using a 525 nm bandpass filter and without automatic gain control in order to preserve the relative intensity of different nematodes' fluorescence. Observations of fluorescence are recorded, and the color images are taken for the documentation of results with Magnafire® software (Olympus, Irving, TX, USA). The lipofuscin levels are measured using ImageJ Software (NIH Image) by determining average pixel intensity in each animal's intestine.

2.9.3 Pharyngeal Pumping

The pharyngeal pumping is also an important endpoint to assess the potential aging process in ENMs exposed nematodes [29, 30]. To assay the pumping rate, nematodes are placed onto NGM plates with food and left undisturbed for 1 h before the measuring. The pharyngeal pumping is counted for 20 s under a digital input control (DIC) optics with a Zeiss axioscope.

2.9.4 Intestinal ROS Production and Locomotion Behavior

Intestinal ROS production and locomotion behavior can be also examined during the aging process in ENMs exposed nematodes to reflect the possible long-term effects of ENMs exposure on aging process of nematodes [26]. With the GO as an example, prolonged exposure to GO caused the more severe decrease in locomotion behavior and induction of intestinal ROS production during the aging process in exposed nematodes compared with controls [26].

2.10 Metabolism

Here, we mainly introduced the related endpoints for fat storage in nematodes. In nematodes, certain ENMs, such as QDs exposure, can dysregulate the fat storage [31].

2.10.1 Sudan Black Staining

To perform the Sudan black staining, the examined nematodes are washed in M9 buffer and fixed in 1% paraformaldehyde in M9 buffer [31]. The nematodes are then subjected to three freeze-thaw cycles and dehydrated through an ethanol series. The nematodes are stained overnight in a 50% saturated solution of Sudan black in 70% ethanol, rehydrated, and then photographed. Thirty nematodes are suggested to be examined per treatment.

2.10.2   Oil Red O Staining

To perform the Oil red O staining, the examined nematodes are fixed with freshly prepared 10% paraformaldehyde solution and immediately frozen in liquid nitrogen [31]. The nematodes are then subjected to two freeze-thaw cycles, after which the nematodes are allowed to settle and the paraformaldehyde solution is removed. One mL of 3 mg/mL oil red diluted with M9 buffer is added to the animal pellet and incubated for 30 min at room temperature, with occasional gentle agitation. The nematodes are allowed to settle and washed with M9 buffer. After removing the staining solution, the nematodes are mounted onto 2% agarose pads for microscopic observation and photography. Thirty nematodes are suggested to be examined per treatment.

2.10.3   Analysis of Triglyceride Content

The triglyceride amount can be measured using an enzymatic kit (Wako Triglyceride E-test, Wako Pure Chemical Ltd., Osaka, Japan) according to the protocol. Ten replicates are suggested to be examined per treatment.

2.10.4   Analysis of Expression Patterns of Genes Encoding Enzymes Required for the Control of Fatty Acid Metabolism

With the CdTe QDs as an example, exposure to CdTe QDs induced increase in fat storage in nematodes [31]. The expression patterns of genes encoding the enzymes required for the synthesis or degradation of fatty acid were further examined in nematodes. *fasn-1* gene encodes a fatty acid synthase, and *pod-2* gene encodes an acetyl-CoA carboxylase. Prolonged exposure to CdTe QDs (20 mg/L) significantly increased the expression levels of *fasn-1* and *pod-2* [31]. *acs-2* gene encodes an Acyl-CoA synthase, and *ech-1* gene encodes a enoyl-CoA hydratase. Prolonged exposure to CdTe QDs (20 mg/L) also significantly decreased the expression levels of *acs-2* and *ech-1* [31]. *C03H5.4* and *C07E3.9* genes encode phospholipases. In contrast, prolonged exposure to CdTe QDs (20 mg/L) did not significantly affect the expression pattern of either *C03H5.4* or *C07E3.9* gene [31].

2.11 Oxidative Stress

2.11.1 Intestinal ROS Production

The intestinal ROS production is a normally used method to detect the activation of oxidative stress in ENMs exposed nematodes [2–4]. The related information has been introduced above.

2.11.2 Oxidative Damage

To determine the oxidative damage, the examined nematodes are washed free of bacteria, pelleted and frozen [32]. The oxidative damage can be analyzed using an Oxyblot assay kit (Millipore, Boston, MA, USA) to detect carbonylated proteins. The carbonyl groups are derivatized with 2,4-dinitrophenylhydrazone (DNP-hydrazone). The quantification of carbonylated proteins is obtained by taking the ratio of DNP staining to tubulin staining according to the manufacturer's protocol.

2.11.3 Oxygen Consumption

The oxygen consumption can be measured using a Clark electrode (COLORLAB1, UK) for a 10 min period [33]. The nematodes are collected, pelleted, and frozen for protein quantification. The proteins are further quantified using a bicinchronic acid protein assay kit (Thermo Scientific, UK) according to the manufacturer's protocol.

2.11.4 Pharmacological Analysis

The pharmacological analysis was used to confirm the direct role of oxidative stress in inducing the toxicity of ENMs in nematodes [33, 34]. Usually, the synchronized examined nematodes (such as the L2-larvae) are treated with 10 mM ascorbate or 5 mM N-acetyl-L-cysteine (NAC) for 24 h and then exposed to certain ENMs when they develop into the adults.

2.11.5 Analysis of Expression Patterns of Genes Required for the Control of Oxidative Stress

In nematodes, the proteins of MEV-1, GAS-1, ISP-1, CLK-1, SKN-1, and GST-4 might play a key role in the primary molecular mechanism for the control of oxidative stress. Meanwhile, *sod* genes or *ctl* genes encode the antioxidative defense system in nematodes. We can further employ the qRT-PCR method to determine the underlying molecular basis for the observed effect of exposure to certain environmental toxicants or stresses in inducing oxidative stress in nematodes [19, 21, 35].

2.11.6 Superoxide Dismutase (SOD) or Catalase (CAT) Activity

The SOD or CAT can also be used to indirectly reflect the activation of oxidative stress in ENMs exposed nematodes [32]. The SOD activity can be measured at 500 nm with a commercially available kit (kit Ransod superoxide dismutase) from Randox Laboratories. To avoid the possible effect of food in solutions, the collected nematodes in solutions are placed on ice first, and then discard the supernatant with most of food in it. The nematodes of at least ten large NGM plates (9 cm) are suggested to be collected. The CAT activity can also be measured a commercial kit. The decrease in absorption is measured spectrophotometrically at 240 nm, and an extinction coefficient of 43.6 M/cm was used to determine the enzyme activity.

2.12 Transgenic Strains Reflecting Stress Response or Oxidative Stress

2.12.1 HSP-16.2::GFP

In nematodes, *hsp-16.2* encodes a heat-shock protein, and HSP-16.2 is an important marker to detect the stress response of nematodes exposed to environmental toxicants. For example, exposure to Al_2O_3-NPs could induce the significant increase in HSP-16.2 expression in nematodes [36].

2.12.2 MTL-2::GFP

In nematodes, *mtl-2* encodes a metallothionein protein, and MTL-2 is also an important marker to detect the stress response of nematodes exposed to environmental toxicants. For example, the MTL-2::GFP expression could be significantly induced by ZnO-NPs exposure in nematodes [6].

2.12.3 SKN-1::GFP and GST-4::GFP

Transgenic strains of SKN-1::GFP and GST-4::GFP are important genetic tools to reflect the activation of oxidative stress in nematodes exposed to certain environmental toxicants or stresses. In nematodes, *skn-1* encodes a Nrf protein having the cellular protective function by acting as a regulator of antioxidant or xenobiotic defense, and *gst-4* encodes a putative glutathione-requiring prostaglandin D synthase acting as a direct targeted gene of *skn-1* during the control of oxidative stress. For example, both SKN-1::GFP and GST-4 could be significantly activated by simulated microgravity treatment, and simulated microgravity treatment could further induce the obvious translocation of SKN-1::GFP into the nucleus in nematodes [35].

2.12.4 PMK-1::GFP

In nematodes, *pmk-1* encodes a MAPK in the p38 signaling pathway, and PMK-1::GFP is an important marker to detect the stress response or the oxidative stress of nematodes exposed to environmental toxicants, such as the ENMs. For example, exposure to GO (100 mg/L) could significantly increase the PMK-1::GFP expression in the intestine [10]. Meanwhile, GO (100 mg/L) further significantly increased the percentage of PMK-1::GFP nucleus localization in intestinal cells [10].

2.12.5 DAF-16::GFP and SOD-3::GFP

In nematodes, *daf-16* encodes a FOXO transcriptional factor in the insulin signaling pathway, and SOD-3 is the direct target of DAF-16 in the regulation of stress response or oxidative stress in nematodes. Both DAF-16::GFP and SOD-3::GFP are important markers to detect the stress response or the oxidative stress of nematodes exposed to environmental toxicants, such as the ENMs. For example, prolonged exposure to GO (100 mg/L) could significantly increase the expressions of both DAF-16::GFP and SOD-3::GFP, and GO exposure also increased the translocation of DAF-16:GFP expression in the nuclei of nematodes [37].

2.13   Perspectives

In this chapter, we introduced the value and the related information in details for the sublethal endpoints associated with development, reproduction, neuronal development and function, intestinal development and function, epidermal development, innate immune response, lifespan, metabolism, oxidative stress, and transgenic strains reflecting stress response or oxidative stress in nematodes. Nevertheless, the well-described cellular and developmental backgrounds of *C. elegans* actually can allow to raise many more useful endpoints with the concerns on different aspects of development and function of targeted organs. We believe that the toxicity assessment endpoints should not be only restricted in these introduced ones, and some more useful toxicity assessment endpoints will be further employed and raised in the future in nematodes.

In the future, besides the secondary targeted organs of reproductive organs and neurons, we suggest to further pay attention to raise the toxicity assessment endpoints associated with other possible secondary targeted organs of ENMs, such as muscle and pharynx. Moreover, the endpoints with potential to reflect the possible toxicity of ENMs on the larvae, the aged animals, and the possible transgenerational toxicity of ENMs also are needed to be further carefully considered and examined.

References

1. Brenner S (1974) The genetics of *Caenorhabditis elegans*. Genetics 77:71–94
2. Leung MC, Williams PL, Benedetto A, Au C, Helmcke KJ, Aschner M, Meyer JN (2008) *Caenorhabditis elegans*: an emerging model in biomedical and environmental toxicology. Toxicol Sci 106:5–28
3. Zhao Y-L, Wu Q-L, Li Y-P, Wang D-Y (2013) Translocation, transfer, and *in vivo* safety evaluation of engineered nanomaterials in the non-mammalian alternative toxicity assay model of nematode *Caenorhabditis elegans*. RSC Adv 3:5741–5757
4. Wang D-Y (2016) Biological effects, translocation, and metabolism of quantum dots in nematode *Caenorhabditis elegans*. Toxicol Res 5:1003–1011
5. Zhao Y-L, Liu Q, Shakoor S, Gong JR, Wang D-Y (2015) Transgenerational safe property of nitrogen-doped graphene quantum dots and the underlying cellular mechanism in *Caenorhabditis elegans*. Toxicol Res 4:270–280
6. Ma H, Bertsch PM, Glenn TC, Kabengi NJ, Williams PL (2009) Toxicity of manufactured zinc oxide nanoparticles in the nematode *Caenorhabditis elegans*. Environ Toxicol Chem 28(6):1324–1330
7. Zhao Y-L, Wang X, Wu Q-L, Li Y-P, Tang M, Wang D-Y (2015) Quantum dots exposure alters both development and function of D-type GABAergic motor neurons in nematode *Caenorhabditis elegans*. Toxicol Res 4:399–408
8. Wang D-Y, Wang Y (2008) Phenotypic and behavioral defects caused by barium exposure in nematode *Caenorhabditis elegans*. Arch Environ Contam Toxicol 54(3):447–453
9. Yang J-N, Zhao Y-L, Wang Y-W, Wang H-F, Wang D-Y (2015) Toxicity evaluation and translocation of carboxyl functionalized graphene in *Caenorhabditis elegans*. Toxicol Res 4:1498–1510

10. Zhao Y-L, Zhi L-T, Wu Q-L, Yu Y-L, Sun Q-Q, Wang D-Y (2016) p38 MAPK-SKN-1/ Nrf signaling cascade is required for intestinal barrier against graphene oxide toxicity in *Caenorhabditis elegans*. Nanotoxicology 10(10):1469–1479
11. Zhao Y-L, Jia R-H, Qiao Y, Wang D-Y (2016) Glycyrrhizic acid, active component from *Glycyrrhizae radix*, prevents toxicity of graphene oxide by influencing functions of microR-NAs in nematode *Caenorhabditis elegans*. Nanomed: Nanotechnol Biol Med 12:735–744
12. Zhi L-T, Ren M-X, Qu M, Zhang H-Y, Wang D-Y (2016) Wnt ligands differentially regulate toxicity and translocation of graphene oxide through different mechanisms in *Caenorhabditis elegans*. Sci Rep 6:39261
13. Zhi L-T, Fu W, Wang X, Wang D-Y (2016) ACS-22, a protein homologous to mammalian fatty acid transport protein 4, is essential for the control of toxicity and translocation of multi-walled carbon nanotubes in *Caenorhabditis elegans*. RSC Adv 6:4151–4159
14. Zhao Y-L, Wu Q-L, Wang D-Y (2016) An epigenetic signal encoded protection mechanism is activated by graphene oxide to inhibit its induced reproductive toxicity in *Caenorhabditis elegans*. Biomaterials 79:15–24
15. Wu Q-L, Zhou X-F, Han X-X, Zhuo Y-Z, Zhu S-T, Zhao Y-L, Wang D-Y (2016) Genome-wide identification and functional analysis of long noncoding RNAs involved in the response to graphene oxide. Biomaterials 102:277–291
16. Liu P-D, He K-W, Li Y-X, Wu Q-L, Yang P, Wang D-Y (2012) Exposure to mercury causes formation of male-specific structural deficits by inducing oxidative damage in nematodes. Ecotoxicol Environ Saf 79:90–100
17. Zhao Y-L, Wang X, Wu Q-L, Li Y-P, Wang D-Y (2015) Translocation and neurotoxicity of CdTe quantum dots in RMEs motor neurons in nematode *Caenorhabditis elegans*. J Hazard Mater 283:480–489
18. Yu X-M, Guan X-M, Wu Q-L, Zhao Y-L, Wang D-Y (2015) Vitamin E ameliorates the neuro-degeneration related phenotypes caused by neurotoxicity of Al$_2$O$_3$-nanoparticles in *C. elegans*. Toxicol Res 4:1269–1281
19. Wu Q-L, Yin L, Li X, Tang M, Zhang T, Wang D-Y (2013) Contributions of altered perme-ability of intestinal barrier and defecation behavior to toxicity formation from graphene oxide in nematode *Caenorhabditis elegans*. Nanoscale 5(20):9934–9943
20. Zhao Y-L, Wu Q-L, Tang M, Wang D-Y (2014) The *in vivo* underlying mechanism for recovery response formation in nano-titanium dioxide exposed *Caenorhabditis elegans* after transfer to the normal condition. Nanomed: Nanotechnol Biol Med 10:89–98
21. Ding X-C, Wang J, Rui Q, Wang D-Y (2018) Long-term exposure to thiolated graphene oxide in the range of μg/L induces toxicity in nematode *Caenorhabditis elegans*. Sci Total Environ 616–617:29–37
22. Gelino S, Chang JT, Kumsta C, She X, Davis A, Nguyen C, Panowski S, Hansen M (2016) Intestinal autophagy improves healthspan and longevity in *C. elegans* during dietary restric-tion. PLoS Genet 12:e1006135
23. Kim SW, Nam S, An Y (2012) Interaction of silver nanoparticles with biological surfaces of *Caenorhabditis elegans*. Ecotoxicol Environ Saf 77:64–70
24. Wu Q-L, Zhao Y-L, Fang J-P, Wang D-Y (2014) Immune response is required for the control of *in vivo* translocation and chronic toxicity of graphene oxide. Nanoscale 6:5894–5906
25. Shakoor S, Sun L-M, Wang D-Y (2016) Multi-walled carbon nanotubes enhanced fungal colo-nization and suppressed innate immune response to fungal infection in nematodes. Toxicol Res 5:492–499
26. Wu Q-L, Zhao Y-L, Zhao G, Wang D-Y (2014) microRNAs control of *in vivo* toxicity from graphene oxide in *Caenorhabditis elegans*. Nanomed: Nanotechnol Biol Med 10:1401–1410
27. Wang D-Y, Cao M, Dinh J, Dong Y-Q (2013) Methods for creating mutations in *C. elegans* that extend lifespan. Method Mol Biol 1048:65–75
28. Qu M, Li Y-H, Wu Q-L, Xia Y-K, Wang D-Y (2017) Neuronal ERK signaling in response to graphene oxide in nematode *Caenorhabditis elegans*. Nanotoxicology 11(4):520–533

29. Scharf A, Piechulek A, von Mikecz A (2013) Effect of nanoparticles on the biochemical and behavioral aging phenotype of the nematode *Caenorhabditis elegans*. ACS Nano 7(12):10695–10703

30. Wu Q-L, Zhao Y-L, Li Y-P, Wang D-Y (2014) Susceptible genes regulate the adverse effects of TiO$_2$-NPs at predicted environmental relevant concentrations on nematode *Caenorhabditis elegans*. Nanomed: Nanotechnol Biol Med 10:1263–1271

31. Wu Q-L, Zhi L-T, Qu Y-Y, Wang D-Y (2016) Quantum dots increased fat storage in intestine of *Caenorhabditis elegans* by influencing molecular basis for fatty acid metabolism. Nanomed: Nanotechnol Biol Med 12(5):1175–1184

32. Li Y-X, Yu S-H, Wu Q-L, Tang M, Pu Y-P, Wang D-Y (2012) Chronic Al$_2$O$_3$-nanoparticle exposure causes neurotoxic effects on locomotion behaviors by inducing severe ROS production and disruption of ROS defense mechanisms in nematode *Caenorhabditis elegans*. J Hazard Mater 219–220:221–230

33. Wu Q-L, Liu P-D, Li Y-X, Du M, Xing X-J, Wang D-Y (2012) Inhibition of ROS elevation and damage on mitochondrial function prevents lead-induced neurotoxic effects on structures and functions of AFD neurons in *Caenorhabditis elegans*. J Environ Sci 24(4):733–742

34. Wu Q-L, Nouara A, Li Y-P, Zhang M, Wang W, Tang M, Ye B-P, Ding J-D, Wang D-Y (2013) Comparison of toxicities from three metal oxide nanoparticles at environmental relevant concentrations in nematode *Caenorhabditis elegans*. Chemosphere 90:1123–1131

35. Zhao L, Rui Q, Wang D-Y (2017) Molecular basis for oxidative stress induced by simulated microgravity in nematode *Caenorhabditis elegans*. Sci Total Environ 607–608:1381–1390

36. Yu S-H, Rui Q, Cai T, Wu Q-L, Li Y-X, Wang D-Y (2011) Close association of intestinal autofluorescence with the formation of severe oxidative damage in intestine of nematodes chronically exposed to Al$_2$O$_3$-nanoparticle. Environ Toxicol Pharmacol 32:233–241

37. Zhao Y-L, Yang R-L, Rui Q, Wang D-Y (2016) Intestinal insulin signaling encodes two different molecular mechanisms for the shortened longevity induced by graphene oxide in *Caenorhabditis elegans*. Sci Rep 6:24024

38. Wu Q-L, Rui Q, He K-W, Shen L-L, Wang D-Y (2010) UNC-64 and RIC-4, the plasma membrane associated SNAREs syntaxin and SNAP-25, regulate fat storage in nematode *Caenorhabditis elegans*. Neurosci Bull 26(2):104–116

39. Nouara A, Wu Q-L, Li Y-X, Tang M, Wang H-F, Zhao Y-L, Wang D-Y (2013) Carboxylic acid functionalization prevents the translocation of multi-walled carbon nanotubes at predicted environmental relevant concentrations into targeted organs of nematode *Caenorhabditis elegans*. Nanoscale 5:6088–6096

Chapter 3
Exposure Routes of Nanomaterials

Abstract The well-described developmental background and the sensitivity to environmental toxicants of *Caenorhabditis elegans* is helpful for designing suitable exposure routes for different engineered nanomaterials (ENMs) with different research aims. We here introduced different exposure routes (acute exposure, prolonged exposure, chronic exposure, one-generation exposure, and transgenerational exposure) for the toxicity assessment of ENMs in nematodes. Moreover, we discussed the value of some exposure routes, such as prolonged exposure and chronic exposure, in detecting the possible toxicity of ENMs at environmentally relevant concentrations on nematodes.

Keywords Exposure route · Toxicity assessment · Environmentally relevant concentration · *Caenorhabditis elegans*

3.1 Introduction

Selection of suitable exposure routes is very important for toxicity assessment of engineered nanomaterials (ENMs) with different research aims. The well-described developmental background and the sensitivity to different environmental toxicants, including the ENMs, of model animal of *Caenorhabditis elegans* are helpful for our designing different exposure routes so as to satisfy different research aims [1–4]. Most of the designed exposure routes belong to the toxicity assessment of ENMs in one generation in nematodes. Some certain exposure routes have also been raised to apply for the toxicity assessment of ENMs transgenerationally in nematodes.

In this chapter, we introduced several different exposure routes of ENMs in nematodes, and these exposure routes contain acute exposure, prolonged exposure, chronic exposure, one-generation exposure, and transgenerational exposure. Besides this, we also introduced and discussed the toxicity assessment of ENMs at environmentally relevant concentrations in nematodes. The information provided in this chapter will provide cues for design suitable exposure routes for different ENMs with certain research aims in nematodes.

© Springer Nature Singapore Pte Ltd. 2018
D. Wang, *Nanotoxicology in* Caenorhabditis elegans,
https://doi.org/10.1007/978-981-13-0233-6_3

3.2 Exposure Routes of ENMs

3.2.1 Acute Exposure

Normally, the acute exposure to environmental toxicants or stresses was performed from L4-larvae stage or young adults for 24 h [5, 6]. For the acute exposure to ENMs from L4-larvae for 24 h, the addition of *E. coli* OP50 as the food source is suggested. With the concern of certain research aims, the acute exposure to ENMs could also be designed from other larval stages, such L1-, L2-, or L3-larvae in the presence of OP50 as the food source. Besides these, with the concern of certain research aims, the acute exposure to ENMs could further be designed from the L4-larvae or the young adults for not more than 24 h. Acute exposure is used to assess the short-term effect of ENMs on nematodes.

With the DMSA coated Fe_2O_3-nanoparticle (DMSA coated Fe_2O_3-NPs) as an example, the possible toxicity of acute exposure to DMSA coated Fe_2O_3-NPs from the L4-larvae for 24 h was investigated in nematodes. After acute exposure, all the examined concentrations (0.5–100 mg/L) of DMSA coated Fe_2O_3-NPs did not affect the survival of nematodes (Fig. 3.1) [5]. After acute exposure, only DMSA coated Fe_2O_3-NPs at the concentration of 100 mg/L could significantly decrease the body length, reduce the brood size, suppress the pumping rate, increase the mean defecation cycle length, and induce the intestinal autofluorescence (Fig. 3.1) [5]. In contrast, after acute exposure, DMSA coated Fe_2O_3-NPs at concentrations more than 50 mg/L could significantly decrease the locomotion behavior as reflected by the endpoints of head thrash and body bend (Fig. 3.1) [5]. These results suggest that acute exposure to DMSA coated Fe_2O_3-NPs at concentrations more than 50 mg/L may have adverse effects on nematodes.

3.2.2 Prolonged Exposure

In nematodes, prolonged exposure to environmental toxicants or stresses was normally performed from L1-larvae to young adults or adult day-1 in the presence of OP50 as the food source [5, 7, 8,]. The prolonged exposure is an exposure route used to assess the long-term effect of ENMs on nematodes.

With the DMSA coated Fe_2O_3-NPs as an example, the possible toxicity of prolonged exposure to DMSA coated Fe_2O_3-NPs from the L1-larvae to adult day-1 was also investigated in nematodes. After prolonged exposure, all the examined concentrations (1–5000 μg/L) of DMSA coated Fe_2O_3-NPs also did not affect the survival of nematodes (Fig. 3.2) [5]. After prolonged exposure, only DMSA coated Fe_2O_3-NPs at the concentration of 5000 μg/L significantly decreased the body length, suppressed the pumping rate, increased the mean defecation cycle length, and induced the intestinal autofluorescence (Fig. 3.2) [5]. In contrast, after prolonged exposure, DMSA coated Fe_2O_3-NPs at concentrations more than 500 μg/L significantly

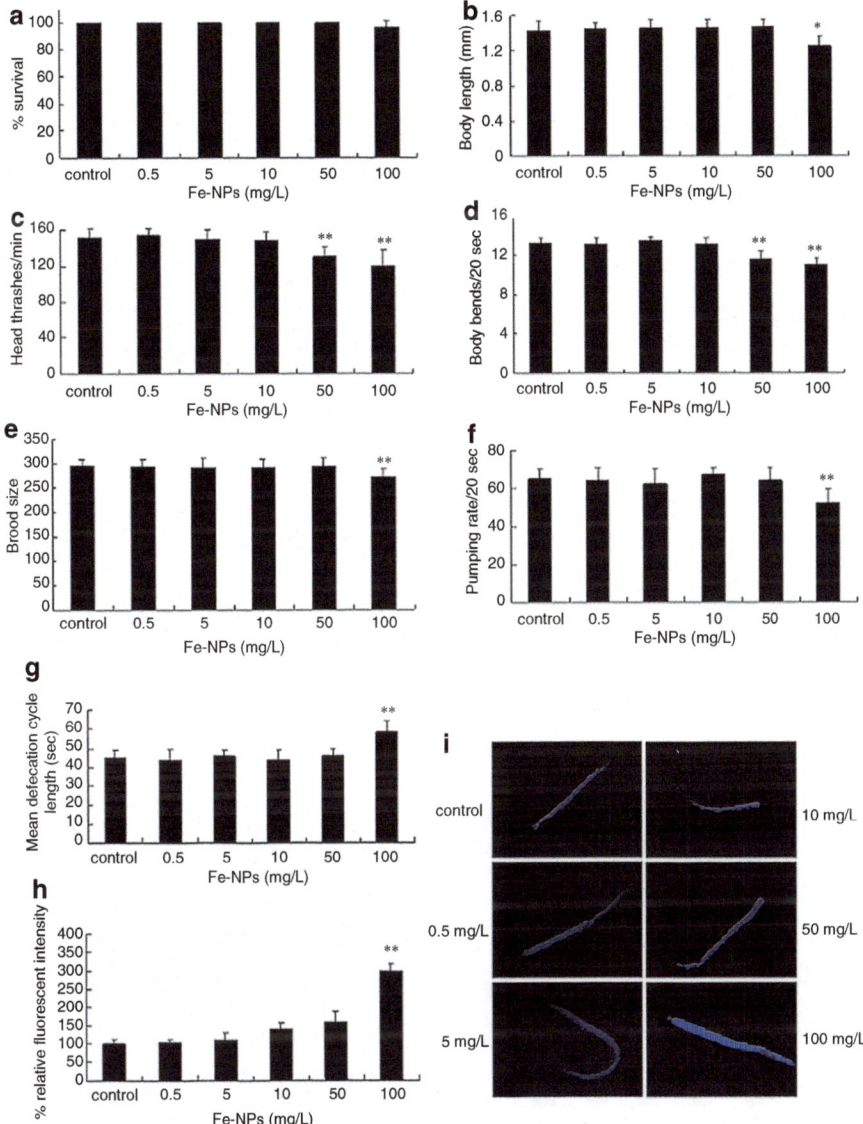

Fig. 3.1 Toxicity assessment of DMSA coated Fe_2O_3-NPs after acute exposure from L4-larvae for 24 h [5]

(**a**) Comparison of lethality in nematodes exposed to different concentrations of DMSA coated Fe_2O_3-NPs. (**b**) Comparison of body length in nematodes exposed to different concentrations of DMSA coated Fe_2O_3-NPs. (**c**) Comparison of head thrash in nematodes exposed to different concentrations of DMSA coated Fe_2O_3-NPs. (**d**) Comparison of body bend in nematodes exposed to different concentrations of DMSA coated Fe_2O_3-NPs. (**e**) Comparison of brood size in nematodes exposed to different concentrations of Fe2O3-nanoparticles. (**f**) Comparison of pumping rate in nematodes exposed to different concentrations of DMSA coated Fe_2O_3-NPs. (**g**) Comparison of mean defecation cycle length in nematodes exposed to different concentrations of DMSA coated Fe_2O_3-NPs. (**h**) Comparison of intestinal autofluorescence in nematodes exposed to different concentrations of DMSA coated Fe_2O_3-NPs. (**i**) Pictures showing the intestinal autofluorescence in nematodes exposed to different concentrations of DMSA coated Fe_2O_3-NPs. Bars represent mean ± SEM. *$p < 0.05$, **$p < 0.01$

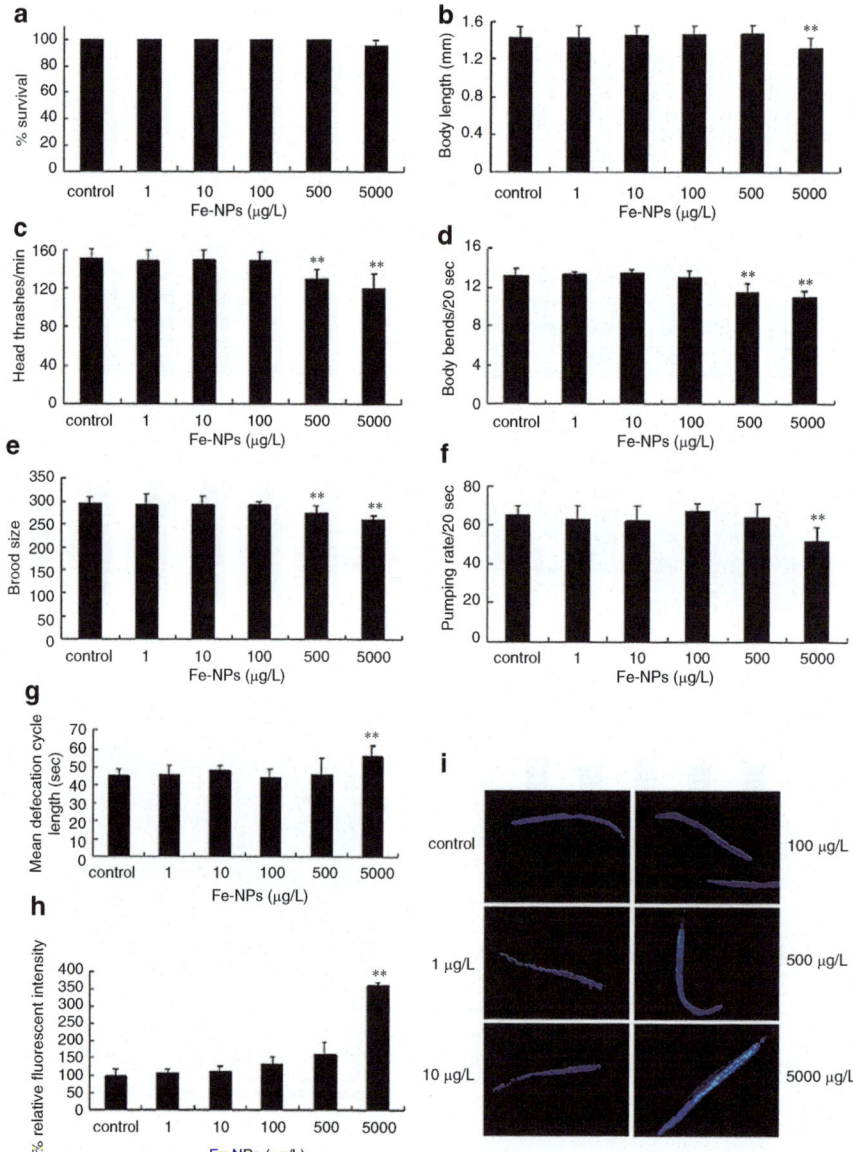

Fig. 3.2 Toxicity assessment of DMSA coated Fe_2O_3-NPs after prolonged exposure from L1-larvae to adult day-1 [5]

(**a**) Comparison of lethality in nematodes exposed to different concentrations of DMSA coated Fe_2O_3-NPs. (**b**) Comparison of body length in nematodes exposed to different concentrations of DMSA coated Fe_2O_3-NPs. (**c**) Comparison of head thrash in nematodes exposed to different concentrations of DMSA coated Fe_2O_3-NPs. (**d**) Comparison of body bend in nematodes exposed to different concentrations of DMSA coated Fe_2O_3-NPs. (**e**) Comparison of brood size in nematodes exposed to different concentrations of DMSA coated Fe_2O_3-NPs. (**f**) Comparison of pumping rate in nematodes exposed to different concentrations of DMSA coated Fe_2O_3-NPs. (**g**) Comparison of mean defecation cycle length in nematodes exposed to different concentrations of DMSA coated Fe_2O_3-NPs. (**h**) Comparison of intestinal autofluorescence in nematodes exposed to different concentrations of DMSA coated Fe_2O_3-NPs. (**i**) Pictures showing the intestinal autofluorescence in nematodes exposed to different concentrations of DMSA coated Fe_2O_3-NPs. Bars represent mean ± SEM. *$p < 0.05$, **$p < 0.01$

decreased the locomotion behavior and reduced the brood size (Fig. 3.2) [5]. Therefore, prolonged exposure to DMSA coated Fe_2O_3-NPs at concentrations more than 500 µg/L may have adverse effects on nematodes.

3.2.3 Chronic Exposure

Chronic exposure to environmental toxicants or stresses can be performed from young adults for 10 days or from adult day-1 to adult day-8 in the presence of OP50 as the food source in nematodes [9, 10]. In nematodes, chronic exposure to environmental toxicants, such as the ENMs, can be further performed from L1-larvae to adult day-8 in the presence of OP50 as the food source [5, 9]. Besides the prolonged exposure, chronic exposure is another usually used exposure route to assess the long-term effect of ENMs on nematodes.

With the DMSA coated Fe_2O_3-NPs as an example, the possible toxicity of chronic exposure to DMSA coated Fe_2O_3-NPs from the L1-larvae to adult day-8 was further investigated in nematodes. After chronic exposure, only DMSA coated Fe_2O_3-NPs at the concentration of 5000 µg/L significantly decreased the survival of nematodes (Fig. 3.3) [5]. In contrast, after chronic exposure, DMSA coated Fe_2O_3-NPs at concentrations more than 500 µg/L could significantly decrease the body length, suppress the pumping rate, and increase the mean defecation cycle length (Fig. 3.3) [5]. Moreover, after chronic exposure, DMSA coated Fe_2O_3-NPs at concentrations more than 100 µg/L significantly decreased the locomotion behavior, reduced the brood size, and induced the intestinal autofluorescence (Fig. 3.3) [5]. Therefore, chronic exposure to DMSA coated Fe_2O_3-NPs at concentrations more than 100 µg/L may have adverse effects on nematodes.

3.2.4 One-Generation Exposure

In nematodes, in order to determine the long-term effects of ENMs or their possible hormesis effects, one-generation exposure can also be designed from L1-larvae till the end of the examined generation in the presence of OP50 as the food source.

With graphene oxide (GO) as an example, the effect of one-generation exposure to GO at different concentrations on lifespan was investigated in nematodes. After one-generation exposure, GO at concentrations of 5–20 mg/L did not significantly alter the lifespan of nematodes (Fig. 3.4) [12], suggesting that the GO at the examined concentrations may exert no adverse effects on the aging process of nematodes under normal conditions.

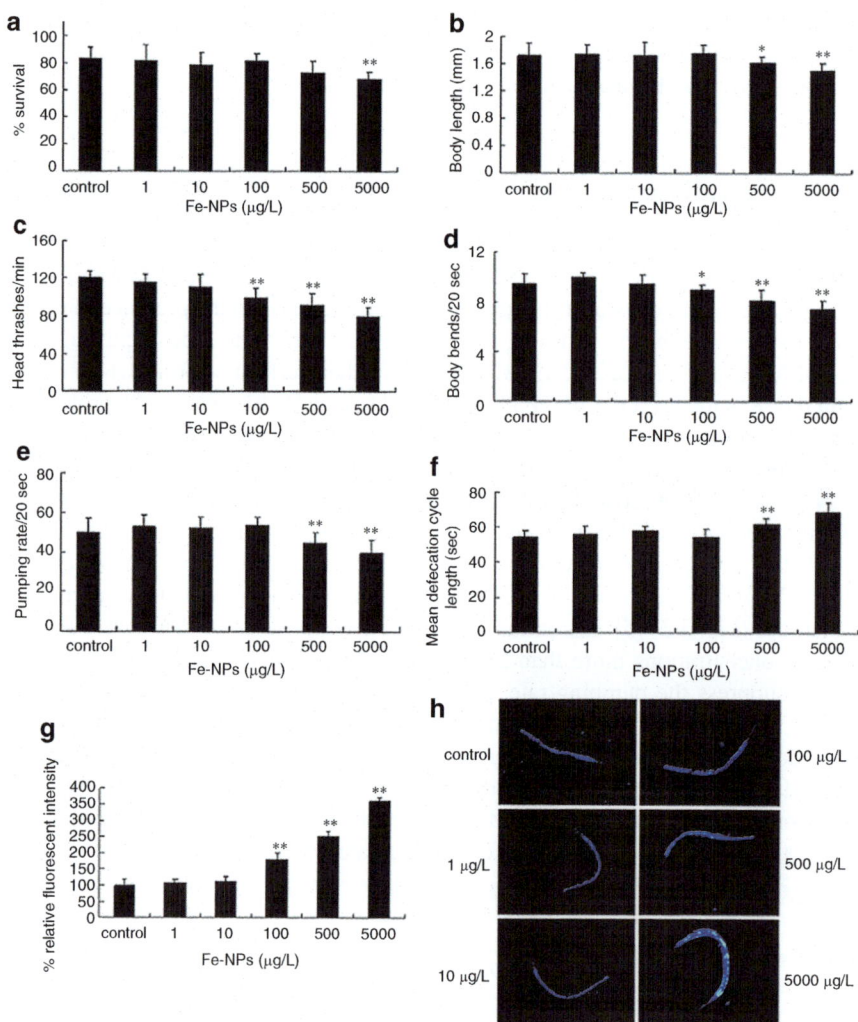

Fig. 3.3 Toxicity assessment of DMSA coated Fe₂O₃-NPs after chronic exposure from L1-larvae to adult day-8 [5]
(**a**) Comparison of lethality in nematodes exposed to different concentrations of DMSA coated Fe₂O₃-NPs. (**b**) Comparison of body length in nematodes exposed to different concentrations of DMSA coated Fe₂O₃-NPs. (**c**) Comparison of head thrash in nematodes exposed to different concentrations of DMSA coated Fe₂O₃-NPs. (**d**) Comparison of body bend in nematodes exposed to different concentrations of DMSA coated Fe₂O₃-NPs. (**e**) Comparison of pumping rate in nematodes exposed to different concentrations of DMSA coated Fe₂O₃-NPs. (**f**) Comparison of mean defecation cycle length in nematodes exposed to different concentrations of DMSA coated Fe₂O₃-NPs. (**g**) Comparison of intestinal autofluorescence in nematodes exposed to different concentrations of DMSA coated Fe₂O₃-NPs. (**h**) Pictures showing the intestinal autofluorescence in nematodes exposed to different concentrations of DMSA coated Fe₂O₃-NPs. Bars represent mean ± SEM. *$p < 0.05$, **$p < 0.01$

Fig. 3.4 Effect of one-generation exposure to GO on lifespan of nematodes [12]

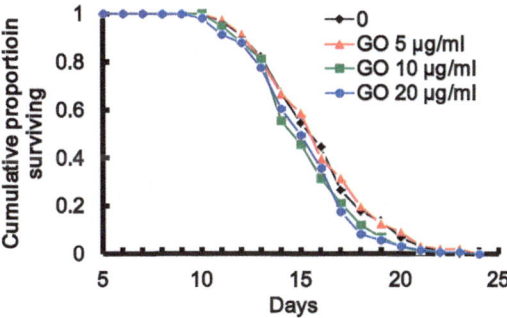

3.2.5 *Transgenerational Exposure*

To examine the transgenerational effect of certain ENMs on nematodes, the exposure is usually performed in the first generation in the presence of OP50 as the food source, and then the nematodes are transferred to normal medium without the addition of ENMs to obtain the progeny of exposed nematodes after the exposure. In the F1 generation, the transgenerational effects of certain ENMs on progeny of exposed nematodes would be examined.

With nitrogen-doped graphene quantum dots (N-GQDs) as an example, the transgenerational effect of N-GQDs was investigated in nematodes after the prolonged exposure from L1-larvae to adult day-1 in the first generation. In the first generation, N-GQDs at concentrations of 0.1–100 mg/L could not significantly affect the lifespan, brood size, and locomotion behavior in nematodes [13]. Additionally, N-GQDs at concentrations of 0.1–100 mg/L could not induce the significant intestinal ROS production in the first generation in nematodes [13]. Moreover, in the second generation, the normal lifespan, brood size, and locomotion behavior were observed, and no significant intestinal ROS production was detected (Fig. 3.5) [13], indicating that no obvious transgenerational toxicity of N-GQDs will be formed in nematodes. Different from the transgenerational effects of N-GQDs, prolonged exposure to nanopolystyrene particles at concentrations more than 10 µg/L significantly decreased the locomotion behavior and reduced the brood size and induced the significant induction of intestinal ROS production in the first generation in nematodes [14]. Moreover, nanopolystyrene particles at concentrations more than 100 µg/L induced the obvious transgenerational toxicity as reflected by the decreased locomotion behavior, the reduced brood size, and the induced significant induction of intestinal ROS production in the second generation in nematodes [14], demonstrating the formation of transgenerational toxicity of nanopolystyrene particles in nematodes.

In nematodes, with the concern of certain research aims, the exposure can be further performed in multiple generations, such as up to six generations [11]. With the $Gd@C_{82}(OH)_{22}$ nanoparticles as an example, $Gd@C_{82}(OH)_{22}$ at concentrations of 0.01–10 mg/L did not obviously affect the lifespan of nematodes after successive exposure to $Gd@C_{82}(OH)_{22}$ for six generations under the normal conditions [11].

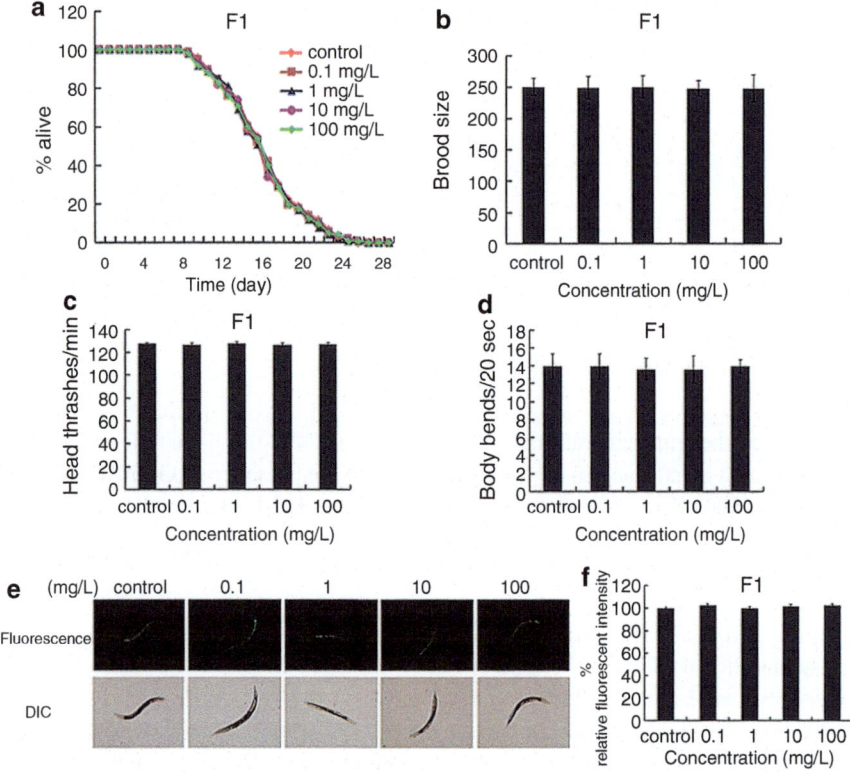

Fig. 3.5 Phenotypic analysis in progeny (F1 generation) of nematodes exposed to N-GQDs [13] (**a**) Lifespan in progeny of nematodes exposed to N-GQDs. (**b**) Brood size in progeny of nematodes exposed to N-GQDs. (**c**) Locomotion behavior in progeny of nematodes exposed to N-GQDs. (**d**) Intestinal ROS production in progeny of nematodes exposed to N-GQDs. Prolonged exposure was performed from L1-larvae to adult day-1. Bars represent means ± SEM

Moreover, under the heat-shock stress, Gd@C$_{82}$(OH)$_{22}$ at concentrations of 0.01–10 mg/L also could not noticeably alter the lifespan of nematodes after successive exposure to Gd@C$_{82}$(OH)$_{22}$ for six generations [11].

3.3 Toxicity Assessment of ENMs at Environmentally Relevant Concentrations

The concentrations of ENMs in the real environment are usually in the range of μg/L or ng/L [15–17]. Due to the sensitivity of nematodes to environmental toxicants, several exposure routes have been used to assess the potential toxicity of ENMs at environmentally relevant concentrations [18–22].

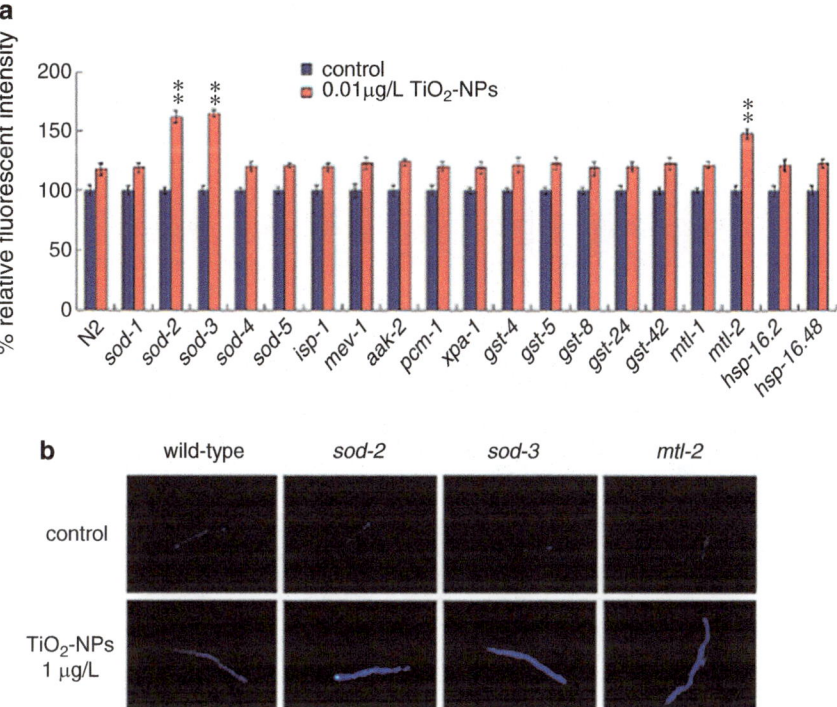

Fig. 3.6 Comparison of intestinal autofluorescence between wild-type and mutants exposed to 1 μg/L of TiO₂-NPs [23]
(**a**) Comparison of intestinal autofluorescence between wild-type and mutant nematodes exposed to TiO₂-NPs. (**b**) Pictures showing the intestinal autofluorescence in wild-type and mutant nematodes. Bars represent means ± SEM. **$P < 0.01$

With the TiO₂-NPs as an example, prolonged exposure from L1-larvae to adult day-1 was employed to determine the long-term effects of TiO₂-NPs at environmentally relevant concentrations on nematodes. After prolonged exposure, TiO₂-NPs at the concentration of 0.01 μg/L could cause the significant reduction in brood size, decrease in locomotion behavior, and induction of intestinal autofluorescence in nematodes (Fig. 3.6) [23]. Using intestinal autofluorescence as the toxicity assessment endpoint, the toxicity of TiO₂-NPs at environmentally relevant concentrations could be even enhanced by *sod-2*, *sod-3*, or *mtl-2* mutation in nematodes (Fig. 3.6) [23].

Further with the TiO₂-NPs as an example, chronic exposure from adult day-1 to adult day-8 was also employed to determine the long-term effects of TiO₂-NPs at environmentally relevant concentrations on nematodes. Using the locomotion behavior as the toxicity assessment endpoint, TiO₂-NPs (4 and 10 nm) at concentrations more than 0.01 μg/L could already significantly decrease the locomotion behavior of nematodes after chronic exposure (Fig. 3.7) [24]. In contrast, after acute

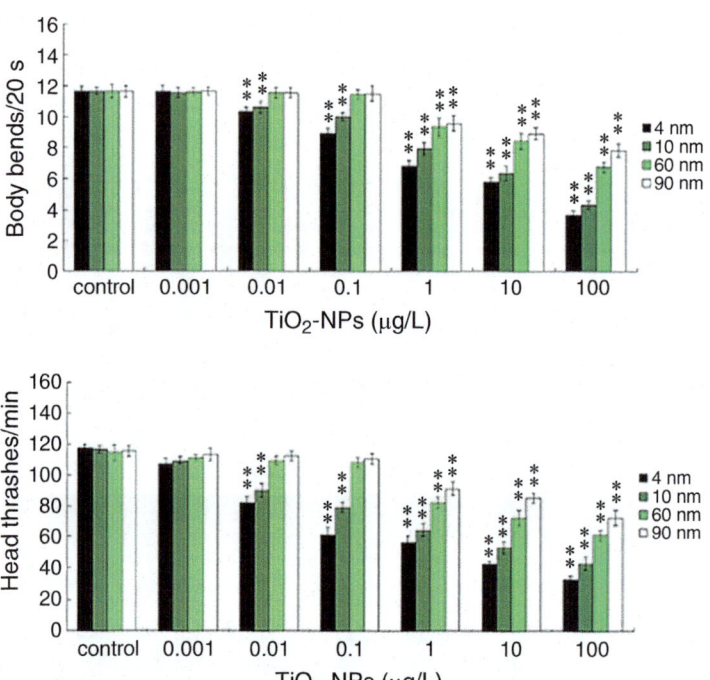

Fig. 3.7 Comparison of locomotion behavior in nematodes chronically exposed to different sizes of TiO$_2$-NPs [24]
Bars represent means ± SEM. **$P < 0.01$

exposure from adult day-1 for 24 h, only TiO$_2$-NPs (4 and 10 nm) at concentrations more than 1 mg/L could significantly decrease the locomotion behavior of nematodes [24].

3.4 Perspectives

We here introduced some useful exposure routes for toxicity assessment of ENMs in nematodes, including acute exposure, prolonged exposure, chronic exposure, one-generation exposure, and transgenerational exposure. Nevertheless, with the concerns on certain research aims, the further modifications of these raised exposure routes are still needed. Additionally, new exposure routes are also welcomed to be further designed in order to satisfy new research aims in nematodes.

In this chapter, we only discussed the value of prolonged exposure and chronic exposure for the toxicity assessment of ENMs at environmentally relevant concentrations in nematodes. Besides these, the possible value of one-generation exposure and multiple-generation exposure for the toxicity assessment of ENMs at environmentally relevant concentrations is also needed to be carefully examined and judged in nematodes. For the use of one-generation exposure or multiple-generation exposure in assessing the possible toxicity of ENMs at environmentally relevant concentrations, both the hormesis effect and the transgenerational effect of ENMs should be carefully considered in nematodes.

References

1. Brenner S (1974) The genetics of *Caenorhabditis elegans*. Genetics 77:71–94
2. Leung MC, Williams PL, Benedetto A, Au C, Helmcke KJ, Aschner M, Meyer JN (2008) *Caenorhabditis elegans*: an emerging model in biomedical and environmental toxicology. Toxicol Sci 106:5–28
3. Zhao Y-L, Wu Q-L, Li Y-P, Wang D-Y (2013) Translocation, transfer, and *in vivo* safety evaluation of engineered nanomaterials in the non-mammalian alternative toxicity assay model of nematode *Caenorhabditis elegans*. RSC Adv 3:5741–5757
4. Wang D-Y (2016) Biological effects, translocation, and metabolism of quantum dots in nematode *Caenorhabditis elegans*. Toxicol Res 5:1003–1011
5. Wu Q-L, Li Y-P, Tang M, Wang D-Y (2012) Evaluation of environmental safety concentrations of DMSA coated Fe_2O_3-NPs using different assay systems in nematode *Caenorhabditis elegans*. PLoS One 7(8):e43729
6. Li W-J, Wang D-Y, Wang D-Y (2018) Regulation of the response of *Caenorhabditis elegans* to simulated microgravity by p38 mitogen-activated protein kinase signaling. Sci Rep 8:857
7. Qu M, Li Y-H, Wu Q-L, Xia Y-K, Wang D-Y (2017) Neuronal ERK signaling in response to graphene oxide in nematode *Caenorhabditis elegans*. Nanotoxicology 11(4):520–533
8. Ding X-C, Wang J, Rui Q, Wang D-Y (2018) Long-term exposure to thiolated graphene oxide in the range of µg/L induces toxicity in nematode *Caenorhabditis elegans*. Sci Total Environ 616–617:29–37
9. Wu Q-L, Zhao Y-L, Fang J-P, Wang D-Y (2014) Immune response is required for the control of *in vivo* translocation and chronic toxicity of graphene oxide. Nanoscale 6:5894–5906
10. Li Y-X, Yu S-H, Wu Q-L, Tang M, Pu Y-P, Wang D-Y (2012) Chronic Al_2O_3-nanoparticle exposure causes neurotoxic effects on locomotion behaviors by inducing severe ROS production and disruption of ROS defense mechanisms in nematode *Caenorhabditis elegans*. J Hazard Mater 219–220:221–230
11. Zhang W, Sun B, Zhang L, Zhao B, Nie G, Zhao Y (2011) Biosafety assessment of Gd@ C82(OH)22 nanoparticles on *Caenorhabditis elegans*. Nanoscale 3:2636–2641
12. Zhang W, Wang C, Li Z, Lu Z, Li Y, Yin J, Zhou Y, Gao X, Fang Y, Nie G, Zhao Y (2012) Unraveling stress-induced toxicity properties of graphene oxide and the underlying mechanism. Adv Mater 24:5391–5397
13. Zhao Y-L, Liu Q, Shakoor S, Gong JR, Wang D-Y (2015) Transgenerational safe property of nitrogen-doped graphene quantum dots and the underlying cellular mechanism in *Caenorhabditis elegans*. Toxicol Res 4:270–280
14. Zhao L, Qu M, Wong G, Wang D-Y (2017) Transgenerational toxicity of nanopolystyrene particles in the range of µg/L in nematode *Caenorhabditis elegans*. Environ Sci Nano 4:2356–2366
15. Tiede K, Hassello¨v M, Breitbarth E, Chaudhry Q, ABA B (2009) Considerations for environmental fate and ecotoxicity testing to support environmental risk assessments for engineered nanoparticles. J Chromatogr A 1216:503–509
16. Mueller N, Nowack B (2008) Exposure modeling of engineered nanoparticles in the environment. Environ Sci Technol 42:4447–4453

17. Gottschalk F, Sonderer T, Scholz RW, Nowack B (2009) Modeled environmental concentrations of engineered nanomaterials (TiO$_2$, ZnO, Ag, CNT, fullerenes) for different regions. Environ Sci Technol 43:9216–9222
18. Zhang H, He X, Zhang Z, Zhang P, Li Y, Ma Y, Kuang Y, Zhao Y, Chai Z (2011) Nano-CeO$_2$ exhibits adverse effects at environmental relevant concentrations. Environ Sci Technol 45(8):3725–3730
19. Li Y-X, Wang W, Wu Q-L, Li Y-P, Tang M, Ye B-P, Wang D-Y (2012) Molecular control of TiO$_2$-NPs toxicity formation at predicted environmental relevant concentrations by Mn-SODs proteins. PLoS One 7(9):e44688
20. Wu Q-L, Nouara A, Li Y-P, Zhang M, Wang W, Tang M, Ye B-P, Ding J-D, Wang D-Y (2013) Comparison of toxicities from three metal oxide nanoparticles at environmental relevant concentrations in nematode *Caenorhabditis elegans*. Chemosphere 90:1123–1131
21. Nouara A, Wu Q-L, Li Y-X, Tang M, Wang H-F, Zhao Y-L, Wang D-Y (2013) Carboxylic acid functionalization prevents the translocation of multi-walled carbon nanotubes at predicted environmental relevant concentrations into targeted organs of nematode *Caenorhabditis elegans*. Nanoscale 5:6088–6096
22. Wu Q-L, Li Y-X, Li Y-P, Zhao Y-L, Ge L, Wang H-F, Wang D-Y (2013) Crucial role of biological barrier at the primary targeted organs in controlling translocation and toxicity of multi-walled carbon nanotubes in nematode *Caenorhabditis elegans*. Nanoscale 5:11166–11178
23. Wu Q-L, Zhao Y-L, Li Y-P, Wang D-Y (2014) Susceptible genes regulate the adverse effects of TiO$_2$-NPs at predicted environmental relevant concentrations on nematode *Caenorhabditis elegans*. Nanomed Nanotechnol Biol Med 10:1263–1271
24. Wu Q-L, Wang W, Li Y-X, Li Y-P, Ye B-P, Tang M, Wang D-Y (2012) Small sizes of TiO$_2$-NPs exhibit adverse effects at predicted environmental relevant concentrations on nematodes in a modified chronic toxicity assay system. J Hazard Mater 243:161–168

Chapter 4
Toxic Effects of Certain Nanomaterials

Abstract Using lethality and some useful sublethal toxicity assessment endpoints and suitable exposure routes, the possible toxic effects of different engineered nanomaterials (ENMs) have been evaluated and determined in *Caenorhabditis elegans*. We here mainly introduce the toxicity assessment of metal nanoparticles, rare earth fluoride nanocrystals, quantum dots, carbon nanotubes, fullerenol, and graphene and its derivatives at different aspects in nematodes. Meanwhile, both the nanotoxicity formation under the oxidative stress condition and the nanotoxicity formation under the susceptible mutation backgrounds in nematodes are also discussed.

Keywords Toxicity assessment · Nanotoxicity · *Caenorhabditis elegans*

4.1 Introduction

In *C. elegans*, some useful sublethal toxicity assessment endpoints and exposure routes have been raised for toxicity assessment and toxicological study of different engineered nanomaterials (ENMs). The values of *C. elegans* in the toxicity assessment and toxicological study of different environmental toxicants are not only due to the typical properties of model animal [1] but also due to the sensitivity to environmental toxicants, especially the toxicants at environmentally relevant concentrations [2, 3].

In this chapter, we will systematically introduce the possible toxic effects of different ENMs on nematodes at different aspects. The mainly introduced ENMs contain metal nanoparticles, rare earth fluoride nanocrystals, quantum dots, carbon nanotubes, fullerenol, and graphene and its derivatives. Moreover, we discussed the nanotoxicity formation under the oxidative stress condition and the nanotoxicity formation under the susceptible mutation backgrounds in nematodes.

© Springer Nature Singapore Pte Ltd. 2018   45
D. Wang, *Nanotoxicology in* Caenorhabditis elegans,
https://doi.org/10.1007/978-981-13-0233-6_4

4.2 Metal Nanoparticles

4.2.1 TiO$_2$-Nanoparticles (TiO$_2$-NPs)

4.2.1.1 TiO$_2$-NPs Toxicity After Acute Exposure

After acute exposure from young adults for 24 h, TiO$_2$-NPs (10 nm) at the concentration of 25 µg/L or 25 mg/L could not affect the survival of nematodes [4].

After acute exposure from young adults for 24 h, TiO$_2$-NPs (10 nm) at the concentration of 25 µg/L or 25 mg/L also could not alter the body length of nematodes [4]. Nevertheless, acute exposure to TiO$_2$-NPs (50 nm) at concentrations more than 47.9 mg/L could significantly decrease the body length of nematodes [6].

After acute exposure from young adults for 24 h, TiO$_2$-NPs (10 nm) at the concentration of 25 mg/L could significantly reduce the brood size in nematodes [4]. Acute exposure to TiO$_2$-NPs (50 nm) at concentrations more than 47.9 mg/L could further significantly reduce both the number of eggs inside the body and the brood size [6]. Moreover, TiO$_2$-NPs (10 nm) at the concentration more than 10 mg/L could significantly decrease the locomotion behavior in nematodes after acute exposure from young adults for 24 h [4, 5].

After acute exposure from young adults for 24 h, TiO$_2$-NPs (10 nm) at the concentration of 25 mg/L or 100 mg/L could further induce the significant induction of intestinal autofluorescence and intestinal reactive oxygen species (ROS) production and reduce the brood size in nematodes [4, 5].

Moreover, after acute exposure from young adults for 24 h, TiO$_2$-NPs (10 nm) at the concentration of 25 mg/L could significantly increase the expressions of *sod-1*, *sod-2*, and *sod-3* encoding the SODs proteins with the functions to counteract the activation of oxidative stress in nematodes [4]. Acute exposure to TiO$_2$-NPs also affect the expression of *cyp35a2*, which may be involved in the fat storage control in nematodes [7].

Besides these, after acute exposure for 24 h, TiO$_2$-NPs (10 nm) at the concentration of 100 mg/L also significantly enhanced the intestinal permeability as indicated by the Nile red labeling [5]. Furthermore, after acute exposure for 24 h, TiO$_2$-NPs (10 nm) at the concentration of 100 mg/L could further significantly increase the mean defecation cycle length and cause the damage on the development of AVL and DVB neurons as reflected by the reduction in size of fluorescent puncta of AVL and DVB neurons [5].

4.2.1.2 TiO$_2$-NPs Toxicity After Prolonged Exposure

After prolonged exposure from L1-larvae to adult day-1, TiO$_2$-NPs (60 and 90 nm) at concentrations of 0.001–10 µg/L did not obviously affect the survival, whereas TiO$_2$-NPs (4 and 10 nm) at concentrations more than 0.1 µg/L and TiO$_2$-NPs (30 nm) at the concentration of 50 µg/L could significantly decrease the survival of nematodes [8, 9].

After prolonged exposure from L1-larvae to adult day-1, TiO_2-NPs (60 and 90 nm) at concentrations of 0.001–10 µg/L did not affect the body length; however, TiO_2-NPs (4 and 10 nm) at concentrations more than 0.1 µg/L and TiO_2-NPs (30 nm) at the concentration of 50 µg/L could significantly decrease the body length of nematodes [8, 9].

After prolonged exposure from L1-larvae to adult day-1, TiO_2-NPs (60 and 90 nm) at concentrations more than 1 µg/L, TiO_2-NPs (30 nm) at concentrations more than 0.5 µg/L, and TiO_2-NPs (4 and 10 nm) at concentrations more than 0.01 µg/L could significantly reduce the brood size in nematodes [8, 9]. Moreover, TiO_2-NPs (60 and 90 nm) at concentrations more than 0.1 µg/L, TiO_2-NPs (30 nm) at concentrations more than 0.05 µg/L, and TiO_2-NPs (4 and 10 nm) at concentrations more than 0.001 µg/L could significantly decrease the locomotion behavior in nematodes after prolonged exposure from L1-larvae to adult day-1 [8, 9].

After prolonged exposure from L1-larvae to adult day-1, TiO_2-NPs (60 and 90 nm) at concentrations more than 10 µg/L and TiO_2-NPs (4 and 10 nm) at concentrations more than 1 µg/L could induce the significant intestinal autofluorescence in nematodes [8]. Moreover, TiO_2-NPs (60 and 90 nm) at concentrations more than 1 µg/L, TiO_2-NPs (30 nm) at concentrations more than 0.05 µg/L, and TiO_2-NPs (4 and 10 nm) at concentrations more than 0.01 µg/L could induce the significant intestinal ROS production in nematodes after prolonged exposure from L1-larvae to adult day-1 [8, 9].

Furthermore, TiO_2-NPs (60 and 90 nm) at concentrations more than 1 µg/L and TiO_2-NPs (4 and 10 nm) at concentrations more than 0.001 µg/L could significantly increase the expressions of *sod-2* and *sod-3* [8].

4.2.1.3 TiO_2-NPs Toxicity After Chronic Exposure

After chronic exposure from adult day-1 to adult day-8, TiO_2-NPs (60 and 90 nm) at concentrations more than 100 µg/L and TiO_2-NPs (4 and 10 nm) at concentrations more than 10 µg/L could significantly decrease the survival of nematodes [10].

After chronic exposure from adult day-1 to adult day-8, TiO_2-NPs (60 and 90 nm) at concentrations more than 100 µg/L and TiO_2-NPs (4 and 10 nm) at concentrations more than 10 µg/L could also significantly decrease the body length of nematodes [10].

TiO_2-NPs (60 and 90 nm) at concentrations more than 1 µg/L and TiO_2-NPs (4 and 10 nm) at concentrations more than 0.01 µg/L could significantly decrease the locomotion behavior in nematodes after chronic exposure from adult day-1 to adult day-8 [10].

After chronic exposure from adult day-1 to adult day-8, TiO_2-NPs (4 and 10 nm) at concentrations more than 1 µg/L could induce the significant intestinal autofluorescence in nematodes [10]. Moreover, TiO_2-NPs (60 and 90 nm) at concentrations more than 1 µg/L and TiO_2-NPs (4 and 10 nm) at concentrations more than 0.01 µg/L also induced the significant intestinal ROS production in nematodes after chronic exposure from adult day-1 to adult day-8 [10].

4.2.2 Silver Nanoparticles (Ag-NPs)

4.2.2.1 Ag-NPs Toxicity After Acute Exposure

After acute exposure for 24 h, Ag-NPs (14 nm) at concentrations of 0.05–0.5 mg/L did not affect the survival and the growth in nematodes [11]. In contrast, acute exposure to Ag-NPs for 24 h at the concentration of 50 mg/L could significantly decrease the body length in nematodes [12]. Additionally, acute exposure to Ag-NPs ($\geq$ 5 mg/L) for 24 h could obviously decrease the survival of nematodes [13, 15]. After acute exposure for 24 h, Ag-NPs at concentrations more than 0.05 mg/L could also significantly reduce the brood size in nematodes [11, 14, 15].

After acute exposure to Ag-NPs for 24 h, the epidermal edema and burst were detected in exposed nematodes, which may be associated with secondary infections in the ecosystems [15].

After acute exposure for 24 h, Ag-NPs (14 nm) at the concentration of 0.5 ppm could significantly increase the expressions of *M162.5*, *mtl-2*, *sod-3*, and *daf-12* in nematodes [11]. Moreover, acute exposure to Ag-NPs (1 mg/L) for 24 h could also significantly increase both the transcriptional expression and the translational expression of *pmk-1* encoding a p38 MAPK in p38 MAPK signaling pathway and the transcriptional expression of *hif-1* encoding a hypoxia-induced factor in nematodes [14].

4.2.2.2 Ag-NPs Toxicity After Prolonged Exposure

After prolonged exposure for 72 h, Ag-NPs (14 nm) at concentrations of 0.05–0.5 mg/L could not alter the survival and the growth in nematodes [11]. In contrast, after prolonged exposure for 72 h, Ag-NPs (14 nm) at concentrations more than 0.1 mg/L could significantly reduce the brood size in nematodes [11]. Prolonged exposure to Ag-NPs ($\geq$3 mg/L) for 72 h could significantly decrease the survival of nematodes [13].

4.2.2.3 Ag-NPs Toxicity After Multigenerational Exposure

After exposure for multiple generations (F1–F3), the smallest Ag-NPs particle, at 2 nm, had a notable impact on brood size in nematodes [16]. In contrast, the largest Ag-NPs particle, at 10 nm, could significantly reduce the lifespan of parent nematodes (P0) by 28.8% and over the span of three generations (F1–F3) [16].

The nematode populations were also initially exposed over six generations, but kept unexposed for the subsequent four generations to allow the recovery from the exposure [17]. Toxicity assessment suggested that continuous exposure to Ag-NPs caused the pronounced sensitization in the F2 generation, which could be sustained until the F10 generation [17].

4.2.3 ZnO-NPs

4.2.3.1 ZnO-NPs Toxicity After Acute Exposure

After acute exposure for 24 h, ZnO-NPs (1.5 nm) at concentrations more than 600 mg/L could significantly decrease the survival of nematodes [18]. After acute exposure for 24 h, ZnO-NPs (20 nm) at concentrations more than 4.1 mg/L significantly decreased the body length in nematodes [6].

After acute exposure for 24 h, ZnO-NPs (1.5 nm) at concentrations more than 30 mg/L could significantly reduce the brood size in nematodes [18]. Additionally, after acute exposure for 24 h, ZnO-NPs (20 nm) at concentrations more than 1.6 mg/L significantly reduced the number of eggs inside the body, and ZnO-NPs (20 nm) at concentrations more than 0.8 mg/L significantly reduced the brood size in nematodes [6].

After acute exposure for 24 h, ZnO-NPs (1.5 nm) at concentrations more than 30 mg/L further significantly increased the expression of MTL-2::GFP in nematodes [18].

4.2.3.2 ZnO-NPs Toxicity After Prolonged Exposure

After prolonged exposure from L1-larvae to adult day-1, ZnO-NPs (30 nm) at the concentration of 50 µg/L could obviously decrease the survival of nematodes [9]. After prolonged exposure from L1-larvae to adult day-1, ZnO-NPs (30 nm) at the concentration of 50 µg/L also significantly decreased the body length in nematodes [9].

After prolonged exposure from L1-larvae to adult day-1, ZnO-NPs (30 nm) at concentrations more than 0.5 µg/L could significantly reduce the brood size [9]. Moreover, after prolonged exposure from L1-larvae to adult day-1, ZnO-NPs (30 nm) at concentrations more than 0.05 µg/L could significantly decrease the locomotion behavior in nematodes [9].

After prolonged exposure from L1-larvae to adult day-1, ZnO-NPs (30 nm) at concentrations more than 0.05 µg/L could induce the significant intestinal ROS production in nematodes [9].

After prolonged exposure from L1-larvae to L4-larvae, ZnO-NPs (30 nm) at the concentration of 50 mg/L could induce the significant increase in expressions of both MTL-2::GFP and PCS-1::GFP in nematodes [19]. After prolonged exposure from L1-larvae to L4-larvae, ZnO-NPs (30 nm) at the concentration of 5 mg/L might also obviously reduce the lifespan in nematodes [19].

4.2.4 Al_2O_3-NPs

4.2.4.1 Al_2O_3-NPs Toxicity After Acute Exposure

After acute exposure from L4-larvae or young adults for 24 h, Al_2O_3-NPs (60 nm) at concentrations more than 12.7 mg/L could significantly decrease the survival of nematodes [20]. In contrast, after acute exposure from L1-larvae for 24 h, Al_2O_3-NPs (60 nm) at concentrations more than 6.3 mg/L could significantly decrease the survival of nematodes [20].

After acute exposure for 24 h, Al_2O_3-NPs (60 nm) at concentrations more than 102 mg/L could significantly decrease the body length in nematodes [6].

After acute exposure for 24 h, Al_2O_3-NPs (60 nm) at concentrations more than 102 mg/L could also significantly decrease the number of eggs inside the body, and Al_2O_3-NPs (60 nm) at concentrations more than 51 mg/L could significantly reduce the brood size in nematodes [6].

After acute exposure from L4-larvae for 24 h, Al_2O_3-NPs (60 nm) at concentrations more than 1 mg/L induced the significant neuronal loss and increase in the number of gaps in ventral cord or dorsal cord in the GABAergic neurons in nematodes [21]. Meanwhile, after acute exposure from L4-larvae for 24 h, Al_2O_3-NPs (60 nm) at concentrations more than 1 mg/L induced the deficit in thermotaxis learning, and Al_2O_3-NPs (60 nm) at concentrations more than 10 mg/L induced the deficits in both the thermotaxis perception and the locomotion behavior [21]. After acute exposure from L4-larvae for 24 h, Al_2O_3-NPs (60 nm) at the concentration of 10 mg/L also caused the damage on the development of AFD sensory neurons and AIY interneurons as reflected by the morphology, fluorescence size, and intensity of cell bodies of AFD and/or AIY neurons and the expressions of genes controlling the cell identity of AFD and AIY neurons [21]. After acute exposure from L4-larvae for 24 h, Al_2O_3-NPs (60 nm) at the concentration of 10 mg/L further induced the deficits in both the presynaptic and the postsynaptic functions as indicated by the significant resistance to both aldicarb and levamisole [21].

After acute exposure from L4-larvae or young adults for 24 h, Al_2O_3-NPs (60 nm) at concentrations more than 12.7 mg/L could induce the significant intestinal autofluorescence in nematodes [20]. In contrast, after acute exposure from L1-larvae for 24 h, Al_2O_3-NPs (60 nm) at concentrations more than 6.3 mg/L could induce the significant intestinal autofluorescence in nematodes [20]. Moreover, after acute exposure from L4-larvae for 24 h, Al_2O_3-NPs (60 nm) at the concentration of 10 mg/L also induced the significant intestinal ROS production [21]. Additionally, after acute exposure from L4-larvae for 24 h, Al_2O_3-NPs (60 nm) at the concentration of 10 mg/L obviously enhanced the intestinal permeability as indicated by Nile red staining in nematodes [21]. More importantly, after acute exposure from L4-larvae for 24 h, the developmental deficits in intestinal microvilli, such as the loss of microvilli, were observed in Al_2O_3-NPs (60 nm) exposed nematodes [21].

After acute exposure from L4-larvae or young adults for 24 h, Al_2O_3-NPs (60 nm) at concentrations more than 51 mg/L could induce the significant increase in expression of HSP-16.2::GFP in nematodes [20]. In contrast, after acute exposure from L1-larvae for 24 h, Al_2O_3-NPs (60 nm) at concentrations more than 6.3 mg/L could induce the significant increase in expression of HSP-16.2::GFP in nematodes [20].

4.2.4.2 Al_2O_3-NPs Toxicity After Prolonged Exposure

After prolonged exposure from L1-larvae to adult day-1, Al_2O_3-NPs (60 nm) at the concentration of 500 µg/L could significantly decrease the survival of nematodes [22].

After prolonged exposure from L1-larvae to adult day-1, Al_2O_3-NPs (60 nm) at concentrations more than 5 µg/L could significantly decrease the locomotion behavior in nematodes [22].

After prolonged exposure from L1-larvae to adult day-1, Al_2O_3-NPs (60 nm) at the concentration of 500 µg/L could further induce the significant intestinal autofluorescence in nematodes [22].

4.2.4.3 Al_2O_3-NPs Toxicity After Chronic Exposure

After chronic exposure from young adults for 10 days, Al_2O_3-NPs (60 nm) at the concentration of 8.1 mg/L could result in the significant decrease in locomotion behavior in nematodes [23].

After chronic exposure from young adults for 10 days, Al_2O_3-NPs (60 nm) at the concentration of 8.1 mg/L could induce the significant intestinal autofluorescence in nematodes [24]. Additionally, after chronic exposure from young adults for 10 days, Al_2O_3-NPs (60 nm) at the concentration of 8.1 mg/L could further induce the significant intestinal ROS production and oxidative damage in nematodes [23, 24]. Moreover, after chronic exposure from young adults for 10 days, Al_2O_3-NPs (60 nm) at the concentration of 8.1 mg/L significantly decreased the SOD activity and the expressions of genes (*sod-2* and *sod-3*) encoding Mn-SODs in nematodes [23].

After chronic exposure from young adults for 10 days, Al_2O_3-NPs (60 nm) at the concentration of 8.1 mg/L caused the significant increase in expression of HSP-16.2::GFP in nematodes [23, 24].

4.2.5 SiO_2-NPs

4.2.5.1 SiO_2-NPs Toxicity After Acute Exposure

After acute exposure, SiO_2-NPs at the concentration of 2.5 g/L could cause the significant reduction in brood size and the obvious formation of bag of worms (BOW) phenotype in nematodes [25]. Besides this, acute exposure to SiO_2-NPs at the concentration of 2.5 g/L further induced an untimely accumulation of insoluble ubiquitinated proteins, nuclear amyloid, and reduction of pharyngeal pumping in nematodes [26].

4.2.5.2 SiO$_2$-NPs Toxicity After Prolonged Exposure

In nematodes, prolonged exposure (from L1-larvae to adult day-1) to SiO$_2$-NPs (30 nm) at concentrations of 0.05–50 µg/L did not significantly affect the survival and the body length in nematodes [9].

In contrast, prolonged exposure (from L1-larvae to adult day-1) to SiO$_2$-NPs (30 nm) at the concentration of 50 µg/L could significantly reduce the brood size [9]. Moreover, prolonged exposure (from L1-larvae to adult day-1) to SiO$_2$-NPs (30 nm) at concentrations more than 5 µg/L could significantly decrease the locomotion behavior in nematodes [9].

In nematodes, after prolonged exposure from L1-larvae to adult day-1, SiO$_2$-NPs (30 nm) at concentrations more than 5 µg/L could induce the significant intestinal ROS production in nematodes [9].

4.2.6 CeO$_2$-NPs

4.2.6.1 CeO$_2$-NPs Toxicity After Acute Exposure

Acute exposure to CeO$_2$-NPs (10–30 nm) for 24 h at concentrations of 1.72–17.21 mg/L could significantly decrease the body length in nematodes [27]. In contrast, acute exposure to CeO$_2$-NPs (10–30 nm) for 24 h at concentrations of 0.17–17.21 mg/L did not obviously affect the lifespan in nematodes [27].

Acute exposure to CeO$_2$-NPs (10–30 nm) for 24 h at concentrations of 0.17–17.21 mg/L could significantly reduce the brood size in nematodes [27].

After acute exposure for 24 h, CeO$_2$-NPs (10–30 nm) at concentrations more than 0.17 mg/L induced the significant increase in expression of HSP-4::GFP, and CeO$_2$-NPs (10–30 nm) at concentrations more than 1.72 mg/L induced the significant increase in expression of GST-4::GFP [27].

More importantly, acute exposure to CeO$_2$-NPs (10–30 nm) for 24 h at concentrations of 1.72–17.21 mg/L could even decrease the thermotolerance of nematodes [27].

4.2.6.2 CeO$_2$-NPs Toxicity After Prolonged Exposure

After prolonged exposure from L1-larvae for 3 days, CeO$_2$-NPs (8.5 nm) at concentrations more than 1 nM could significantly reduce the lifespan in nematodes [28].

After prolonged exposure from L1-larvae for 3 days, CeO_2-NPs (8.5 nm) at concentrations more than 5 nM induced the significant intestinal autofluorescence in nematodes [28].

After prolonged exposure from L1-larvae for 3 days, CeO_2-NPs (8.5 nm) at concentrations more than 1 nM even decreased the thermotolerance of nematodes [28].

4.2.7 DMSA Coated Fe_2O_3-NPs

4.2.7.1 DMSA Coated Fe_2O_3-NPs Toxicity After Acute Exposure

After acute exposure from L4-larvae for 24 h, DMSA coated Fe_2O_3-NPs at concentrations of 0.5–100 mg/L did not obviously affect the survival of nematodes [29]. After acute exposure from L4-larvae for 24 h, DMSA coated Fe_2O_3-NPs only at the concentration of 100 mg/L significantly decreased the body length in nematodes [29].

After acute exposure from L4-larvae for 24 h, DMSA coated Fe_2O_3-NPs at the concentration of 100 mg/L could significantly reduce the brood size in nematodes [29]. Additionally, after acute exposure from L4-larvae for 24 h, DMSA coated Fe_2O_3-NPs at concentrations more than 50 mg/L could significantly decrease the locomotion behavior in nematodes [29].

Moreover, after acute exposure from L4-larvae for 24 h, DMSA coated Fe_2O_3-NPs at the concentration of 100 mg/L induced the significant intestinal autofluorescence in nematodes [29]. After acute exposure from L4-larvae for 24 h, DMSA coated Fe_2O_3-NPs at concentrations more than 50 mg/L induced the significant intestinal ROS production in nematodes [29]. After acute exposure from L4-larvae for 24 h, DMSA coated Fe_2O_3-NPs at the concentration of 100 mg/L could further significantly decrease the pumping rate and increase the mean defecation cycle length in nematodes [29].

4.2.7.2 DMSA Coated Fe_2O_3-NPs Toxicity After Prolonged Exposure

After prolonged exposure from L1-larvae to adult day-1, DMSA coated Fe_2O_3-NPs at concentrations of 1–5000 µg/L did not obviously affect the survival of nematodes [29]. After prolonged exposure from L1-larvae to adult day-1, DMSA coated Fe_2O_3-NPs only at the concentration of 5000 µg/L significantly decreased the body length in nematodes [29].

After prolonged exposure from L1-larvae to adult day-1, DMSA coated Fe_2O_3-NPs at concentrations more than 500 µg/L could significantly reduce the brood size in nematodes [29]. Additionally, after prolonged exposure from L1-larvae to adult day-1, DMSA coated Fe_2O_3-NPs at concentrations more than 500 µg/L could also significantly decrease the locomotion behavior in nematodes [29].

Moreover, after prolonged exposure from L1-larvae to adult day-1, DMSA coated Fe_2O_3-NPs at the concentration of 5000 µg/L induced the significant intestinal autofluorescence in nematodes [29]. After prolonged exposure from L1-larvae to adult day-1, DMSA coated Fe_2O_3-NPs at concentrations more than 500 µg/L induced the significant intestinal ROS production in nematodes [29]. After prolonged exposure from L1-larvae to adult day-1, DMSA coated Fe_2O_3-NPs at the concentration of 5000 µg/L could further significantly decrease the pumping rate and increase the mean defecation cycle length in nematodes [29].

4.2.7.3 DMSA Coated Fe_2O_3-NPs Toxicity After Chronic Exposure

After chronic exposure from L1-larvae to adult day-8, DMSA coated Fe_2O_3-NPs at the concentration of 5000 µg/L could obviously decrease the survival of nematodes [29]. After chronic exposure from L1-larvae to adult day-8, DMSA coated Fe_2O_3-NPs only at concentrations more than 500 µg/L significantly decreased the body length in nematodes [29].

After chronic exposure from L1-larvae to adult day-8, DMSA coated Fe_2O_3-NPs at concentrations more than 100 µg/L could significantly decrease the locomotion behavior in nematodes [29].

Moreover, after chronic exposure from L1-larvae to adult day-8, DMSA coated Fe_2O_3-NPs at concentrations more than 100 µg/L induced the significant intestinal autofluorescence in nematodes [29]. After chronic exposure from L1-larvae to adult day-8, DMSA coated Fe_2O_3-NPs at concentrations more than 100 µg/L induced the significant intestinal ROS production in nematodes [29]. After chronic exposure from L1-larvae to adult day-8, DMSA coated Fe_2O_3-NPs at concentrations more than 500 µg/L could further significantly decrease the pumping rate and increase the mean defecation cycle length in nematodes [29].

4.2.8 Au-NPs

4.2.8.1 Toxicity of Au-NPs After Acute Exposure

After acute exposure from L3-larvae for 12 h, citrate-coated Au-NPs (4 nm) at least at concentrations of 10 mg/L significantly decreased the survival of nematodes [30].

After acute exposure from L3-larvae for 12 h, citrate-coated Au-NPs (4 nm) at the concentration of 5.9 mg/L could further significantly increase expressions of *abu-11*, *dyn-1*, *act-5*, *apl-1*, and *hsp-4* in nematodes [30].

4.2.8.2   Transgenerational Effect of Au-NPs on Nematodes

The acute exposure to Au-NP (10 nm) was performed for 12 h in P0 generation, and then the nematodes were transferred to new NGM plates without the addition of Au-NPs [31]. In nematodes, no significant alteration in the survival rate under all generations was induced by maternal exposure to P0 generation [31]. In contrast, the reproduction rate was obviously affected in the F1 generation, and it was gradually recovered in the F2 and the F3 generations [31], which might be largely due to the germline transfer and the possible damage on the gonad and the embryo germ cells during their development after maternal exposure of Au-NPs.

4.3   Rare Earth Fluoride Nanocrystals (NCs)

After acute exposure from L4-larvae or young adults for 6 h, NaYF4:Yb,Tm NCs might be not able to affect the ELT-2::GFP expression, lifespan, egg production, egg viability, and growth rate in nematodes [32]. In contrast, after acute exposure for 24 h, NaYF4:Yb at concentrations more than 2.5 mg/L significantly decreased the survival of nematodes [33].

4.4   Quantum Dots (QDs)

4.4.1   Toxicity of QDs After Acute Exposure

After acute exposure for 24 h, MPA-modified CdTe QDs (5–6 nm) at the concentration of 200 nM could obviously decrease the body length in nematodes [34].

After acute exposure for 24 h, MPA-modified CdTe QDs (5–6 nm) at the concentration of 200 nM could also significantly reduce the brood size in nematodes [34].

After acute exposure for 24 h, MPA-modified CdTe QDs (5–6 nm) at the concentration of 200 nM could further significantly reduce the lifespan in nematodes [34].

Moreover, after acute exposure from L2-larvae for 24–h, CdTe QDs invasion could disrupt key subcellular processes, such as the endocytic recycling, the nutrition storage, and the lysosome formation in the intestinal cells of nematodes [35].

4.4.2   Toxicity of QDs After Prolonged Exposure

After prolonged exposure from L1-larvae to young adults, CdTe QDs (3.7 nm) at concentrations more than 10 μg/L could significantly decrease the survival of nematodes [36]. After prolonged exposure from L1-larvae to young adults, CdTe QDs with different sizes could obviously decrease the body length in nematodes [35].

After prolonged exposure from L1-larvae to young adults, CdTe QDs (3.7 nm) at the concentration of 100 μg/L induced the formation of obvious paralysis in nematodes [36]. After prolonged exposure from L1-larvae to young adults, CdTe QDs (3.7 nm) at concentrations more than 0.1 μg/L caused the damage on D-type motor neurons in GABAergic neurons as reflected by the relative fluorescence intensity of cell body, the neuronal loss, and the number of gaps on both ventral cord and dorsal cord, and CdTe QDs (3.7 nm) at concentrations more than 0.1 μg/L induced the formation of shrinker behavior in nematodes [36]. Similarly, after prolonged exposure from L1-larvae to young adults, CdTe QDs (3.7 nm) at concentrations more than 0.1 μg/L caused the damage on RMEs motor neurons in GABAergic neurons as reflected by the relative fluorescence intensity and the relative fluorescence size of cell body, and CdTe QDs (3.7 nm) at concentrations more than 0.1 μg/L induced the formation of abnormal foraging behavior in nematodes [37]. Additionally, after prolonged exposure from L1-larvae to young adults, CdTe QDs (3.7 nm) at the concentration of 1 μg/L caused the damage on AVL and DVB neurons in GABAergic neurons as also reflected by the relative fluorescence intensity and the relative fluorescence size of cell body, and CdTe QDs (3.7 nm) at concentrations more than 0.1 μg/L induced the significant increase in mean defection cycle length in nematodes [37]. Prolonged exposure to CdTe QDs (1 μg/L) further significantly decreased the expression of *unc-30*, *unc-25*, and *unc-47*, which regulate the synthesis and transport of GABA neurotransmitter, as well as the cell identity of GABAergic neurons [36]. Besides these, after prolonged exposure from L1-larvae to young adults, CdTe QDs (3.7 nm) at least at concentrations more than 2.5 mg/L resulted the significant decrease in locomotion behavior in nematodes [38].

Furthermore, after prolonged exposure from L1-larvae to young adults, CdTe QDs (3.7 nm) at least at concentrations more than 2.5 mg/L induced the significant intestinal ROS production in nematodes [38].

After prolonged exposure from L1-larvae to young adults, CdTe QDs with different sizes could also cause the significant reduction in lifespan in nematodes [35].

4.4.3   Transgenerational Toxicity of QDs in Nematodes

After prolonged exposure from L1-larvae to young adults, the significant decrease in the locomotion behavior could be observed in the progeny of CdTe QDs (10 or 20 mg/L) exposed nematodes [38]. Similarly, after prolonged exposure from L1-larvae to young adults, the significant induction of intestinal ROS production could also be detected in the progeny of CdTe QDs (10 or 20 mg/L) exposed nematodes [38], which implies that the CdTe QDs at high concentrations may induce the transgenerational toxicity in nematodes.

4.4.4 Toxicity of QDs After Multigenerational Exposure

After four-generation exposure, the biological effects of CdSe QDs (3.4 nm) at the highest concentrations for the first generation were commonly similar in magnitude to those found in future generations using development, reproduction, and locomotion as the toxicity assessment endpoints in nematodes [39], which implies that the nematodes may have the potential to adapt or acclimate to the presence of low levels of trace metals.

4.5 Carbon Nanotubes

4.5.1 Single-Walled Carbon Nanotubes (SWCNTs)

After exposure from L1-larvae for 48 h, α-SWCNTs at concentrations more than 250 mg/L could significantly decrease the survival rate of nematodes [40]. After exposure from L1-larvae for 48 h, α-SWCNTs at the concentration of 500 mg/L could significantly decrease the body length in nematodes [40].

After exposure from L1-larvae for 48 h, α-SWCNTs at the concentration of 500 mg/L also significantly reduced the brood size in nematodes [40].

After exposure from L1-larvae for 48 h, α-SWCNTs at the concentration of 500 mg/L could cause the significant reduction in oxygen consumption and nuclear translocation of DAF-16::GFP in nematodes [40]. Besides this, after exposure from L1-larvae for 48 h, α-SWCNTs at the concentration of 500 mg/L could also induce the defective endocytosis in nematodes [40].

After exposure from L1-larvae for 48 h, α-SWCNTs at the concentration of 500 mg/L could further significantly reduce the lifespan and induce the BOW phenotype in nematodes [40].

4.5.2 Multiwalled Carbon Nanotubes (MWCNTs)

4.5.2.1 Toxicity of MWCNTs After Acute Exposure

After acute exposure from L4-larvae for 24 h, MWCNTs at concentrations more than 1 mg/L could significantly reduce the brood size in nematodes [41]. Additionally, after acute exposure from L4-larvae for 24 h, MWCNTs at concentrations more than 1 mg/L could also significantly decrease the locomotion behavior in nematodes [41].

After acute exposure from L4-larvae for 24 h, MWCNTs at concentrations more than 1 mg/L could induce the significant intestinal ROS production in nematodes [41].

Moreover, after acute exposure from L4-larvae for 24 h, MWCNTs at concentrations more than 1 mg/L could obviously reduce the lifespan in nematodes [41].

4.5.2.2   Toxicity of MWCNTs After Prolonged Exposure

After prolonged exposure from L1-larvae to adult day-1, MWCNTs at concentrations more than 0.1 µg/L could cause the significant reduction in brood size in nematodes [42, 43]. Moreover, after prolonged exposure from L1-larvae to adult day-1, MWCNTs at concentrations more than 0.01 µg/L significantly decreased the locomotion behavior in nematodes [42, 43].

After prolonged exposure from L1-larvae to adult day-1, MWCNTs at concentrations more than 1 µg/L could induce the significant intestinal autofluorescence in nematodes [42, 43]. Additionally, after prolonged exposure from L1-larvae to adult day-1, MWCNTs at concentrations more than 0.1 µg/L could induce the significant intestinal ROS production in nematodes [42, 43].

In nematodes, after prolonged exposure from L1-larvae to adult day-1, MWCNTs at the concentration of 1 mg/L could obviously enhance the intestinal permeability as indicated by the Nile red staining signals in nematodes [42, 43].

4.6   Fullerenol

After exposure from L4-larvae, the hydroxylated fullerene (fullerol, $C_{60}(OH)19$–24) at concentrations more than 1 mg/L could obviously decrease the survival of nematodes [44]. After exposure from L4-larvae, the hydroxylated fullerene (fullerol, $C_{60}(OH)19$–24) at the concentration of 100 mg/L could cause the significant decrease in body length in nematodes [44]. After exposure from L4-larvae, the hydroxylated fullerene (fullerol, $C_{60}(OH)19$–24) at the concentration of 100 mg/L could also significantly reduce the lifespan of nematodes [44]. In contrast, after exposure from L1-larvae for 3 days, the polyhydroxylated fullerene at concentrations of 0.01–100 µM could not affect the survival of nematodes [45]. After exposure from L1-larvae for 3 days, the polyhydroxylated fullerene at concentrations of 0.01–100 µM could also not influence the body length in nematodes [45].

After exposure from L4-larvae, the hydroxylated fullerene (fullerol, $C_{60}(OH)19$–24) at the concentration of 100 mg/L caused the significant reduction in the brood size in nematodes [44]. After exposure from L1-larvae for 3 days, the polyhydroxylated fullerene at the concentration of 1 µM could significantly reduce the brood size and decrease the locomotion behavior in nematodes [45].

After exposure from L4-larvae, the hydroxylated fullerene (fullerol, $C_{60}(OH)19$–24) at the concentration of 100 mg/L could induce the malfunction in digestive system in nematodes [44]. After exposure from L1-larvae for 3 days, the polyhydroxylated fullerene at the concentration of 1 µM could obviously affect the pumping rate in nematodes [45].

After exposure from L4-larvae, the hydroxylated fullerene (fullerol, C60(OH)19–24) at the concentration of 100 mg/L could further induce the apoptotic cell corpses in the body [44].

4.7   Graphene and Its Derivatives

4.7.1   Graphene Oxide (GO)

4.7.1.1   Toxicity of GO After Acute Exposure

After acute exposure from L4-larvae for 24 h, GO at concentrations more than 100 mg/L could significantly reduce the brood size in nematodes [46]. Meanwhile, after acute exposure from L4-larvae for 24 h, GO at concentrations more than 10 mg/L could significantly decrease the locomotion behavior in nematodes [46].

In nematodes, after acute exposure from L4-larvae for 24 h, GO at concentrations more than 100 mg/L could induce the significant intestinal autofluorescence in nematodes [46]. Moreover, after acute exposure from L4-larvae for 24 h, GO at concentrations more than 10 mg/L could induce the significant intestinal ROS production in nematodes [46].

4.7.1.2   Toxicity of GO After Prolonged Exposure

After prolonged exposure from L1-larvae to adult day-1, GO at concentrations more than 1 mg/L could significantly reduce the brood size in nematodes [46]. Meanwhile, after prolonged exposure from L1-larvae to adult day-1, GO at concentrations more than 0.5 mg/L could significantly decrease the locomotion behavior in nematodes [46].

In nematodes, after prolonged exposure from L1-larvae to adult day-1, GO at concentrations more than 1 mg/L could induce the significant intestinal autofluorescence [46]. Moreover, after prolonged exposure from L1-larvae to adult day-1, GO at concentrations more than 0.5 mg/L could induce the significant intestinal ROS production in nematodes [46]. Prolonged exposure to GO (100 mg/L) could further result in the disrupted ultrastructure of microvilli and the loss of many microvilli on intestinal cells and enhanced intestinal permeability as indicated by Nile red staining signals in nematodes [46]. Furthermore, prolonged exposure to GO (100 mg/L) could also significantly increase the mean defecation cycle length in nematodes [46].

4.7.1.3   Toxicity of GO After Chronic Exposure

After chronic exposure from adult day-1 to adult day-8, GO at concentrations of 0.001–1 mg/L could not obviously affect the survival of nematodes [47]. In contrast, after chronic exposure from L1-larvae to adult day-8, GO at concentrations more than 1 mg/L could significantly decrease the survival of nematodes [47].

After chronic exposure from adult day-1 to adult day-8, GO at concentrations more than 1 mg/L could significantly decrease the locomotion behavior in nematodes [47]. In contrast, after chronic exposure from L1-larvae to adult day-8, GO at concentrations more than 0.01 mg/L could significantly decrease the locomotion behavior in nematodes [47].

After chronic exposure from adult day-1 to adult day-8, GO at concentrations more than 1 mg/L could induce the significant intestinal autofluorescence and intestinal ROS production in nematodes [47]. In contrast, after chronic exposure from L1-larvae to adult day-8, GO at concentrations more than 0.01 mg/L could induce the significant intestinal autofluorescence and intestinal ROS production in nematodes [47].

After chronic exposure from L1-larvae to adult day-8, GO at the concentration of 1 mg/L could significantly decrease the expressions of *F08G5.6*, *pqm-1*, *K11D12.5*, *prx-11*, *spp-1*, *lys-7*, *lys-2*, *abf-2*, *acdh-1*, and *lys-8*, as well as the expression of P*nlp-29::GFP*, and these genes encode the antimicrobial peptides in nematodes [47], which implies the suppression of innate immune response in nematodes after chronic exposure to GO.

Moreover, after chronic exposure from L1-larvae to adult day-8, GO (1 mg/L) could cause the significant increase in mean defecation cycle length and the damage on the development of AVL and DVB neurons controlling the defecation behavior as reflected by the reduced fluorescence size of cell bodies of neurons in nematodes [47].

4.7.2   Thiolated GO (GO-SH)

After prolonged exposure from L1-larvae to adult day-1, GO-SH concentrations more than 0.1 mg/L could significantly reduce the brood size in nematodes [48]. Meanwhile, after prolonged exposure from L1-larvae to adult day-1, GO-SH at concentrations more than 0.1 mg/L could also significantly decrease the locomotion behavior in nematodes [48].

In nematodes, after prolonged exposure from L1-larvae to adult day-1, GO-SH at concentrations more than 0.1 mg/L could induce the significant intestinal ROS production in nematodes [48]. Prolonged exposure to GO-SH at concentrations more than 0.01 mg/L could further enhance the intestinal permeability as indicated by Nile red staining signals in nematodes [48].

4.8 Nanotoxicity Formation Under the Oxidative Stress Condition

So far, at least two examples have been raised to explain the related information. With the CeO_2-NPs (8.5 nm) as the example, prolonged exposure (from L1-larvae for 3 days) to CeO_2-NPs at concentrations more than 5 nM significantly reduced the lifespan in nematodes under juglone-induced oxidative stress [28]. Meanwhile, after prolonged exposure, CeO_2-NPs at concentrations more than 5 nM induced the significant ROS accumulation in nematodes under juglone-induced oxidative stress [28].

Additionally, although the PEG-modified GO (GO-PEG) are relatively safe under normal conditions, GO-PEG could still significantly reduce the lifespan under juglone-induced oxidative stress [49]. For this observed toxicity, it has been supposed that the GO-PEG may enhance and accelerate the electron transfer among juglone/O_2, H_2O_2/$^\bullet OH$, and cyt c/H_2O_2 to impair the inherent antioxidant defense system, which will eventually cause the dramatic toxicity on nematodes [49].

4.9 Nanotoxicity Formation Under the Susceptible Mutation Backgrounds

In nematodes, mutation of many genes can induce the susceptibility to the toxicity of ENMs, which provide important basis for the examination of ENMs toxicity under susceptible genetic backgrounds.

With TiO_2-NPs as an example, using reproduction and locomotion behavior as toxicity assessment endpoints, mutation of *sod-2*, *sod-3*, *mtl-2*, or *hsp-16.48* could cause the susceptibility to the toxicity of TiO_2-NPs (1 µg/L) after prolonged exposure from L1-larvae to young adults (Fig. 4.1) [50]. Using survival and intestinal development as toxicity assessment endpoints, mutation of *sod-2*, *sod-3*, or *mtl-2* could cause the susceptibility to the toxicity of TiO_2-NPs (1 µg/L) after prolonged exposure from L1-larvae to young adults (Fig. 4.1) [50]. Using development as the toxicity assessment endpoint, only mutation of *mtl-2* could cause the susceptibility to the toxicity of TiO_2-NPs (1 µg/L) after prolonged exposure from L1-larvae to young adults (Fig. 4.1) [50].

After acute exposure from young adults for 24 h, mutation of *sod-2*, *sod-3*, *mtl-2*, or *hsp-16.48* could also result in the susceptibility to the toxicity of TiO_2-NPs (25 mg/L) using locomotion behavior, oxidative stress, and stress response as the toxicity assessment endpoints in nematodes (Fig. 4.2) [4].

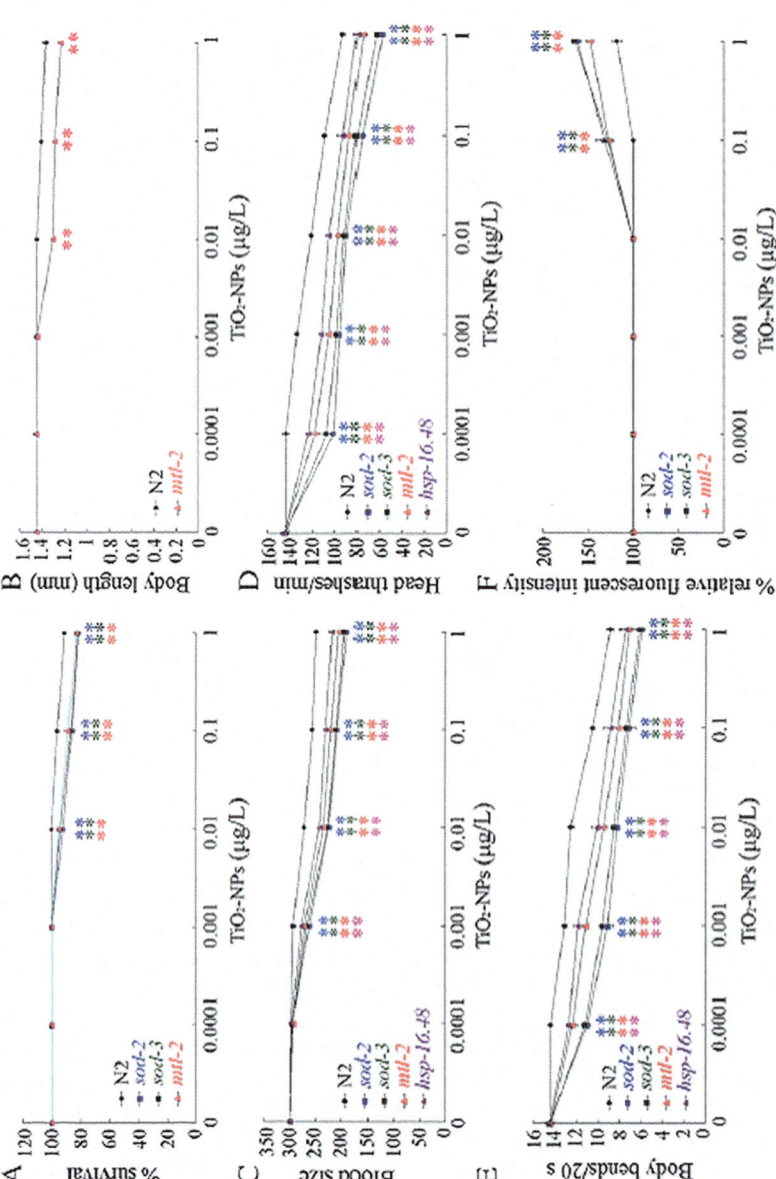

Fig. 4.1 Toxicity assessment of TiO₂-NPs using the strains with mutations of susceptible genes [50] (**a**) Toxicity assessment with the aid of lethality as endpoint. (**b**) Toxicity assessment with the aid of brood size as endpoint. (**c**) Toxicity assessment with the aid of body length as endpoint. (**d**) Toxicity assessment with the aid of head thrash as endpoint. (**e**) Toxicity assessment with the aid of body bend as endpoint. (**f**) Toxicity assessment with the aid of intestinal autofluorescence as endpoint. Bars represent means ± SEM. $^{**}P < 0.01$

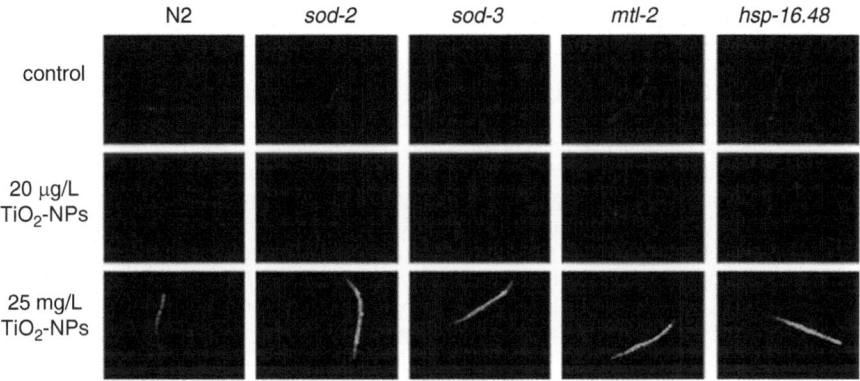

Fig. 4.2 Comparison of intestinal ROS production between wild-type and mutant nematodes acutely exposed to TiO$_2$-NPs [4]

Acute exposures to TiO$_2$-NPs were performed from young adult for 24 h

4.10 Perspectives

We here systematically introduced the possible toxic effects of metal nanoparticles, rare earth fluoride nanocrystals, quantum dots, carbon nanotubes, fullerenol, and graphene and its derivatives in nematodes. In contrast, so far, there are little or very limited data on toxicity assessment of different nanocomposites in organisms, which should be paid more attention to in the future research in this field. Moreover, most of the obtained data on nanotoxicity assessment in nematodes does not reflect the possible effect of certain ENMs at environmentally relevant concentrations. With the aid of more suitable exposure routes, the possible toxic effects of different ENMs at environmentally relevant concentrations on organisms including the nematodes should be carefully determined.

In this chapter, we discussed the values of the nanotoxicity formation under the oxidative stress condition and the nanotoxicity formation under the susceptible mutation backgrounds in nematodes. Nevertheless, the nanotoxicity formation under the oxidative stress condition or under the certain pathological conditions and the underlying mechanisms still need to be further deeply and systematically elucidated.

References

1. Brenner S (1974) The genetics of *Caenorhabditis elegans*. Genetics 77:71–94
2. Leung MC, Williams PL, Benedetto A, Au C, Helmcke KJ, Aschner M, Meyer JN (2008) *Caenorhabditis elegans*: An emerging model in biomedical and environmental toxicology. Toxicol Sci 106:5–28

3. Zhao Y-L, Wu Q-L, Li Y-P, Wang D-Y (2013) Translocation, transfer, and *in vivo* safety evaluation of engineered nanomaterials in the non-mammalian alternative toxicity assay model of nematode *Caenorhabditis elegans*. RSC Adv 3:5741–5757

4. Rui Q, Zhao Y-L, Wu Q-L, Tang M, Wang D-Y (2013) Biosafety assessment of titanium dioxide nanoparticles in acutely exposed nematode *Caenorhabditis elegans* with mutations of genes required for oxidative stress or stress response. Chemosphere 93(10):2289–2296

5. Zhao Y-L, Wu Q-L, Tang M, Wang D-Y (2014) The *in vivo* underlying mechanism for recovery response formation in nano-titanium dioxide exposed *Caenorhabditis elegans* after transfer to the normal condition. Nanomed: Nanotechnol Biol Med 10:89–98

6. Wang H, Wick RL, Xing B (2009) Toxicity of nanoparticulate and bulk ZnO, Al₂O₃ and TiO₂ to the nematode *Caenorhabditis elegans*. Environ Pollut 157:1171–1177

7. Roh J, Park Y, Park K, Choi J (2010) Ecotoxicological investigation of CeO₂ and TiO₂ nanoparticles on the soil nematode *Caenorhabditis elegans* using gene expression, growth, fertility, and survival as endpoints. Environ Toxicol Pharmacol 29:167–172

8. Li Y-X, Wang W, Wu Q-L, Li Y-P, Tang M, Ye B-P, Wang D-Y (2012) Molecular control of TiO₂-NPs toxicity formation at predicted environmental relevant concentrations by Mn-SODs proteins. PLoS One 7(9):e44688

9. Wu Q-L, Nouara A, Li Y-P, Zhang M, Wang W, Tang M, Ye B-P, Ding J-D, Wang D-Y (2013) Comparison of toxicities from three metal oxide nanoparticles at environmental relevant concentrations in nematode *Caenorhabditis elegans*. Chemosphere 90:1123–1131

10. Wu Q-L, Wang W, Li Y-X, Li Y-P, Ye B-P, Tang M, Wang D-Y (2012) Small sizes of TiO₂-NPs exhibit adverse effects at predicted environmental relevant concentrations on nematodes in a modified chronic toxicity assay system. J Hazard Mater 243:161–168

11. Roh J, Sim SJ, Yi J, Park K, Chung KH, Ryu D, Choi J (2009) Ecotoxicity of silver nanoparticles on the soil nematode *Caenorhabditis elegans* using functional ecotoxicogenomics. Environ Sci Technol 43:3933–3940

12. Meyer JN, Lord CA, Yang XY, Turner EA, Badireddy AR, Marinakos SM, Chilkoti A, Wiesner MR, Auffan M (2010) Intracellular uptake and associated toxicity of silver nanoparticles in *Caenorhabditis elegans*. Aquat Toxicol 100(2):140–150

13. Ellegaard-Jensen L, Jensen KA, Johansen A (2012) Nano-silver induces dose-response effects on the nematode *Caenorhabditis elegans*. Ecotoxicol Environ Saf 80:216–223

14. Lim D, Roh J, Eom H, Choi J, Hyun J, Choi J (2012) Oxidative stress-related PMK-1 p38 MAPK activation as a mechanism for toxicity of silver nanoparticles to reproduction in the nematode *Caenorhabditis elegans*. Environ Toxicol Chem 31:585–592

15. Kim SW, Nam S, An Y (2012) Interaction of silver nanoparticles with biological surfaces of *Caenorhabditis elegans*. Ecotoxicol Environ Saf 77:64–70

16. Contreras EQ, Puppala HL, Escalera G, Zhong W, Colvin VL (2014) Size-dependent impacts of silver nanoparticles on the lifespan, fertility, growth, and locomotion of *Caenorhabditis elegans*. Environ Toxicol Chem 33:2716–2723

17. Schultz CL, Wamucho A, Tsyusko OV, Unrine JM, Crossley A, Svendsen C, Spurgeon DJ (2016) Multigenerational exposure to silver ions and silver nanoparticles reveals heightened sensitivity and epigenetic memory in *Caenorhabditis elegans*. Proc R Soc B 283. https://doi.org/10.1098/rspb.2015.2911. 20152911

18. Ma H, Bertsch PM, Glenn TC, Kabengi NJ, Williams PL (2009) Toxicity of manufactured zinc oxide nanoparticles in the nematode *Caenorhabditis elegans*. Environ Toxicol Chem 28(6):1324–1330

19. Polak N, Read DS, Jurkschat K, Matzke M, Kelly FJ, Spurgeon DJ, Stürzenbaum SR (2014) Metalloproteins and phytochelatin synthase may confer protection against zinc oxide nanoparticle induced toxicity in *Caenorhabditis elegans*. Compara Biochem Physiol Part C 160:75–85

20. Wu S, Lu J-H, Rui Q, Yu S-H, Cai T, Wang D-Y (2011) Aluminum nanoparticle exposure in L1 larvae results in more severe lethality toxicity than in L4 larvae or young adults by strengthening the formation of stress response and intestinal lipofuscin accumulation in nematodes. Environ Toxicol Pharmacol 31:179–188

21. Yu X-M, Guan X-M, Wu Q-L, Zhao Y-L, Wang D-Y (2015) Vitamin E ameliorates the neuro-degeneration related phenotypes caused by neurotoxicity of Al_2O_3-nanoparticles in *C. elegans*. Toxicol Res 4:1269–1281

22. Li Y-X, Yu S-H, Wu Q-L, Tang M, Wang D-Y (2013) Transmissions of serotonin, dopamine and glutamate are required for the formation of neurotoxicity from Al_2O_3-NPs in nematode *Caenorhabditis elegans*. Nanotoxicology 7(5):1004–1013

23. Li Y-X, Yu S-H, Wu Q-L, Tang M, Pu Y-P, Wang D-Y (2012) Chronic Al_2O_3-nanoparticle exposure causes neurotoxic effects on locomotion behaviors by inducing severe ROS production and disruption of ROS defense mechanisms in nematode *Caenorhabditis elegans*. J Hazard Mater 219–220:221–230

24. Yu S-H, Rui Q, Cai T, Wu Q-L, Li Y-X, Wang D-Y (2011) Close association of intestinal autofluorescence with the formation of severe oxidative damage in intestine of nematodes chronically exposed to Al_2O_3-nanoparticle. Environ Toxicol Pharmacol 32:233–241

25. Pluskota A, Horzowski E, Bossinger O, von Mikecz A (2009) In *Caenorhabditis elegans* nanoparticle-bio-interactions become transparent: silica-nanoparticles induce reproductive senescence. PLoS One 4(8):e6622

26. Scharf A, Piechulek A, von Mikecz A (2013) Effect of nanoparticles on the biochemical and behavioral aging phenotype of the nematode *Caenorhabditis elegans*. ACS Nano 7(12):10695–10703

27. Rogers S, Rice KM, Manne NDPK, Shokuhfar T, He K, Selvaraj V, Blough ER (2015) Cerium oxide nanoparticle aggregates affect stress response and function in *Caenorhabditis elegans*. SAGE Open Med 3. https://doi.org/10.1177/2050312115575387. 2050312115575387

28. Zhang H, He X, Zhang Z, Zhang P, Li Y, Ma Y, Kuang Y, Zhao Y, Chai Z (2011) Nano-CeO_2 exhibits adverse effects at environmental relevant concentrations. Environ Sci Technol 45(8):3725–3730

29. Wu Q-L, Li Y-P, Tang M, Wang D-Y (2012) Evaluation of environmental safety concentrations of DMSA coated Fe_2O_3-NPs using different assay systems in nematode *Caenorhabditis elegans*. PLoS One 7(8):e43729

30. Tsyusko OV, Unrine JM, Spurgeon D, Blalock E, Starnes D, Tseng M, Joice G, Bertsch PM (2012) Toxicogenomic responses of the model organism *Caenorhabditis elegans* to gold nanoparticles. Environ Sci Technol 46:4115–4124

31. Kim SW, Kwak JI, An Y (2013) Multigenerational study of gold nanoparticles in *Caenorhabditis elegans*: transgenerational effect of maternal exposure. Environ Sci Technol 47:5393–5399

32. Zhou J, Yang Z, Dong W, Tang R, Sun L, Chun-Hua Yan C (2011) Bioimaging and toxicity assessments of near-infrared upconversion luminescent NaYF4:Yb,Tm nanocrystals. Biomaterials 32:9059–9067

33. Chen J, Guo C, Wang M, Huang L, Wang L, Mi C, Li J, Fang X, Mao C, Xu S (2011) Controllable synthesis of NaYF4: Yb,Er upconversion nanophosphors and their application to *in vivo* imaging of *Caenorhabditis elegans*. J Mater Chem 21(8):2632

34. Qu Y, Li W, Zhou Y, Liu X, Zhang L, Wang L, Li Y, Iida A, Tang Z, Zhao Y, Chai Z, Chen C (2011) Full assessment of fate and physiological behavior of quantum dots utilizing *Caenorhabditis elegans* as a model organism. Nano Lett 11:3174–3183

35. Wang Q, Zhou Y, Song B, Zhong Y, Wu S, Cui R, Cong H, Su Y, Zhang H, He Y (2016) Linking subcellular disturbance to physiological behavior and toxicity induced by quantum dots in *Caenorhabditis elegans*. Small 23:3143–3154

36. Zhao Y-L, Wang X, Wu Q-L, Li Y-P, Tang M, Wang D-Y (2015) Quantum dots exposure alters both development and function of D-type GABAergic motor neurons in nematode *Caenorhabditis elegans*. Toxicol Res 4:399–408

37. Zhao Y-L, Wang X, Wu Q-L, Li Y-P, Wang D-Y (2015) Translocation and neurotoxicity of CdTe quantum dots in RMEs motor neurons in nematode *Caenorhabditis elegans*. J Hazard Mater 283:480–489

38. Liu Z-F, Zhou X-F, Wu Q-L, Zhao Y-L, Wang D-Y (2015) Crucial role of intestinal barrier in the formation of transgenerational toxicity in quantum dots exposed nematodes *Caenorhabditis elegans*. RSC Adv 5:94257–94266

39. Contreras EQ, Cho M, Zhu H, Puppala HL, Escalera G, Zhong W, Colvin VL (2013) Toxicity of quantum dots and cadmium salt to *Caenorhabditis elegans* after multigenerational exposure. Environ Sci Technol 47:1148–1154

40. Chen P, Hsiao K, Chou C (2013) Molecular characterization of toxicity mechanism of single-walled carbon nanotubes. Biomaterials 34:5661–5669

41. Shu C-J, Yu X-M, Wu Q-L, Zhuang Z-H, Zhang W-M, Wang D-Y (2015) Pretreatment with paeonol prevents the adverse effects and alters the translocation of multi-walled carbon nanotubes in nematode *Caenorhabditis elegans*. RSC Adv 5:8942–8951

42. Nouara A, Wu Q-L, Li Y-X, Tang M, Wang H-F, Zhao Y-L, Wang D-Y (2013) Carboxylic acid functionalization prevents the translocation of multi-walled carbon nanotubes at predicted environmental relevant concentrations into targeted organs of nematode *Caenorhabditis elegans*. Nanoscale 5:6088–6096

43. Wu Q-L, Li Y-X, Li Y-P, Zhao Y-L, Ge L, Wang H-F, Wang D-Y (2013) Crucial role of biological barrier at the primary targeted organs in controlling translocation and toxicity of multi-walled carbon nanotubes in nematode *Caenorhabditis elegans*. Nanoscale 5:11166–11178

44. Cha YJ, Lee J, Choi SS (2012) Apoptosis-mediated in vivo toxicity of hydroxylated fullerene nanoparticles in soil nematode *Caenorhabditis elegans*. Chemosphere 87:49–54

45. Cong W, Wang P, Qu Y, Tang J, Bai R, Zhao Y, Chen C, Bi X (2015) Evaluation of the influence of fullerenol on aging and stress resistance using *Caenorhabditis elegans*. Biomaterials 42L:78–86

46. Wu Q-L, Yin L, Li X, Tang M, Zhang T, Wang D-Y (2013) Contributions of altered permeability of intestinal barrier and defecation behavior to toxicity formation from graphene oxide in nematode *Caenorhabditis elegans*. Nanoscale 5(20):9934–9943

47. Wu Q-L, Zhao Y-L, Fang J-P, Wang D-Y (2014) Immune response is required for the control of *in vivo* translocation and chronic toxicity of graphene oxide. Nanoscale 6:5894–5906

48. Ding X-C, Wang J, Rui Q, Wang D-Y (2018) Long-term exposure to thiolated graphene oxide in the range of μg/L induces toxicity in nematode *Caenorhabditis elegans*. Sci Total Environ 616–617:29–37

49. Zhang W, Wang C, Li Z, Lu Z, Li Y, Yin J, Zhou Y, Gao X, Fang Y, Nie G, Zhao Y (2012) Unraveling stress-induced toxicity properties of graphene oxide and the underlying mechanism. Adv Mater 24:5391–5397

50. Wu Q-L, Zhao Y-L, Li Y-P, Wang D-Y (2014) Susceptible genes regulate the adverse effects of TiO$_2$-NPs at predicted environmental relevant concentrations on nematode *Caenorhabditis elegans*. Nanomed: Nanotechnol Biol Med 10:1263–1271

Chapter 5
Physicochemical Basis for Nanotoxicity Formation

Abstract The unique physicochemical properties of different engineered nanomaterials (ENMs) not only allow their potential industrial or medical applications at different aspects but also lead to different biological effects on organisms and interactions with targeted cells or tissues in organisms. The well-described cellular, developmental, molecular, and genetic backgrounds and the sensitivity to toxicity of environmental toxicants or stresses of *Caenorhabditis elegans* provide a powerful in vivo model system to determine the roles of physicochemical properties of ENMs in the toxicity formation of different ENMs in organisms. We here systematically introduce the important roles of different physicochemical properties of ENMs, mainly including size, surface charge, shape, surface groups, and impurity, in the toxicity formation of ENMs, which provides the underlying important chemical basis for the observed toxicity of ENMs in nematodes.

Keywords Physicochemical property · Chemical basis · Nanotoxicology · *Caenorhabditis elegans*

5.1 Introduction

Different engineered nanomaterials (ENMs) have different unique physicochemical properties, which allow their potentials for different industrial or medical applications. The different physicochemical properties of ENMs may cause the various biological effects on organisms. Meanwhile, the different physicochemical properties of ENMs may lead to the formation of different interaction between ENMs and targeted cells or tissues in the body of organisms. The classic model animal of *Caenorhabditis elegans* has well-described cellular, developmental, molecular, and genetic backgrounds [1]. More importantly, with the rapid increase in application of *C. elegans* in toxicity assessment of different environmental toxicants or environmental stresses, *C. elegans* has been shown to be very sensitive to the toxicity of environmental toxicants or environmental stresses [2–4]. These research and knowledge backgrounds provide us the solid foundation to determine the roles of physicochemical properties of ENMs in the toxicity formation of different ENMs in organisms.

© Springer Nature Singapore Pte Ltd. 2018
D. Wang, *Nanotoxicology in* Caenorhabditis elegans,
https://doi.org/10.1007/978-981-13-0233-6_5

In this chapter, we systematically introduce the roles of physicochemical properties of ENMs in the nanotoxicity formation. The introduced physicochemical properties of ENMs mainly include size, surface charge, shape, surface groups, and impurity. We also introduced the underlying chemical mechanism for the oxidative stress induced by ENMs in nematodes. The introduced information in this chapter provides an important physicochemical basis for our attempt to understand and to elucidate the underlying cellular and physiological mechanisms of nanotoxicity formation.

5.2 Size

5.2.1 Toxicity Comparison of ENMs Between Nanosize and Bulk Size

Size is normally considered as a key factor for the induction of toxicity of ENMs in organisms. The size of ENMs plays a pivotal role in how the body responds to, distributes, and eliminates. In organisms, metal-nanoparticles (metal-NPs) or metal oxide-NPs usually have more severe toxicity than the corresponding bulk metal or bulk metal oxide. For example, although both NPs and their bulk counterparts were toxic, the 24-h LC_{50} values were significantly different between TiO_2-NPs and bulk TiO_2, as well as between Al_2O_3-NPs-NPs and bulk Al_2O_3 exposure [5]. Besides this, after acute exposure, the stress response and the intestinal lipofuscin autofluorescence were also significantly different between Al_2O_3-NPs-NPs and bulk Al_2O_3 exposed nematodes [6]. After chronic exposure, the significant differences of intestinal autofluorescence, stress response, and intestinal reactive oxygen species (ROS) production of Al_2O_3-NPs exposed nematodes from those in bulk Al_2O_3 exposed nematodes were further detected at the examined concentrations [7].

5.2.2 Toxicity Comparison of ENMs with Different Nanosizes

Moreover, the toxicities of ENMs with different nanosize may be very different. With titanium oxide-NPs (TiO_2-NPs) as an example, it has been demonstrated that the small sizes (4 nm and 10 nm) of TiO_2-NPs could induce more severe toxicities than large sizes (60 nm and 90 nm) of TiO_2-NPs on animals using lethality, growth, reproduction, and locomotion behavior as endpoints after prolonged exposure from L1-larvae to adult day-1 [8]. After prolonged exposure, TiO_2-NPs (60 nm and 90 nm) at concentrations of 0.001–10 mg/L and TiO_2-NPs (4 nm and 10 nm) at concentrations of 0.001–0.01 mg/L did not obviously influence the survival of nematodes, whereas TiO_2-NPs (4 nm and 10 nm) at concentrations of 0.1–10 mg/L could significantly increase the mortality of nematodes (Fig. 5.1) [8]. Additionally,

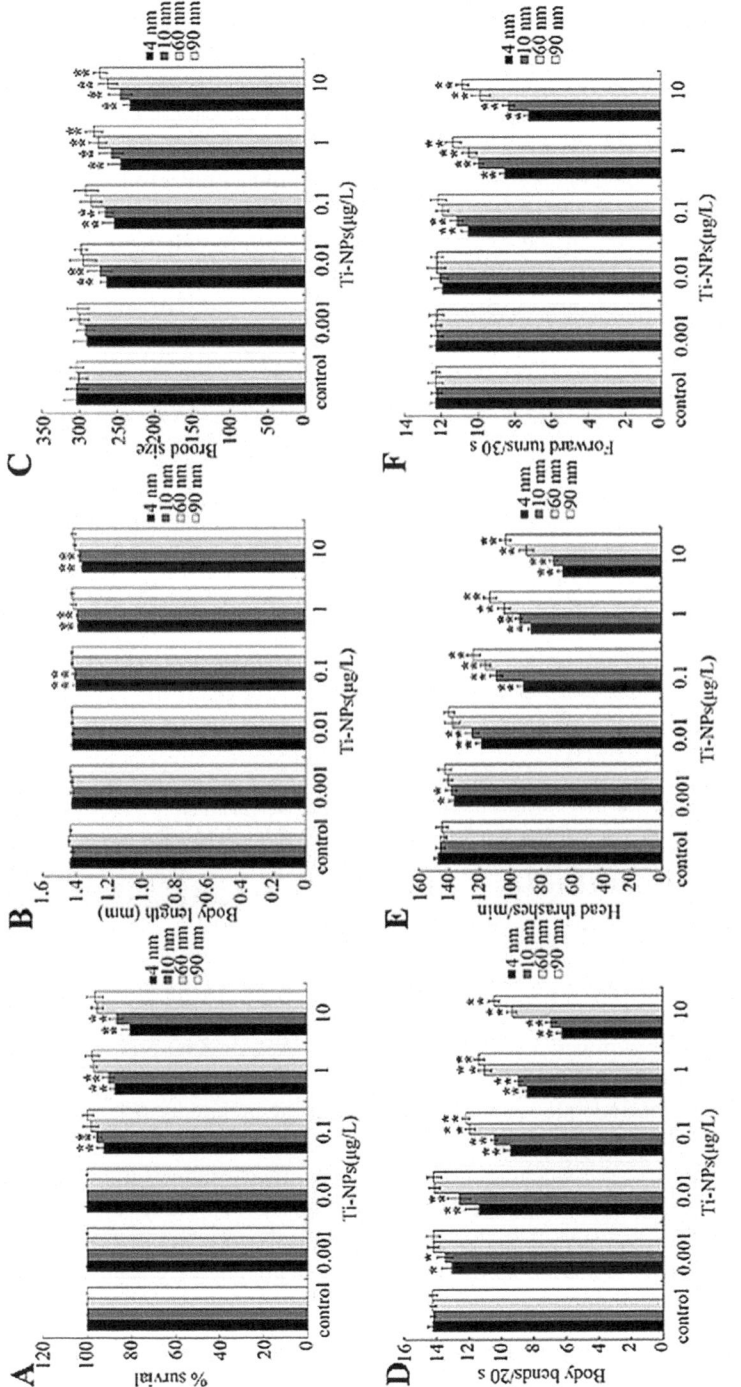

Fig. 5.1 Effects of different sizes of TiO$_2$-NPs on survival, growth, reproduction, and locomotion behavior in nematodes [8] (**a**) Effects of different sizes of TiO$_2$-NPs on survival. (**b**) Effects of different sizes of TiO$_2$-NPs on growth. (**c**) Effects of different sizes of TiO$_2$-NPs on reproduction. (**d**, **e**, and **f**) Effects of different sizes of TiO$_2$-NPs on locomotion behavior. Exposure of TiO$_2$-NPs was performed from L1-larvae, and the endpoints were examined when nematodes developed into the adult day-1. Ti-NPs, TiO$_2$-NPs. The bars represent mean ± SEM. *$p < 0.05$, **$p < 0.01$

all the examined sizes of TiO_2-NPs at concentrations of 0.001–0.1 mg/L and TiO_2-NPs (60 nm and 90 nm) at concentrations of 1–10 mg/L did not obviously affect the growth as reflected by the endpoint of body length; however, the body length of nematodes was significantly reduced by TiO_2-NPs (4 nm or 10 nm) at concentrations of 1 and 10 mg/L (Fig. 5.1) [8]. Moreover, all the examined TiO_2-NPs at the concentration of 0.001 mg/L and TiO_2-NPs (60 nm and 90 nm) at concentrations of 0.01 and 0.1 mg/L could not cause alterations in the reproduction as reflected by the endpoint of brood size, whereas TiO_2-NPs (4 nm and 10 nm) at concentrations of 0.01 and 0.1 mg/L and all the examined TiO_2-NPs at concentrations of 1 and 10 mg/L could significantly reduce the brood size (Fig. 5.1) [8]. Besides these, TiO_2-NPs (60 nm and 90 nm) at concentrations of 0.001 and 0.01 mg/L did not significantly alter the locomotion behavior as reflected by the endpoints of head thrash and body bend, and TiO_2-NPs (4 nm and 10 nm) at the concentration of 0.001 mg/L only moderately but significantly decreased the locomotion behavior (Fig. 5.1) [8]. In contrast, TiO_2-NPs (4 nm and 10 nm) at the concentration of 0.01 mg/L and all the examined TiO_2-NPs at the concentrations of 0.1–10 mg/L could significantly decrease the locomotion behavior (Fig. 5.1) [8]. Different from the observations on head thrash and body bend, the endpoint of forward turn was less sensitive for assessing the neurotoxicity of TiO_2-NPs. Exposure to all the examined TiO_2-NPs at concentrations of 0.001–0.01 mg/L and TiO_2-NPs (60 nm and 90 nm) at the concentration of 0.1 mg/L could not significantly alter the forward turns; however, TiO_2-NPs (4 nm and 10 nm) at concentrations of 0.1–10 mg/L and TiO_2-NPs (60 nm and 90 nm) at concentrations of 1–10 mg/L could significantly decrease the forward turns (Fig. 5.1) [8].

5.2.3 Toxicity Comparison of Different ENMs with the Same Nanosize

Wu et al. (2013) compared the toxicities from three different metal oxide-NPs (TiO_2-NPs, ZnO-NPs, and SiO_2-NPs) with the same nanosize (30 nm) at predicted environmental relevant concentrations in nematodes. It has been found that the toxicity could be detected in nematodes exposed to TiO_2-NPs or ZnO-NPs at concentrations more than 0.05 μg/L and SiO_2-NPs at concentrations more than 5 μg/L with locomotion behavior and intestinal ROS production as endpoints (Fig. 5.2) [9]. Using the growth, the locomotion behavior, the reproduction, and the intestinal ROS production as toxicity assessment endpoints, toxicity order for the examined metal oxide-NPs was ZnO-NPs > TiO_2-NPs > SiO_2-NPs (Fig. 5.2) [9]. These results imply that, besides the particle size, the other physicochemical properties of ENMs may also affect the toxicity formation of ENMs to a great degree in organisms.

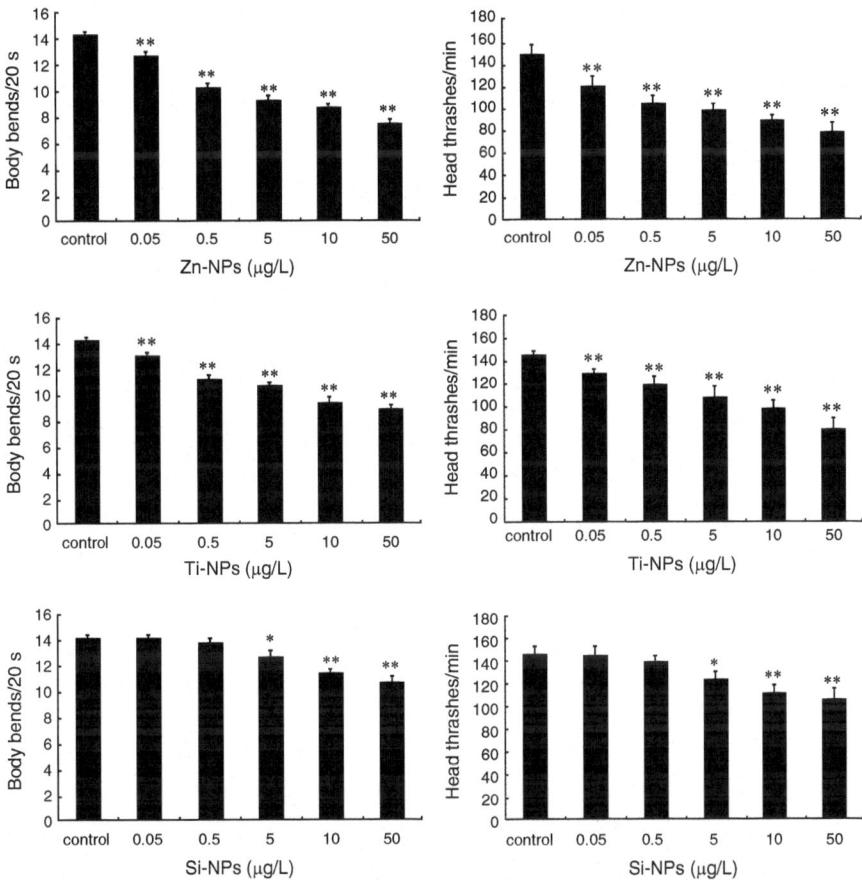

Fig. 5.2 Effects of different metal oxide nanoparticles (NPs) on locomotion behavior in nematodes [9]

The locomotion behavior was evaluated by body bend and head thrash. Exposure was performed from the L1-larvae, and the endpoint was examined when nematodes developed into the adult day-1. Bars represent mean ± SEM. *$P < 0.05$; **$P < 0.01$

5.2.4   Toxicity Comparison of Metal Oxide-NPs with the Corresponding Metal Ion

It was also reported that exposure to specific metal oxide-NPs with a very small size can even induce the similar toxicity on nematodes to those from exposure to the corresponding metal ion. For example, ZnO-NPs showed no significant differences from $ZnCl_2$ in inducing the toxicity on nematodes using lethality, reproduction, behavior, and expression of P*mtl-2::gfp* as the toxicity assessment endpoints [10]. One of the possible explanations is that the intracellular biotransformation of

ZnO-NPs might occur. Another possible explanation is that ZnO-NPs might have been dissolved to Zn^{2+} to enact the toxicity on nematodes.

5.3   Surface Charge

With CeO_2-NPs as an example, the effects of different surface charges on the toxicity formation of ENMs in nematodes were examined. The surface of CeO_2-NPs was functionalized with diethylaminoethyl groups to confer either a net positive charge (diethylaminoethyl-dextran; DEAE-CeO_2(+)) or with carboxymethyl groups to confer a net negative charge (carboxymethyl-dextran; CM-CeO_2(−)). With the aid of these ENMs, it has been shown that the positively charged CeO_2-NPs (DEAE-CeO_2(+)) exhibited the more toxic effects on nematodes than the neutral or the negatively charged CeO_2-NPs (CM-CeO_2(−)) (Fig. 5.3) [11]. Meanwhile, more positively charged CeO_2-NPs (DEAE-CeO_2(+)) was accumulated in the body of nematodes than the neutral or the negatively charged CeO_2-NPs (CM-CeO_2(−)) [11]. The chemical analysis further demonstrates that the surface charge of CeO_2-NPs affected the oxidation state of Ce in the body of nematodes after the uptake [11]. That is, a greater reduction of Ce from Ce (IV) to Ce (III) was detected in the body of nematodes when exposed to the neutral or the negatively charged CeO_2-NPs (CM-CeO_2(−)) relative to the positively charged CeO_2-NPs (DEAE-CeO_2(+)) [11].

5.4   Shape

Three silver nanomaterials (Ag-NMs) containing silver-NPs (Ag-NPs), silver nanocubes (Ag-NCs), and silver nanowires (Ag-NWs) with different shape were selected to determine the possible effects of shape for ENMs on the toxicity formation in organisms. The examined AgPs and AgCs were approximately the same size [12]. In plant species *Lolium multiflorum*, Ag-NCs showed a lower toxicity compared to quasi-spherical (Ag-NPs) and Ag-NWs [12]. However, all these three examined Ag-NMs exhibited the similar toxicity on nematodes, as well as on *Danio rerio* [12], suggesting that the shape-based toxicity of Ag-NMs may only be important in plants. Nevertheless, we cannot exclude the possibility that the shape-based toxicity of other ENMs exists in nematodes.

5.5   Surface Groups

The chemical interactions of ENMs are largely defined or determined by chemical compositions or groups at the surface of ENMs. The surface of ENMs is the direct contact with the body or tissues and influences the ENMs bioavailability to targeted

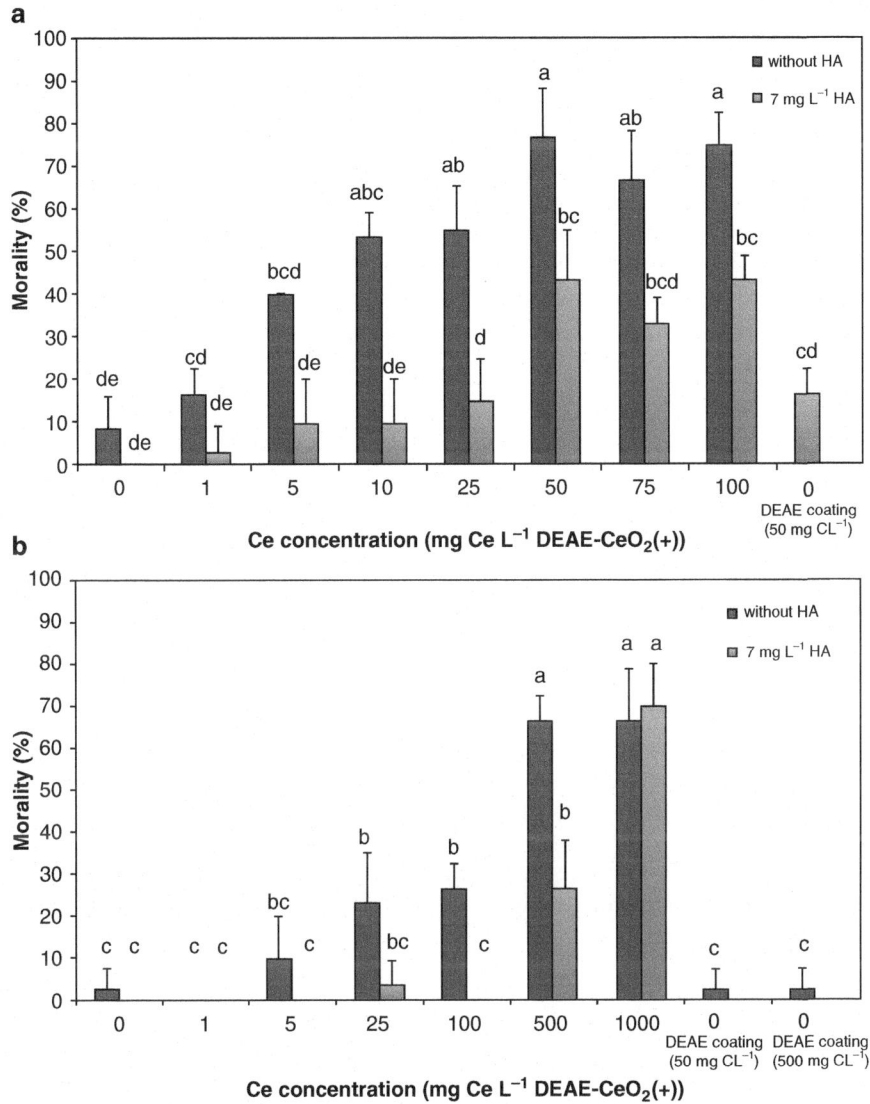

Fig. 5.3 Mortality of nematodes exposed to positively charged diethylaminoethyl-dextran coated CeO$_2$-NPs (DEAE-CeO$_2$(+)) at L1 (**a**) and L3 (**b**) stages for 48 h [11]

organs. So far, various surface modifications have been developed for different ENMs with the aim of increasing blood circulation, making them more biocompatible, and targeted therapy. Among these surface chemical modifications, some surface design can noticeably affect the toxicity formation of certain ENMs. For example, LC$_{50}$ for PVP-coated Ag-NPs (28 nm) with more toxicity on nematodes was lower than that for Ag-NPs (1 nm) [13]. Nevertheless, both the Ag-NPs and the

PVP-coated Ag-NPs showed adverse dose-response effects and mortality in nematodes [13].

Different from this, the citrate-coated Ag-NPs were less toxic for nematodes than PVP-coated Ag-NPs [14]. The supernatant from citrate-coated Ag-NPs showed no toxicity on nematodes, whereas the toxicity from supernatant of PVP-coated Ag-NPs existed for nematodes [14], implying that the citrate-coated Ag-NPs may release less silver than the PVP-coated Ag-NPs in nematodes.

Besides these, some important surface modifications or coatings have been successfully designed to reduce the toxicity of ENMs, such as surface PEG modification and surface FBS or ZnS coating. We will introduce the related information in details in Chap. 10.

5.6   Impurity

During the preparation, some ENMs usually have the property of impurity. For example, the prepared carbon nanotubes (CNTs) usually contain a certain amount of heavy metals. This aspect of property has been considered for multiwalled CNTs (MWCNTs) toxicity formation in nematodes [15, 16]. In the prepared MWCNTs, the presence of Ni and Fe impurities at concentrations not more than 0.1% as determined by elemental inductively coupled plasma mass spectrometry (ICP-MS) was detected [15, 16]. After prolonged exposure from L1-larvae to young adults, MWCNTs in the range of μg/L could cause the damage on the functions of primary targeted organs, such as the intestine, and secondary targeted organs, such as the neurons and the reproductive organs, in nematodes [15, 16]. After prolonged exposure, it has been shown that both 0.077% Ni and 0.017% Fe in MWCNTs (1000 mg/L) could not induce the significant intestinal ROS production (Fig. 5.4) [16], implying the observed toxicity of MWCNTs on nematodes may be not due to the impurity of Ni or Fe in MWCNTs.

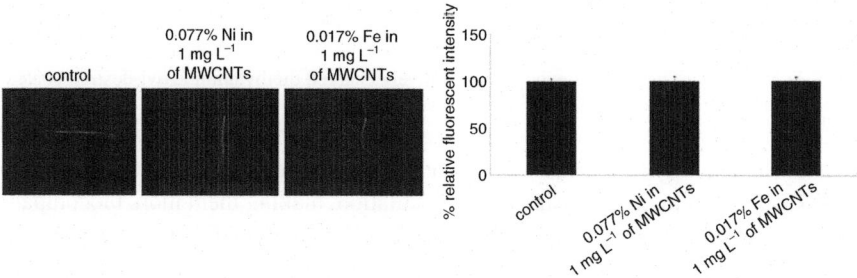

Fig. 5.4 Effects of Ni or Fe in MWCNTs on intestinal ROS production [16]
Prolonged exposure was performed from L1-larvae to young adults. Bars represent mean ± SEM

5.7   The Underlying Chemical Mechanism for the Oxidative Stress Induced by ENMs

Graphene oxide (GO) is a widely studied carbon-based ENMs. In nematodes, it has been demonstrated that GO exposure could cause the adverse effects on the functions of both primary targeted organs, such as the intestine, and the secondary targeted organs, such as the reproductive organs and the neurons [17, 18]. Moreover, GO exposure could induce the activation of oxidative stress as indicated by the significant induction of ROS production in nematodes [17, 18]. Under the normal conditions, surface chemical PEG modification could prevent the activation of oxidative stress, as well as the toxicity on the functions of both primary and secondary targeted organs, induced by GO exposure in nematodes [19, 20]. That is, the nematodes may be very tolerant toward the surface chemical PEG-modified GO (GO/PP) GO/PP under normal conditions.

Using a simple model of H_2O_2 decomposition by the ESR method with the aim to examine the possible chemical role of GO/PP on ROS formation and conversion, three lines comprising the inherent ESR signal of GO/PP (nitrogen-centered radicals) were detected in the spectrum of GO/PP, and GO/PP alone produced no ·OH species [17]. Meanwhile, typical ESR signal of DEPMPO/·OH could dominate the spectrum observed in H_2O_2 solution with GO/PP and overlap with that of GO/PP [17]. This increase in the ESR signal suggests that the GO/PP may potentially facilitate the production of ·OH from H_2O_2 and mediate the electron transfer reaction.

In organisms, GO may have the potential to act as a catalyst to facilitate the generation of ROS from H_2O_2 [17]. After comparison of formation of ·OH from H_2O_2 with and without GO using the density functional theory method, both paths (H_2O_2 (path 1) and GO (path 2, the coronene-aided decomposition of H_2O_2)) yield two ·OH ions from H_2O_2 [17]. Path 1 was a highly endothermic process; however, path 2 consists of three steps, and path 2 avoids the highly endothermic reaction step of path 1 [17], which suggests that this coronene can behave as a catalyst to make the generation of ·OH easier along the path 2. Meanwhile, the hydroxyl groups can also possibly be detached from the GO sheets to form ·OH [17]. Therefore, the GO/PP may also potentially promote the ROS production from H_2O_2, since the catalytic effects of functionalized GO sheets largely depend on the uncovered GO surface (Fig. 5.5) [17]. Considering the fact that H_2O_2 generation is precisely controlled at a basic level to ensure its role in redox signaling to contribute to the normal cell function under normal conditions, such a restricted source of H_2O_2 implies that the facilitating ability of GO/PP on the H_2O_2 decomposition to ·OH might only occur in a controlled manner and thus cannot cause toxicity on nematodes under normal conditions (Fig. 5.5) [17].

In nematodes, exposure to GO/PP could still cause the toxic effects on nematodes under the oxidative stress condition induced by juglone, a widely used peroxidant to induce oxidative stress [17]. It has been further raised that the accelerating effects of GO/PP on ·OH formation may be enhanced due to H_2O_2 overproduction triggered by oxidative stress (Fig. 5.5) [17]. Meanwhile, the GO/PP may facilitate

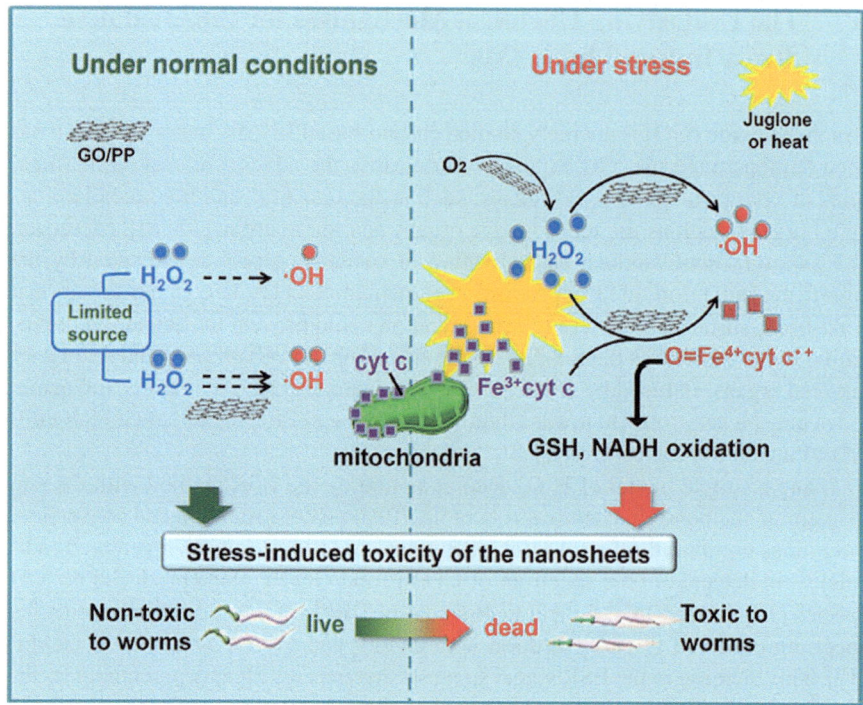

Fig. 5.5 Schematic diagram of the proposed mechanism for the stress-induced toxicity of GO/ PEG on nematodes [17]

the electron transfer of cyt c to H_2O_2 and generate active free radicals under the oxidative stress condition (Fig. 5.5) [17]. That is, the cyt c released from mitochondria to the cytoplasm provides a chance for GO/PP to come into contact with it, and the GO/PP may enhance the electron transfer among juglone/O_2, H_2O_2/·OH, and cyt c/H_2O_2 to cause the damage on the inherent antioxidant defense system and eventually lead to the toxicity on nematodes (Fig. 5.5) [17]. In contrast, under normal conditions, cyt c may be only restricted on the inner membrane of mitochondria, and thus the GO/PP will not be able to directly in contact with the cyt c (Fig. 5.5) [17].

5.8 Perspectives

For the roles of physicochemical properties in influencing the toxicity formation of ENMs in nematodes, we here mainly introduced the size, the surface charge, the shape, the surface groups, and the impurity of ENMs. Actually, so far, the published data at this aspect are still very limited with the concern on the in vivo model animal of *C. elegans*. On the one hand, besides the size of ENMs, most of the introduced physicochemical properties of ENMs have only been examined in one or limited number of ENMs. On the other hand, the roles of some other important physico-chemical properties of ENMs, such as the surface area, in the toxicity formation of ENMs have not been determined in nematodes in details. More importantly, more and more different forms of nanocomposites have been developed with the different application potentials. In these nanocomposites, the roles of various combinational physicochemical properties for the assembly between ENM(s) and ENM(s), or between ENM(s) and chemical(s), in toxicity induction in organisms are really largely unclear.

For the underlying chemical basis of the observed toxicity of ENMs in nema-todes, we here mainly introduced the chemical mechanism for the oxidative stress induced by ENMs with the GO as an example. In Chap. 8, we also introduced the chemical metabolism of certain ENMs in the body of nematodes with the quantum dots (QDs) as an example. Nevertheless, more efforts are needed to be conducted to elucidate the underlying chemical mechanisms for the observed toxicity of ENMs in organisms, especially the underlying chemical basis for how ENMs interact with the biological barrier, the different tissues, the different cells, and even the different organelle in organisms.

References

1. Brenner S (1974) The genetics of *Caenorhabditis elegans*. Genetics 77:71–94
2. Leung MC, Williams PL, Benedetto A, Au C, Helmcke KJ, Aschner M, Meyer JN (2008) *Caenorhabditis elegans*: an emerging model in biomedical and environmental toxicology. Toxicol Sci 106:5–28
3. Zhao Y-L, Wu Q-L, Li Y-P, Wang D-Y (2013) Translocation, transfer, and *in vivo* safety evaluation of engineered nanomaterials in the non-mammalian alternative toxicity assay model of nematode *Caenorhabditis elegans*. RSC Adv 3:5741–5757
4. Wang D-Y (2016) Biological effects, translocation, and metabolism of quantum dots in nematode *Caenorhabditis elegans*. Toxicol Res 5:1003–1011
5. Wang H, Wick RL, Xing B (2009) Toxicity of nanoparticulate and bulk ZnO, Al_2O_3 and TiO_2 to the nematode *Caenorhabditis elegans*. Environ Pollut 157:1171–1177
6. Wu S, Lu J-H, Rui Q, Yu S-H, Cai T, Wang D-Y (2011) Aluminum nanoparticle exposure in L1 larvae results in more severe lethality toxicity than in L4 larvae or young adults by strengthening the formation of stress response and intestinal lipofuscin accumulation in nematodes. Environ Toxicol Pharmacol 31:179–188
7. Yu S-H, Rui Q, Cai T, Wu Q-L, Li Y-X, Wang D-Y (2011) Close association of intestinal autofluorescence with the formation of severe oxidative damage in intestine of nematodes chronically exposed to Al_2O_3-nanoparticle. Environ Toxicol Pharmacol 32:233–241
8. Li Y-X, Wang W, Wu Q-L, Li Y-P, Tang M, Ye B-P, Wang D-Y (2012) Molecular control of TiO_2-NPs toxicity formation at predicted environmental relevant concentrations by Mn-SODs proteins. PLoS One 7(9):e44688
9. Wu Q-L, Nouara A, Li Y-P, Zhang M, Wang W, Tang M, Ye B-P, Ding J-D, Wang D-Y (2013) Comparison of toxicities from three metal oxide nanoparticles at environmental relevant concentrations in nematode *Caenorhabditis elegans*. Chemosphere 90:1123–1131
10. Ma H, Bertsch PM, Glenn TC, Kabengi NJ, Williams PL (2009) Toxicity of manufactured zinc oxide nanoparticles in the nematode *Caenorhabditis elegans*. Environ Toxicol Chem 28(6):1324–1330

11. Collin B, Oostveen E, Tsyusko OV, Unrine JM (2014) Influence of natural organic matter and surface charge on the toxicity and bioaccumulation of functionalized ceria nanoparticles in *Caenorhabditis elegans*. Environ Sci Technol 48:1280–1289
12. Gorka DE, Osterberg JS, Gwin CA, Colman BP, Meyer JN, Bernhardt ES, Gunsch CK, DiGulio RT, Liu J (2015) Reducing environmental toxicity of silver nanoparticles through shape control. Environ Sci Technol 49:10093–10098
13. Ellegaard-Jensen L, Jensen KA, Johansen A (2012) Nano-silver induces dose-response effects on the nematode *Caenorhabditis elegans*. Ecotoxicol Environ Saf 80:216–223
14. Meyer JN, Lord CA, Yang XY, Turner EA, Badireddy AR, Marinakos SM, Chilkoti A, Wiesner MR, Auffan M (2010) Intracel006Cular uptake and associated toxicity of silver nanoparticles in *Caenorhabditis elegans*. Aquat Toxicol 100(2):140–150
15. Nouara A, Wu Q-L, Li Y-X, Tang M, Wang H-F, Zhao Y-L, Wang D-Y (2013) Carboxylic acid functionalization prevents the translocation of multi-walled carbon nanotubes at predicted environmental relevant concentrations into targeted organs of nematode *Caenorhabditis elegans*. Nanoscale 5:6088–6096
16. Wu Q-L, Li Y-X, Li Y-P, Zhao Y-L, Ge L, Wang H-F, Wang D-Y (2013) Crucial role of biological barrier at the primary targeted organs in controlling translocation and toxicity of multi-walled carbon nanotubes in nematode *Caenorhabditis elegans*. Nanoscale 5:11166–11178
17. Zhang W, Wang C, Li Z, Lu Z, Li Y, Yin J, Zhou Y, Gao X, Fang Y, Nie G, Zhao Y (2012) Unraveling stress-induced toxicity properties of graphene oxide and the underlying mechanism. Adv Mater 24:5391–5397
18. Wu Q-L, Yin L, Li X, Tang M, Zhang T, Wang D-Y (2013) Contributions of altered permeability of intestinal barrier and defecation behavior to toxicity formation from graphene oxide in nematode *Caenorhabditis elegans*. Nanoscale 5(20):9934–9943
19. Wu Q-L, Zhao Y-L, Fang J-P, Wang D-Y (2014) Immune response is required for the control of *in vivo* translocation and chronic toxicity of graphene oxide. Nanoscale 6:5894–5906
20. Wu Q-L, Zhou X-F, Han X-X, Zhuo Y-Z, Zhu S-T, Zhao Y-L, Wang D-Y (2016) Genome-wide identification and functional analysis of long noncoding RNAs involved in the response to graphene oxide. Biomaterials 102:277–291

Chapter 6
Cellular and Physiological Mechanisms of Nanotoxicity Formation

Abstract In organisms, understanding the cellular and physiological mechanisms can provide the fundamental insights in the toxicity formation or behavior of engineered nanomaterials (ENMs). The well-described cellular and developmental basis of *Caenorhabditis elegans* provides us a powerful tool to determine the possible cellular and physiological mechanisms of nanotoxicity formation in organisms. We here systematically introduce several aspects of cellular and physiological mechanisms of nanotoxicity formation in nematodes. The raised cellular and physiological mechanisms of toxicity formation of ENMs in the nematodes provide valuable information on the interaction between the ENMs and their targeted organs or tissues in organisms.

Keywords Cellular mechanism · Physiological mechanism · Nanotoxicology · Nanomaterials · *Caenorhabditis elegans*

6.1 Introduction

Cellular and physiological mechanisms are important aspects for our understanding of the toxicity formation or behavior of engineered nanomaterials (ENMs) in organisms. *Caenorhabditis elegans* is a classic model animal [1], and its cellular and developmental basis has been well described. Meanwhile, it has been widely used in the toxicity assessment and toxicological study of different environmental toxicants, including the ENMs [2–4]. The cellular and physiological mechanisms of toxicity formation of ENMs provide important information on the interaction between the ENMs and the organs or the tissues in nematodes. Additionally, the cellular and physiological mechanisms of toxicity formation of ENMs are helpful for explaining the observed toxicity of ENMs on different organs or tissues in nematodes. Moreover, the examination of cellular and physiological mechanisms of toxicity formation of ENMs provides important basis or clues for further elucidation of the underlying molecular mechanisms of toxicity formation of ENMs.

In this chapter, we systematically introduce several aspects of cellular mechanisms of nanotoxicity formation in nematodes, including release of metal ion, oxidative

© Springer Nature Singapore Pte Ltd. 2018
D. Wang, *Nanotoxicology in* Caenorhabditis elegans,
https://doi.org/10.1007/978-981-13-0233-6_6

stress, intestinal permeability, defecation behavior, bioavailability to targeted organs, acceleration in aging process, innate immune response, mitochondrial damage and DNA damage, developmental fate, and deficit in cellular endocytosis in intestinal cells. Moreover, we also introduce several aspects of physiological mechanisms of nanotoxicity formation in nematodes, including environmental factors, exposure, physiological state of nematodes, developmental stages, and hormesis of nematodes.

6.2   Cellular Mechanisms for Nanotoxicity Formation

6.2.1   Release of Metal Ions Is an Important Toxicological Mechanism for Toxicity Formation of some Metal Nanoparticles (Metal-NPs) or Quantum Dots (QDs)

6.2.1.1   Role of the Release of Metal Ions in Toxicity Formation of Metal-NPs

Dissolution of ENMs can provide an important basis for the toxicity formation of metal-NPs or metal oxide-NPs. The toxicity of some metal-NPs or metal oxide-NPs may be at least partially mediated by the release of ions from the chemical composition, although their toxicity could not be adequately explained by the dissolution of the particles alone. With ZnO-NPs as an example, the toxicity of ZnO-NPs was similar to that of $ZnCl_2$ (Fig. 6.1) [5], implying that the dissolved Zn^{2+} may play an important role in inducing the toxicity of ZnO-NPs in nematodes. Similarly, Ag ion and Ag-NPs had the similar toxicity on reproduction, implying that the dissolved Ag^+ may also play a key role in inducing the toxicity of Ag-NPs [6]. The linear correlation between the Ag-NPs toxicity and the dissolved silver ion was further detected in nematodes [7]. Nevertheless, some Ag-NPs (typically less soluble due to

Fig. 6.1 Concentration-survival curves of L1 larvae for 24 h exposure to various toxicants in the presence of *E. coli* OP50 as a food [5]
ZnO NP (▲), bulk ZnO (●), and $ZnCl_2$ (■)

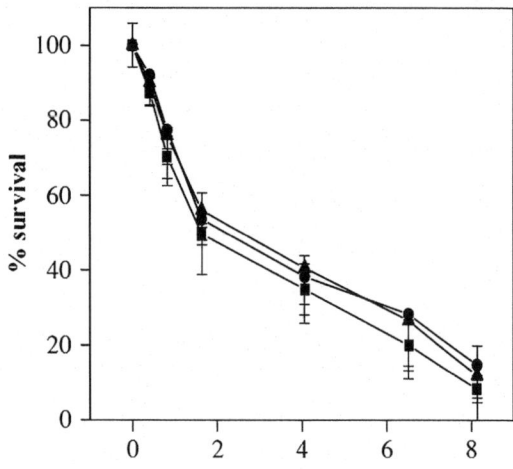

size or coating) still acted via the activation of oxidative stress in nematodes [7], implying that the toxicity formation from Ag-NPs may be more exactly due to the combinational effects from Ag-solubility, coating, and higher uptake rates. Therefore, although some reports suggest that the toxicity of ZnO-NPs on organisms, such as freshwater microalga, may be attributable to the dissolved zinc [8], an increasing evidence has implied that the dissolved metal ion can only partially explain the observed toxicity of metal-NPs or metal oxide-NPs. That is, this mechanism is potentially used to explain the toxicity formation of metal-NPs and metal oxide-NPs to a certain degree.

In nematodes, it has been further observed that the supernatant after the centrifugation was lower than the suspension of Al_2O_3-NPs [5], indicating the existence of NPs-specific toxic mechanism. More importantly, some metal-NPs or metal oxide-NPs may potentially release only very limited metal ion. For example, TiO_2-NPs only dissolved approximately 0.02 mg Ti/L after 96 h exposure [5].

6.2.1.2 Role of Release of Metal Ions in Toxicity Formation of QDs

In nematodes, both the deficit in egg-laying behavior and the reduction in lifespan were detected in QDs-MSA-exposed nematodes [9]. However, cadmium could cause a reduction in lifespan, but a low incidence of deficit in egg-laying behavior could be observed in exposed nematodes [9]. Therefore, the release of metal ion may explain the toxicity of QDs-MSA in reducing lifespan to a certain degree. Meanwhile, the Cd and the QDs-MSA may affect the nematodes by different mechanisms.

In nematodes, it has been further found that the observed toxicity of QDs at environmentally relevant concentrations on nematodes may not due to the Cd ion release. Since the molecular weight of CdTe QDs is 240 g/mol, 0.0042 μM of Cd^{2+} would be released from 1 μg/L of CdTe QDs with the assumption that the maximum amount of Cd^{2+} would be released from the CdTe QDs. The effects of this possible release of Cd^{2+} from the CdTe QDs on development and function of RMEs motor neurons were examined in nematodes. Prolonged exposure to CdTe QDs (1 μg/L) resulted in the deficits in the development of RMEs motor neurons as reflected by the altered fluorescent intensity of cell bodies and fluorescent size of RMEs motor neurons [10]. Prolonged exposure to CdTe QDs (1 μg/L) could further cause the deficit in the function of RMEs motor neurons as reflected by the induction of abnormal foraging behavior [10]. In contrast, the possible release of Cd^{2+} from CdTe QDs (1 μg/L) could not induce abnormal foraging behavior (Fig. 6.2) [10]. Additionally, the possible release of Cd^{2+} from CdTe QDs (1 μg/L) could not obviously affect the morphology, the relative fluorescent intensity, and the relative fluorescent size of cell bodies of RMEs neurons in exposed nematodes (Fig. 6.2) [10]. These data imply that, for the QDs at environmentally relevant concentrations, the observed toxicity of QDs on environmental organisms may be not mainly due to the possible release of Cd^{2+}.

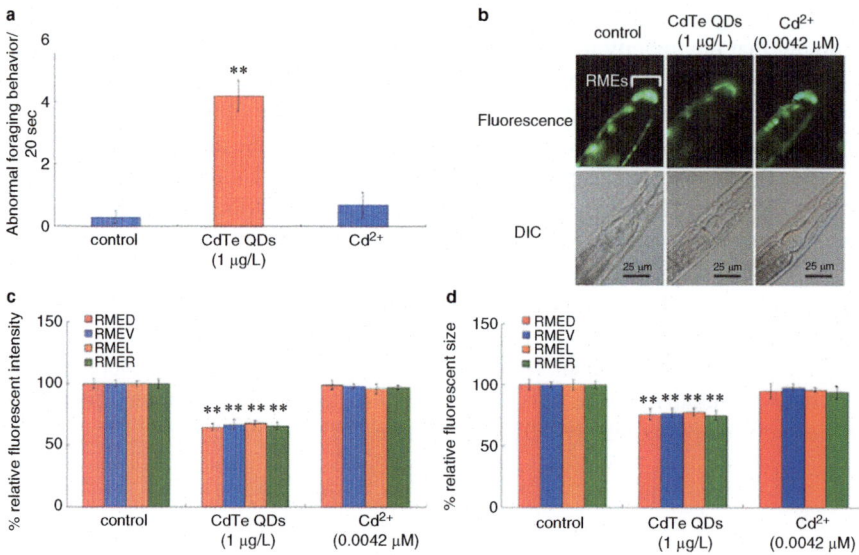

Fig. 6.2 Effect of the possible released Cd ion on neurotoxicity of CdTe QDs on development and function of RMEs motor neurons [10]
(**a**) Effect of the possible released Cd ion from CdTe QDs on foraging behavior. (**b**) Effect of the possible released Cd ion from CdTe QDs on morphology of RMEs motor neurons. (**c**) Effect of the possible released Cd ion from CdTe QDs on fluorescent intensity of cell bodies in RMEs motor neurons. (**d**) Effect of the possible released Cd ion from CdTe QDs on fluorescent size of cell bodies in RMEs motor neurons. CdTe QDs was exposed from L1-larvae to young adult. Bars represent mean ± SEM. **$P < 0.01$

6.2.2   Oxidative Stress

6.2.2.1   Induction of Reactive Oxygen Species (ROS) Production Acts As an Important Crucial Cellular Mechanism for Nanotoxicity Formation

The ROS production in nematodes can be determined at least by directly labeling the nematodes with molecular probe of CM-H$_2$DCFDA. After chronic exposure, Al$_2$O$_3$-NPs at concentrations of 8.1–30.6 mg/L led to the significant induction of intestinal ROS production in nematodes (Fig. 6.3) [11], and this observed induction of intestinal ROS production was further correlated with the induction of intestinal autofluorescence in Al$_2$O$_3$-NPs-exposed nematodes [11]. In contrast, the induction of intestinal ROS production could only be detected in nematodes exposed to bulk Al$_2$O$_3$ at concentrations more than 23.1 mg/L, and the induction of intestinal ROS production in Al$_2$O$_3$-NPs-exposed nematodes was higher than that in bulk Al$_2$O$_3$-exposed nematodes at all the examined concentrations [11]. The ROS production can also be assessed by the examination of carbonylated protein or GST activation in nematodes. For example, Ag-NPs exposure could result in the obvious GST activation in nematodes [12].

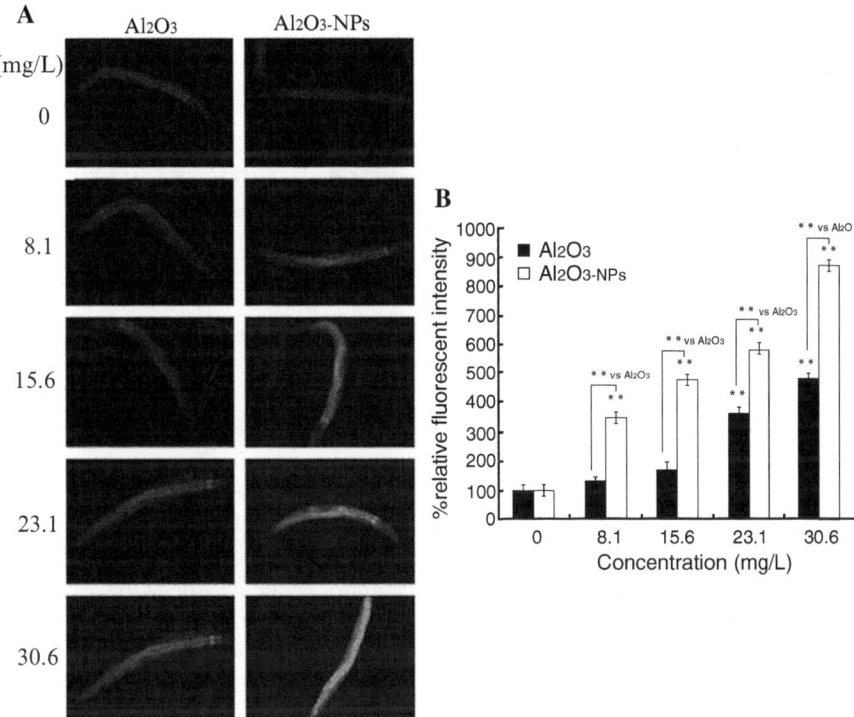

Fig. 6.3 Intestinal ROS production in wild-type nematodes exposed to different concentrations of Al_2O_3-NPs or bulk Al_2O_3 [11]

(**a**) Pictures showing the intestinal ROS production in nematodes exposed to 8.1, 15.6, 23.1, and 30.6 mg/L of Al_2O_3-NPs or bulk Al_2O_3. (**b**) Comparison of intestinal ROS production in nematodes exposed to different concentrations of Al_2O_3-NPs or bulk Al_2O_3. Chronic exposure was performed from L1-larvae for 10 days. Bars represent mean ± SEM. $**P < 0.01$ vs. 0 mg/L (if not otherwise indicated)

6.2.2.2 Some Evidence Raised to Prove the Important Role of Oxidative Stress in Nanotoxicity Induction in Nematodes

One of the important evidence to prove the role of oxidative stress in toxicity induction of ENMs in nematodes is from the pharmacological analysis. In nematodes, prolonged exposure to graphene oxide (GO) (100 mg/L) from L1-larvae to young adults could decrease lifespan, reduce brood size, and decrease locomotion behavior [13, 14]. Meanwhile, prolonged exposure to GO (100 mg/L) induced the significant intestinal ROS production in nematodes [13, 14]. Paraquat is a ROS-generating drug, and ascorbate is a classic antioxidant for organisms. It has been found that both the GO exposure and the treatment with paraquat (2 mM) induced a significant induction of ROS production in nematodes (Fig. 6.4) [13]. In contrast, treatment with ascorbate (10 mM) could significantly suppress the induction of ROS production in GO (100 mg/L)-exposed nematodes (Fig. 6.4) [13].

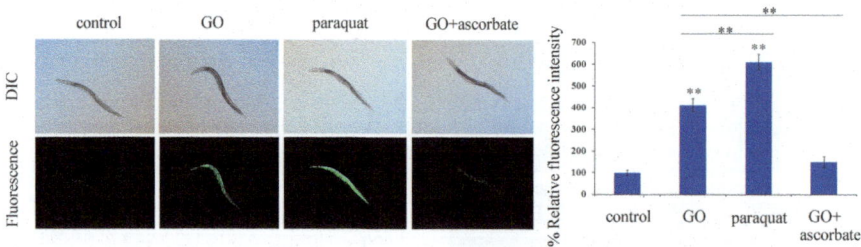

Fig. 6.4 Pharmacological assay [13]
L4-larval nematodes were treated with 2 mM paraquat for 12 h in the 12-well sterile tissue culture plates. Nematodes were exposed to GO (100 mg/L) first from L1-larvae to L4-larvae and then treated with 10 mM ascorbate for 24 h. Bars represent mean ± SEM. **$P < 0.01$

An indirect evidence indicating the important role of ROS production in the induction of toxicity of ENMs is the altered expression levels of genes required for control of oxidative stress. In nematodes, small sizes (4 nm and 10 nm) of TiO_2-NPs could induce more severe intestinal ROS production than large sizes (60 nm and 90 nm) of TiO_2-NPs after prolonged exposure from L1-larvae to adult day-1 [15]. Meanwhile, expression patterns of *sod-2* and *sod-3* encoding Mn-SODs were significantly dysregulated in nematodes exposed to different sizes (4, 10, 60, or 90 nm) of TiO_2-NPs, and the increase in *sod-2* or *sod-3* expression was higher in TiO_2-NPs (4 and 10 nm)-exposed nematodes than that in TiO_2-NPs (60 and 90 nm)-exposed nematodes [15]. The expressions of *sod-2* and *sod-3* were closely correlated with the toxicity of TiO_2-NPs in nematodes as reflected by the endpoints of lethality, growth, reproduction, locomotion behavior, intestinal autofluorescence, and ROS production [15]. Besides these, expression levels of *sod-1* and *sod-3* encoding superoxide dismutases were significantly altered in Ag-NPs-exposed nematodes, and the *sod-3* expression was correlated with the observed toxicity on reproduction in nematodes by a Pearson correlation test [6].

Based on the study on Al_2O_3-NPs toxicology, it has been suggested that the activation of oxidative stress in nematodes exposed to Al_2O_3-NPs was actually due to both the increase in intestinal ROS production and the suppression of ROS defense mechanisms [16], which is the normally observed underlying mechanism for the induction of oxidative stress in nematodes exposed to environmental toxicants, including ENMs.

6.2.2.3 Functional Evidence for the Important Role of Oxidative Stress in Nanotoxicity Induction in Nematodes

In nematodes, Mn-SODs encoded by *sod-2* and *sod-3* genes are key antioxidant enzymes with the function to defend against the activation of oxidative stress. Moreover, a functional evidence for the important role of oxidative stress in nanotoxicity induction was raised with the aid of nematodes overexpressing *sod-2* or *sod-3*. Under normal conditions, overexpression of *sod-2* or *sod-3* in RMEs motor neurons cannot noticeably affect the development of RMEs motor neurons and induce the

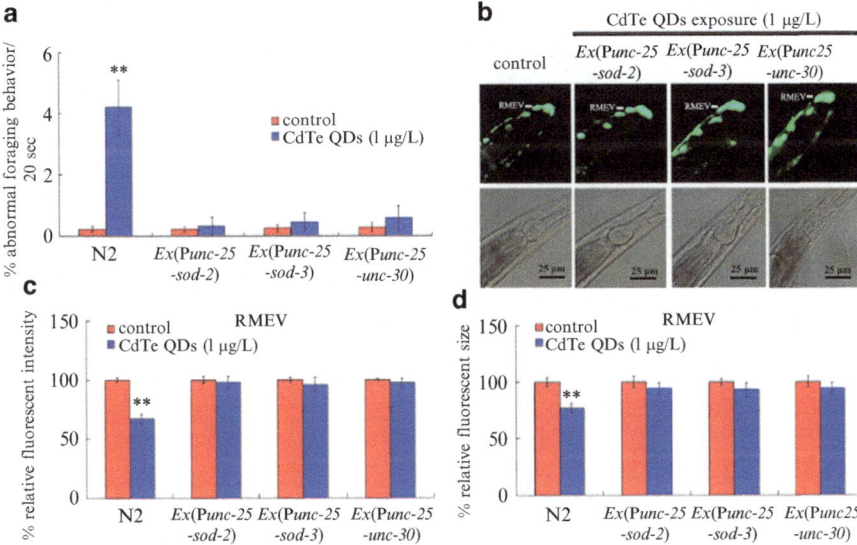

Fig. 6.5 Contribution of oxidative stress and cell identity to formation of CdTe QDS neurotoxicity on development and function of RMEs motor neurons [10]
(**a**) Effects of overexpressing *sod-2*, *sod-3*, or *unc-30* in RMEs motor neurons on foraging behavior in CdTe QDs-exposed nematodes. (**b**) Effects of overexpressing *sod-2*, *sod-3*, or *unc-30* in RMEs motor neurons on development of RMEs motor neurons in CdTe QDs-exposed nematodes. (**c**) Effects of overexpressing *sod-2*, *sod-3*, or *unc-30* in RMEs motor neurons on fluorescent intensity of cell body of RMEV motor neuron in CdTe QDs-exposed nematodes. (**d**) Effects of overexpressing *sod-2*, *sod-3*, or *unc-30* in RMEs motor neurons on fluorescent size of cell body of RMEV motor neuron in CdTe QDs-exposed nematodes. CdTe QDs was exposed from L1-larvae to young adult. Bars represent mean ± SEM. **$P < 0.01$

abnormal foraging behavior (Fig. 6.5) [10]. After prolonged exposure to CdTe QDs (1 µg/L), the formation of abnormal foraging behavior still could not be observed in nematodes overexpressing *sod-2* or *sod-3* in the RMEs motor neurons (Fig. 6.5) [10]. Additionally, the deficits in development of RMEs motor neurons could also not be detected in nematodes overexpressing *sod-2* or *sod-3* in the RMEs motor neurons after prolonged exposure to CdTe QDs (1 µg/L) (Fig. 6.5) [10]. Similarly, ectopically expression of human or nematode Mn-SODs genes also could effectively prevent the induction of ROS production and the toxicity formation of TiO$_2$-NPs [15].

6.2.3 Intestinal Permeability

Enhancement in intestinal permeability provides an important cellular basis for the bioavailability of ENMs to targeted tissues or organs in nematodes through the intestinal barrier. One of the strategies to assess the intestinal permeability is by labeling the examined nematodes with lipophilic fluorescent dye of Nile red. In nematodes, prolonged exposure to GO-SH at concentrations ≥100 µg/L could cause the toxicity on functions of both primary targeted organs, such as the intestine, and

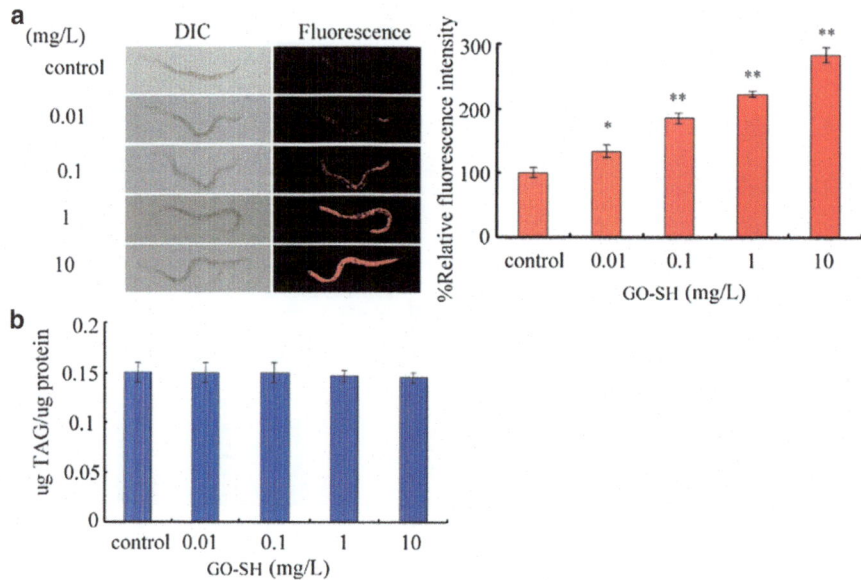

Fig. 6.6 Enhancement in intestinal permeability in GO-SH-exposed nematodes [17]
(**a**) Nile red staining results. The left-hand side shows pictures, and the right-hand side compares the relative fluorescence intensity of Nile red signals. (**b**) Comparison of triglyceride content. Prolonged exposure was performed from L1-larvae to adult day-1. Bars represent mean ± SD. *$P < 0.05$ vs. control, **$P < 0.01$ vs. control

secondary targeted organs, such as the neurons and the reproductive organs [17]. More importantly, prolonged exposure to GO-SH at concentrations ≥10 µg/L significantly increased the relative fluorescence intensity of Nile red signals in the body of nematodes (Fig. 6.6) [17]. Some of the fluorescent signals of Nile red could be detected in the secondary targeted organs, such as the reproductive organs, and even in the embryos in the body of GO-SH-exposed nematodes (Fig. 6.6) [17]. Meanwhile, although Nile red is also a normally used molecular probe to label the fat storage [18], prolonged exposure to GO-SH at all the examined concentrations could not affect the triglyceride content in nematodes (Fig. 6.6) [17], which implies that prolonged exposure to GO-SH may potentially enhance the intestinal permeability in nematodes.

6.2.4   Defecation Behavior

Besides the enhancement in intestinal permeability, prolonged defecation cycle length is another important cellular basis for the severe accumulation of ENMs in the body and the bioavailability of ENMs to targeted tissues or organs of nematode. With the CdTe QDs as an example, prolonged exposure to CdTe QDs (0.1–1 µg/L)

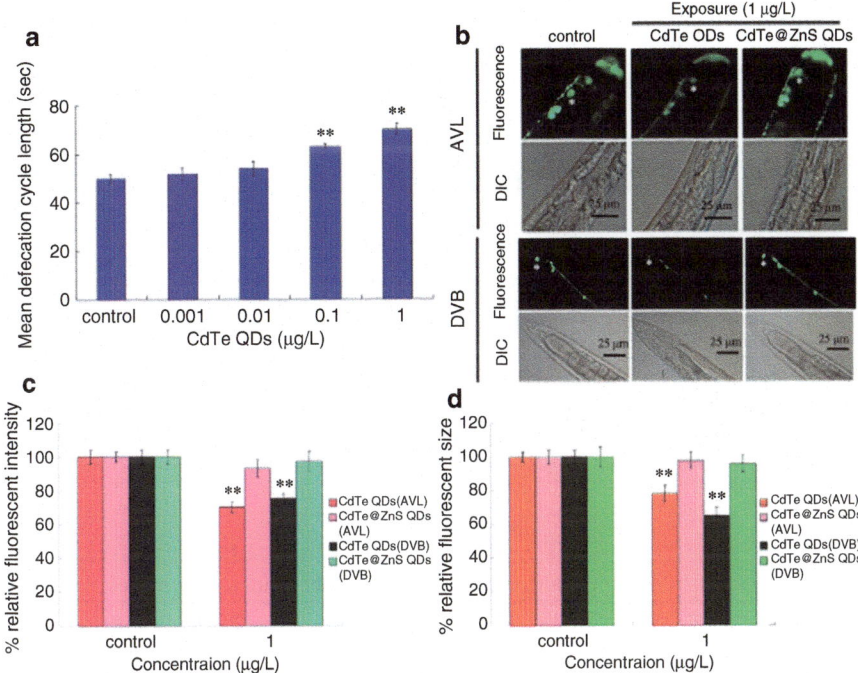

Fig. 6.7 Alterations of defecation behavior and damage on the neurons controlling defecation behavior in CdTe QDs-exposed nematodes [10]
(**a**) Effects of CdTe QDs on mean defecation cycle length. (**b, c,** and **d**) Effects of CdTe QDs or CdTe@ZnS QDs (1 μg/L) exposure on development of AVL and DVB neurons. Bars represent mean ± SEM. **$P < 0.01$

could significantly increase the mean defecation cycle length in nematodes (Fig. 6.7) [10]. Moreover, prolonged exposure to CdTe QDs (1 μg/L) could also significantly decrease both the relative fluorescent intensity and the relative fluorescent size of cell bodies of AVL and DVB neurons controlling the defecation behavior (Fig. 6.7) [10], which provide the corresponding important cellular basis for the observed increase in defecation cycle length in CdTe QDs-exposed nematodes. In contrast, surface ZnS coating could obviously prevent the neurotoxicity of CdTe QDs exposure in increasing mean defecation cycle length and the damage of CdTe QDs exposure on the development of AVL and DVB neurons (Fig. 6.7) [10].

In nematodes, prolonged exposure to GO (1–100 mg/L) could also significantly increase the mean defecation cycle length [14]. Meanwhile, prolonged exposure to GO (100 mg/L) further significantly decreased the expressions of *unc-101*, *itr-1*, *smp-1*, *iri-1*, and *cab-1*, and increased the expressions of *fat-3*, *isp-1*, *unc-44*, *fat-2*, *clk-1*, *hlh-8*, *gat-1*, *lim-6*, *unc-93*, *mlg-2*, *ced-10*, and *egl-30* [14]. These genes have been shown to be required for the control of defecation behavior in nematodes at different aspects [19]. These results imply that the altered defecation behavior may be associated with the functional changes of UNC-101, ITR-1, SMP-1, IRI-1, CAB-

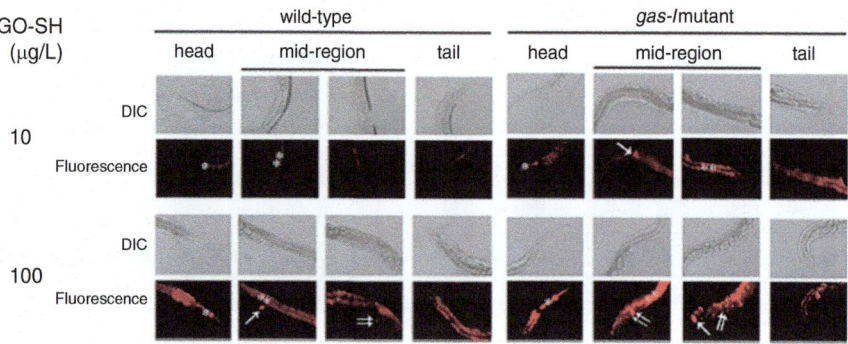

Fig. 6.8 Comparison of translocation and distribution of GO-SH/Rho B in wild-type and *gas-1* mutant nematodes [17]

Arrowheads indicate the spermatheca (single arrowheads) and the embryos (double arrowheads). The pharynx (*) and intestine (**) are also indicated. Prolonged exposure was performed from L1-larvae to adult day-1

1, FAT-3, ISP-1, UNC-44, FAT-2, CLK-1, HLH-8, GAT-1, LIM-6, UNC-93, MLG-2, CED-10, and EGL-30 proteins, which provides the underlying molecular basis for the observed increase in mean defecation cycle length in GO-exposed nematodes.

6.2.5   Bioavailability to Targeted Organs

Bioavailability to targeted organs plays a pivotal role in the toxicity induction of ENMs in nematodes. With the GO-SH as an example, the biodistribution and the translocation of GO-SH in nematodes were investigated by loading Rhodamine B (Rho B) on GO-SH. After prolonged exposure to GO-SH/Rho B (10 μg/L), only a very weak accumulation of GO-SH/Rho B was observed in the pharynx and the intestine in nematodes (Fig. 6.8) [17]. Moreover, after prolonged exposure to GO-SH/Rho B (100 μg/L), a large amount of GO-SH/Rho B was accumulated in the pharynx and the intestine, as well as in the reproductive organs, such as the spermatheca, in nematodes (Fig. 6.8) [17]. In contrast, treatment with Rho B alone caused the relatively equal distribution of fluorescence signals in the body of nematodes [17]. These results well explained the observed toxicity of GO-SH on the functions of both primary and secondary targeted organs in nematodes [17]. After prolonged exposure to GO-SH/Rho B (100 μg/L), GO-SH/Rho B could be even detected in the embryos of exposed nematodes (Fig. 6.8) [17], implying the possible transgenerational toxicity of GO-SH after long-term exposure.

In nematodes, the biodistribution and the translocation of GO-SH are under the control of *gas-1* encoding a subunit of mitochondrial complex I. Mutation of *gas-1* could induce a susceptibility of nematodes to GO-SH toxicity as indicated by the endpoints of locomotion behavior, brood size, and intestinal ROS production [17].

Meanwhile, after prolonged exposure to GO-SH/Rho B (10 µg/L), mutation of *gas-1* caused the accumulation of GO-SH/Rho B in the reproductive organs, such as the spermatheca, and the more severe accumulation of GO-SH/Rho B in the pharynx and in the intestine (Fig. 6.8) [17]. Moreover, after prolonged exposure to GO-SH/Rho B (100 µg/L), mutation of *gas-1* further resulted in the more severe accumulation of GO-SH/Rho B in the pharynx, the intestine, the spermatheca, and the embryos in nematodes (Fig. 6.8) [17]. One important underlying cellular basis for the function of GAS-1 in regulating the biodistribution and the translocation of GO-SH is that mutation of *gas-1* could obviously enhance the intestinal permeability in GO-SH-exposed nematodes [17].

In nematodes, after prolonged exposure, the toxicity of nanopolystyrene particles at concentrations higher than 10 µg/L could be found in exposed nematodes, and the transgenerational toxicity of nanopolystyrene particles at concentrations higher than 100 µg/L could be further detected from the P0 generation to F1 generation [20]. Meanwhile, after prolonged exposure to nanopolystyrene particles at concentrations (10 µg/L), a large amount of nanopolystyrene particles were observed to be accumulated in the intestine (especially in the middle and in the posterior regions of the intestine) and in the tail, and a limited amount of nanopolystyrene particles were observed to be accumulated in the pharynx in nematodes (Fig. 6.9) [20]. More importantly, the obvious accumulation of nanopolystyrene particles was detected in both side arms of the gonads of exposed nematodes (Fig. 6.9) [20]. In contrast, only a certain amount of nanopolystyrene particles in the intestine was detected in the progeny of nanopolystyrene particles (10 µg/L)-exposed nematodes (Fig. 6.9) [20], which is corresponding to the observations of no significant induction of intestinal ROS production, decrease in locomotion behavior, and reduction in brood size in the progeny of nanopolystyrene particles (10 µg/L)-exposed nematodes [20].

6.2.6   Acceleration in Aging Process

In nematodes, lifespan and aging-related phenotypes are important endpoints to detect the long-term adverse effects of environmental toxicants on animals. After prolonged exposure from L1 larvae to adult day-1, although GO at concentrations of 0.1–1 mg/L could not significantly affect the lifespan of nematodes, GO at concentrations more than 10 mg/L significantly reduced the lifespan in nematodes (Fig. 6.10) [21]. More importantly, prolonged exposure to GO further induced the alterations in aging-related phenotypes. Lipofuscin, an endogenous intralysosomal autofluorescent marker, can accumulate in aging nematodes. After prolonged exposure, GO at concentrations more than 1 mg/L could induce the noticeable intestinal autofluorescence and intestinal ROS production in nematodes at the stage of adult day-10 (Fig. 6.10) [21]. Furthermore, prolonged exposure to GO (10 mg/L) could significantly decrease the locomotion behavior of nematodes at the stages from adult day-4 to adult day-12 (Fig. 6.10) [21].

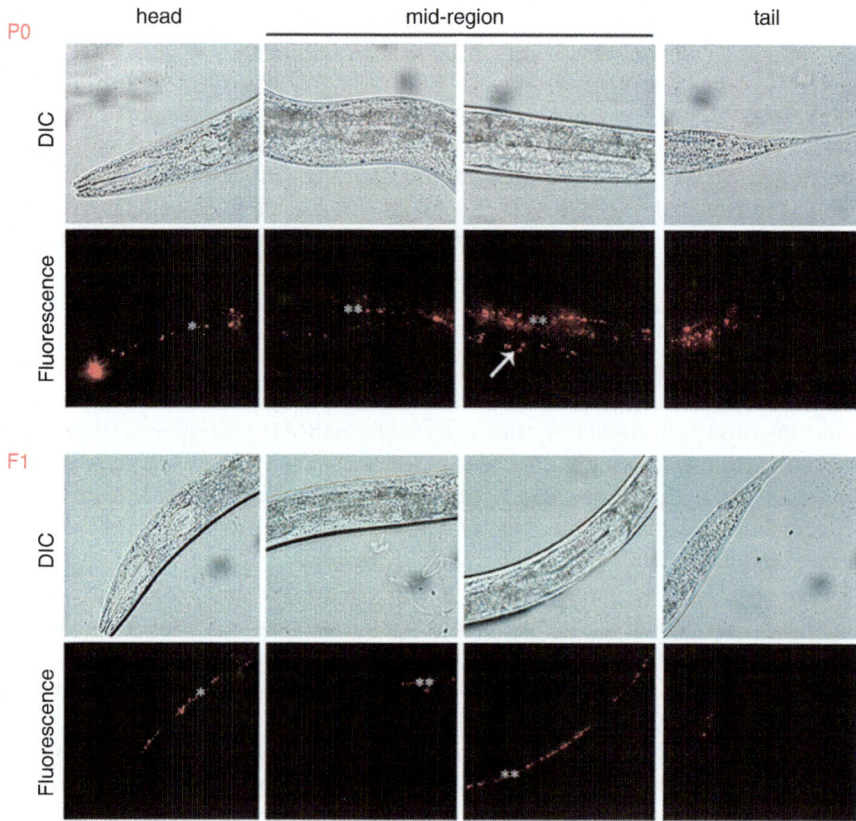

Fig. 6.9 Distribution and translocation of nanopolystyrene particles in nematodes [20]
The pharynx (*) and the intestine (**) are indicated by asterisks. The single arrowhead indicates
the gonad. Exposure concentration of nanopolystyrene particles was 10 μg/L. Prolonged exposure
was performed from L1-larvae to adult day-1

In nematodes, the acceleration in aging process can be also reflected by the formation of reproductive senescence. In silica-NPs-exposed nematodes, although the lifespan has not been significantly affected, a reduction in the progeny production was already detected in silica-NPs-exposed nematodes [22]. This reduction in progeny production was accompanied by an increase in "bag of worms" phenotype that is characterized by failed egg laying and usually formed in aged nematodes [22]. Based on the experimental design, the possibility that the formation of "bag of worms" phenotype is due to the deficit in vulva development or inadequate motility of the vulval muscles in silica-NPs-exposed nematodes has been excluded.

In nematodes, another important endpoint to reflect the acceleration in aging process is the premature reduction in pharyngeal pumping rate. Pumping rate represents a sensitive endpoint to help us to detect the aging processes that are characterized by a linear rate reduction over time in nematodes. A normal, age-related decline

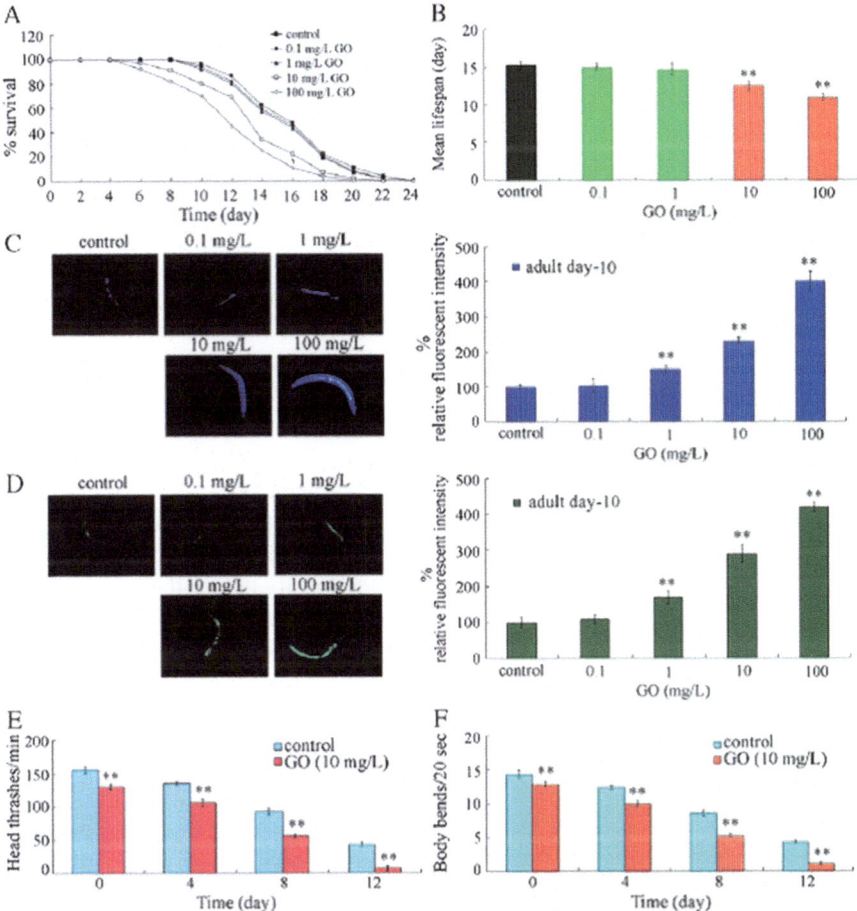

Fig. 6.10 Toxicity assessment of GO on lifespan and aging-related phenotypes in nematodes [21] (**a**) Effects of GO exposure on lifespan. (**b**) Comparison of mean lifespans. (**c**) Effects of GO exposure on intestinal autofluorescence. (**d**) Effects of GO exposure on intestinal ROS production. (**e** and **f**) Effects of GO exposure on locomotion behavior as indicated by endpoints of head thrash and body bend. Bars represent mean ± SEM. **P < 0.01

of pumping values is observed in silica-NP-s treated nematodes until day-3 after silica-NPs treatment; however, between the day-3 and the day-4, the pharyngeal pumping was significantly reduced in silica-NPs-exposed nematodes [23]. An underlying cellular basis has been raised that the silica-NPs accumulated in the pharynx may have the potential to interfere with pharyngeal function, which in turn induces the untimely premature induction of age-related decline of pumping in nematodes [23].

Increase in insolubility and aggregation of endogenous proteins is also a prominent aging-related molecular phenotype in aged nematodes. The induction of an insoluble, SDS-resistant fraction of ubiquitinated proteins associated with the acceleration in aging process was further found in silica-NPs-exposed nematodes. Using the filter retardation technique to trap the insoluble, SDS-resistant proteins on a cellulose-acetate membrane, it has been shown that the adult nematodes exposed to silica-NPs accumulated the severe insoluble, ubiquitinated proteins with the two to threefold increase in insoluble, ubiquitinated proteins compared with control after 24 h exposure to silica-NPs (Fig. 6.11) [23]. Considering the threefold increase in insoluble, ubiquitinated proteins from young to middle-aged nematodes and fourfold increase in insoluble, ubiquitinated proteins from young to old nematodes (Fig. 6.11) [23], exposure to silica-NPs for 24 h may induce a fraction of insoluble, SDS-resistant, ubiquitinated proteins that resembles the increase in the insoluble, endogenous proteins in old nematodes. Because the bulk-silica particles could not trap the SDS-resistant, ubiquitinated proteins on the filters, the induction of protein insolubility may relatively be a specific effect of silica at the nanosize.

The aberrant protein aggregation as reflected by the formation of amyloid-like structures was further identified to exist locally in nematodes. After labeling the adult nematodes with amyloid-dye Congo red, the intestinal cells formed the amyloid-like structures in cell nucleoli after silica-NPs exposure (Fig. 6.11) [23], implying that silica-NPs may have the potential to locally induce the fibrillation of endogenous proteins to amyloid-like aggregates in the specific microenvironments. Moreover, after gel electrophoresis analysis, the biochemical fractionation of insoluble proteins showed an increase in band intensities in silica-NPs-exposed nematodes [23], indicating that silica-NPs may selectively induce the insoluble proteins. Therefore, at least in silica-NPs-exposed nematodes, alteration in protein homeostasis toward protein insolubility can act as a hallmark of in vivo interactions between ENMs and the animals.

6.2.7   Innate Immune Response

The nematodes can normally survive after feeding with E. coli OP50 as the food. However, after chronic exposure to GO (1 mg/L) from L1-larvae to adult day-4 or adult day-8, the severe intestinal accumulation of E. coli OP50 was observed in GO-exposed nematodes (Fig. 6.12) [24]. To examine the potential association of GO with the accumulation of food OP50, the possible colocalization of GO with OP50 was investigated in GO-exposed nematodes. In day-8 adults, an obvious colocalization of GO with OP50 was detected in GO-exposed nematodes (Fig. 6.12) [24]. The colocalization of GO with OP50 was mainly formed in the anterior region of the intestine in adults day-6, and the colocalization of GO with OP50 could be detected at both the anterior region and the posterior region of intestine in adult day-8 [24]. In day-6 adults, the accumulation of both GO and OP50 in the tail region was seldom observed in GO-exposed nematodes; however, in day-8 adults, the colocalization of GO with OP50 could be formed at the region of the defecation

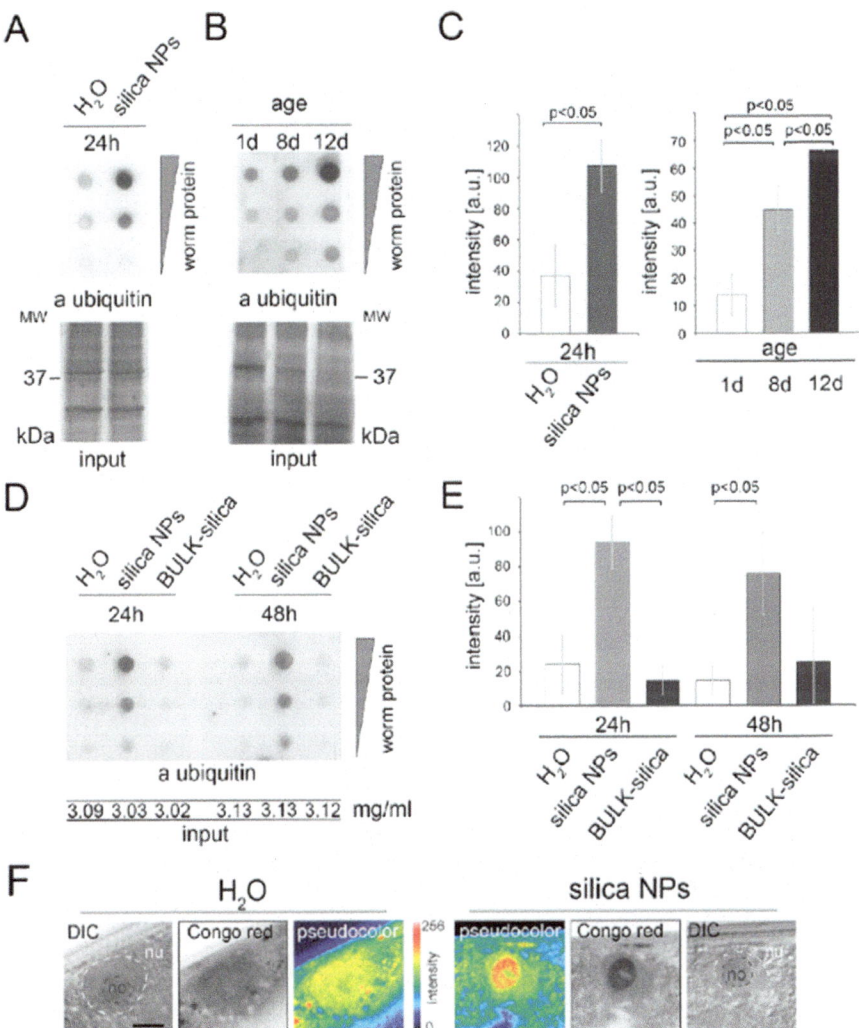

Fig. 6.11 Accumulation of SDS-insoluble, ubiquitinated proteins in silica-NPs-exposed nematodes [23]

Filter retardation assays show an accumulation of SDS-insoluble ubiquitinated proteins in nematodes (a) treated with 2.5 mg/mL silica-NPs for 24 h in comparison with H_2O as a mock control (b) with age. (c) Respective densitometric quantification of filter retardation assays. (d) Filter retardation assay showing accumulation of SDS-insoluble ubiquitinated proteins in nematodes incubated for 24 and 48 h with H_2O, 2.5 mg/mL silica-NPs, or 2.5 mg/mL bulk-silica particles (500 nm diameter). (e) Respective densitometric quantification of the filter retardation assays in (d). (f) Fluorescent micrographs of representative anteriormost intestine nuclei of 4-day-old, adult nematodes that were mock treated (H_2O) or treated with 2.5 mg/mL silica-NPs from day-1 of adulthood, fixed and stained with the amyloid-dye Congo red. a, anti; a.u., arbitrary units; d, days of adulthood; DIC, differential interference contrast; h, hours; kDa, kilodalton; MW, molecular weight; no, nucleolus; nu, nucleus. Bar, 7.5 μm

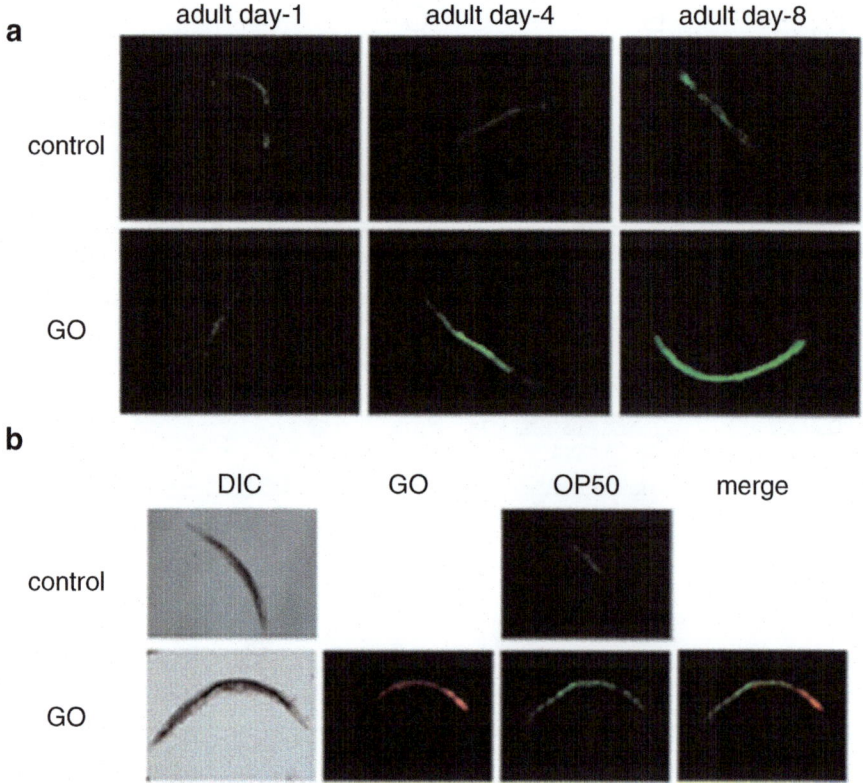

Fig. 6.12 Chronic exposure to GO (1 mg/L) induced the intestinal accumulation of *E. coli* OP50 (**a**) and colocalization of GO with *E. coli* OP50 in the intestine (**b**) in nematodes [24]

structure in the tail [24], implying the potential damage on the defecation behavior by GO exposure at this developmental stage.

In nematodes, GO at the examined concentrations could not obviously affect the survival of OP50 [24]. In contrast, chronic exposure to GO (1 mg/L) could significantly decrease the expressions of genes (*F08G5.6*, *pqm-1*, *K11D12.5*, *prx-11*, *spp-1*, *lys-7*, *lys-2*, *abf-2*, *acdh-1*, and *lys-8*) encoding antimicrobial peptides [24]. The *nlp-29* gene encodes another antimicrobial peptide, and chronic exposure to GO (1 mg/L) also significantly decreased the expression of P*nlp-29::GFP* in nematodes [24]. Moreover, chronic exposure to GO (1 mg/L) significantly decreased the expressions of *nsy-1*, *sek-1*, and *pmk-1*, which encode the core signaling cascade of p38 MAP kinase signaling, one of the key signaling pathways controlling the innate immune response to bacterial infection in nematodes [24]. Meanwhile, chronic exposure to GO (1 mg/L) caused the severe toxicity on nematodes, such as the significant reduction in lifespan in GO-exposed nematodes [24]. Therefore, the decrease in innate immune response may act as a crucial cellular mechanism for the observed toxicity induced by chronic exposure to GO in nematodes.

6.2.8   Mitochondrial Damage and DNA Damage

It has been shown that the toxicity formation of Ag-NPs may be largely due to the induction of ROS production and only partially due to dissolved Ag ions [25]. The mitochondrial damage can be measured by the mitochondrial membrane potential (Dwm) with TMRE staining using the fluorescence microscopy and the COPAS™ Select. In CCCP (a positive control)-exposed nematodes, a significant decrease in mitochondrial membrane potential was detected after TMRE staining (Fig. 6.13) [25]. At the low concentration (LC10), no change in mitochondrial membrane potential was found in any of examined Ag-NPs- and AgNO$_3$-exposed nematodes, whereas the decreased fluorescence could be detected in nematodes exposed to PVP8-Ag-NPs and AgNO$_3$ at the high concentration (LC50) (Fig. 6.13) [25], which suggests that both Ag-NPs and AgNO$_3$ have the potential to alter the mitochondrial membrane permeability in nematodes.

Besides the mitochondrial damage, it has been further demonstrated that exposure to Ag-NPs or AgNO$_3$ caused the obvious oxidative DNA damage as indicated by a significant increase in 8-OHdG levels in exposed nematodes (Fig. 6.14) [25]. The DNA damage was analyzed using two complementary methods, detection of oxidative base modification and polymerase-inhibiting lesions. Guanine base modification 8-OHdG was measured to assess the oxidative DNA damage, and DNA polymerase-inhibiting DNA lesions were measured by qPCR to detect a wider range of DNA damage, including strand breaks, adducts, dimers, crosslinks, etc. qPCR assay was performed on both the mitochondria and the nuclear genomes, because the mitochondrial genome is normally considered to be more sensitive to genotoxic chemicals than the nuclear genome. Additionally, PVP8-Ag-NPs induced an increase in the 8-OHdG levels only at high concentration (LC50), and no significant increase was observed after PVP38-Ag-NPs exposure (Fig. 6.14) [25]. Based on the qPCR analysis, neither any of the Ag-NPs nor AgNO$_3$ could result in both the nuclear DNA (nDNA) damage and the mitochondrial DNA (mtDNA) damage (Fig. 6.14) [25]. Similarly, neither any of the Ag-NPs nor AgNO$_3$ could cause the significant alteration in the mtDNA:nDNA ratio (Fig. 6.14) [25]. The *cep-1(gk138)* mutant lacks the *C. elegans* homolog of p53 protein and has the sensitivity to DNA damage. Mutation of *cep-1* also did not affect the phenotypes of DNA lesions either in nDNA or mtDNA in Ag-NPs- or AgNO$_3$-exposed nematodes (Fig. 6.14) [25]. A hypothesis has been raised that the Ag-NPs inside the mitochondria and nucleus may lead to the damage on mitochondrial respiratory chain, the induction of ROS production, and the interruption of ATP synthesis, in turn causing the DNA damage [25]. Therefore, besides the oxidative stress, the potential mitochondrial toxicity or genotoxicity of ENMs also needs the great attention in organisms.

Fig. 6.13 Analysis of mitochondrial membrane potential (Dwm) [25] (**a**) Pictures showing the Dwm assay. (**b**) The fluorescence intensity was quantified. Nematodes were incubated with tetramethylrhodamine ethyl ester (TMRE) for 3 h, and carbonyl cyanide 3-chlorophenyl hydrazone (CCCP) was used as a positive control. The result of fluorescence intensity was expressed as the mean value compared to control

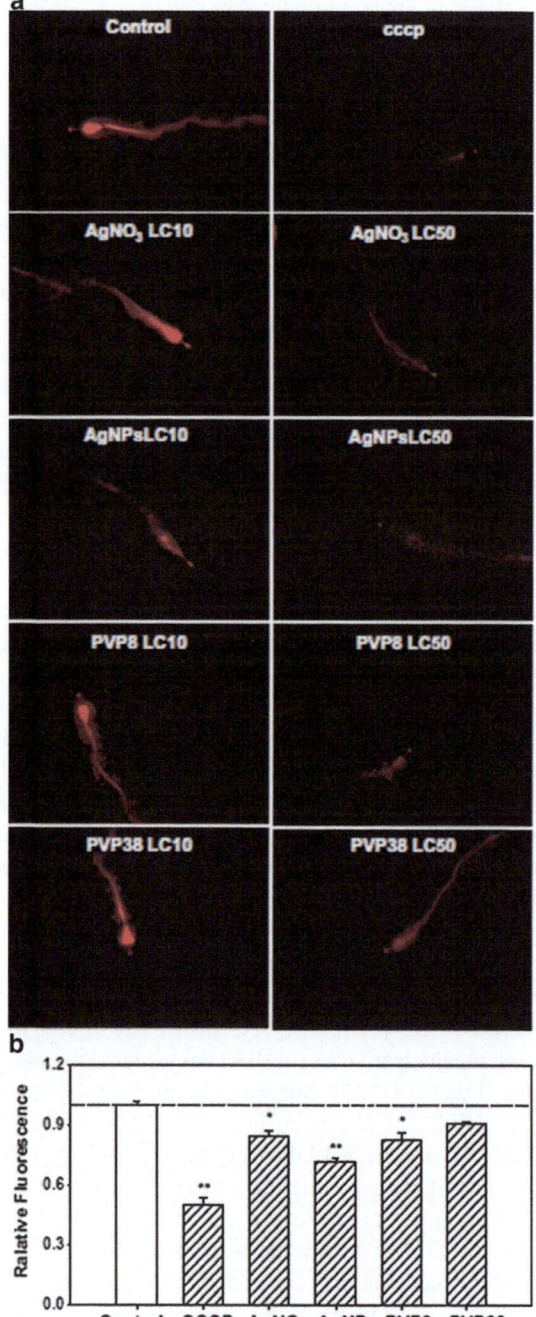

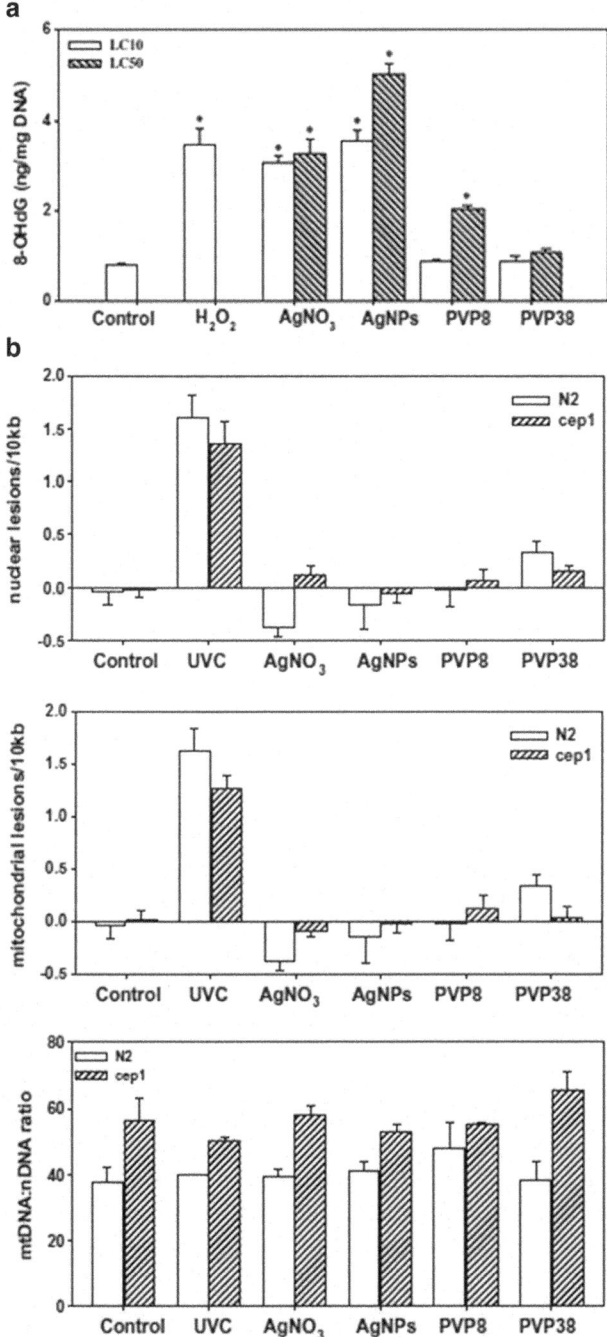

Fig. 6.14 Measurement of DNA damage based on 8-OHdG levels [25]
(**a**) Analysis of 8-OHdG level. (**b**) qPCR assay to assess nuclear and mitochondrial DNA damage. Polymerase-inhibiting DNA lesions were measured in wild-type and *cep-1(gk138)* mutant nematodes exposed to AgNO₃ or Ag-NPs using qPCR in the nuclear genome and/or mitochondrial genome. The mitochondrial-nuclear DNA ratio was also measured in the same nematodes. UVC was used as the positive control.

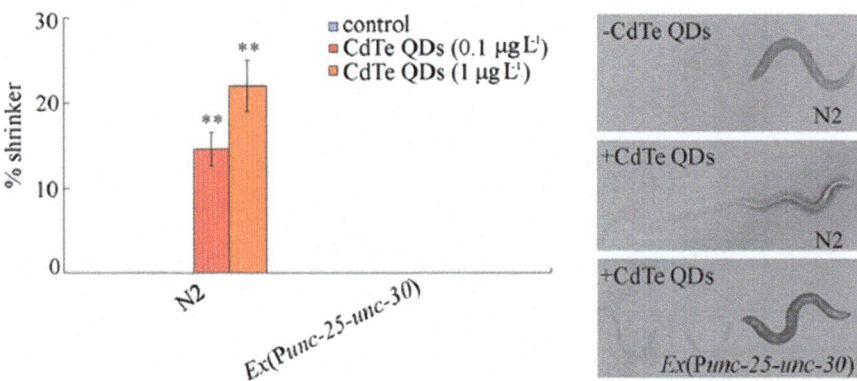

Fig. 6.15 Formation of shrinking behavior in CdTe QDs exposed wild-type or *Ex(Punc-25-unc-30)* transgenic strain [26]
Control, without CdTe QDs exposure. CdTe QDs was exposed from L1-larvae to young adults. Bars represent mean ± SEM. **$P < 0.01$

6.2.9 Developmental Fate

In *C. elegans*, *unc-30* encoding a homeodomain transcription factor is involved in maintaining cell identity of GABAergic neurons, such as the RMEs motor neurons. To determine the possible role of cell identity in toxicity formation of CdTe QDs, *unc-30* was overexpressed in GABAergic neurons to strengthen the cell identity of RMEs motor neurons. Under normal conditions, overexpression of *unc-30* cannot induce the formation of shrinking behavior or abnormal foraging behavior and cause the damage on the development of GABAergic neurons, such as the RMEs motor neurons and the D-type motor neurons [10, 26]. After prolonged exposure to CdTe QDs (1 µg/L), the shrinking behavior or abnormal foraging behavior still could not be detected in nematodes with the overexpression of *unc-30* in GABAergic neurons (Figs. 6.5 and 6.15) [10, 26]. Moreover, after prolonged exposure to CdTe QDs (1 µg/L), the damage on the development of RMEs or D-type motor neurons as reflected by the endpoints of relative fluorescent intensity and relative fluorescent size of cell body of RME or D-type neurons also could not be observed in nematodes with the overexpression of *unc-30* in GABAergic neurons (Fig. 6.5) [10, 26]. These two lines of evidence suggest that strengthening the identity of RMEs or D-type motor neurons may induce a resistance to the neurotoxicity of CdTe QDs on both the development and the function of RMEs or D-type motor neurons in nematodes.

6.2.10 Deficit in Cellular Endocytosis in Intestinal Cells

With QDs as an example, QDs could be localized in the intestinal lysosomes based on the observation on the colocalization between QDs and lysosomes in nematodes after acute exposure for 12 h [27]. In nematodes, the intestinal endocytosis process is involved in the control of nutrition uptake. mCherry::RAB-11 was employed as a genetic tool to label the apical endocytic vesicles. Under the normal conditions, the mCherry::RAB-11 display the evenly distributed and small punctuate structures in intestinal cells (Fig. 6.16) [27]. However, after QDs exposure, the mCherry::RAB-11 signals tending to form large aggregates were detected in nematodes (Fig. 6.16) [27], suggesting the disturbed endocytosis process in QDs-exposed nematodes. This disruption of endocytosis process was in accordance with the growth inhibition and lifespan reduction caused by QDs exposure [27], implying that the QDs toxicity may be contributed by the disruption in endocytosis process. Because the $CdCl_2$ treatment resulted in only very little endocytic vesicle aggregation, the observed subcellular toxicity of QDs may be mainly due to their nanoeffects rather than the released Cd ion in nematodes [27]. Moreover, it has been shown that the different toxic effects caused by different QDs might be more likely due to their difference in the size and/or in the surface properties [27]. Additionally, the possibility that QDs may induce the severe endosome dysfunction has also been excluded [27].

To investigate the effects of QDs on lysosome function, mRFP::RME-1, a marker for basolateral recycling endosomes, was further employed. After QDs exposure, it has been found that the abnormal aggregation of mRFP::RME-1 was sufficient to induce the disruption of basal endosome recycling, and the RME-1 was mainly aggregated in the apical side in intestinal cells [27], implying that the subcellular toxicity induced by QDs may be more restricted to the apical side in intestinal cells. Moreover, the ingested nutrition is mainly stored as the form of fat in intestinal cells, and QDs exposure could cause the severe nutrition deprivation [27], which may be one of the important reasons for explaining the observed growth inhibition in QDs-exposed nematodes.

In nematodes, hTAC is a marker for clathrin-independent endocytosis, and hTfR is a marker for clathrin-dependent endocytosis. Under normal conditions, hTfR::GFP and hTAC::GFP are both evenly distributed and enriched at the basolateral membranes in intestinal cells [27]. However, after QDs exposure, both hTAC::GFP and hTfR::GFP exhibited the loss of basolateral enrichment and the abnormal aggregation at the apical side in intestinal cells [27]. These results demonstrate that both clathrin-independent endocytosis and clathrin-dependent endocytosis are involved in the toxicity formation of QDs in nematodes.

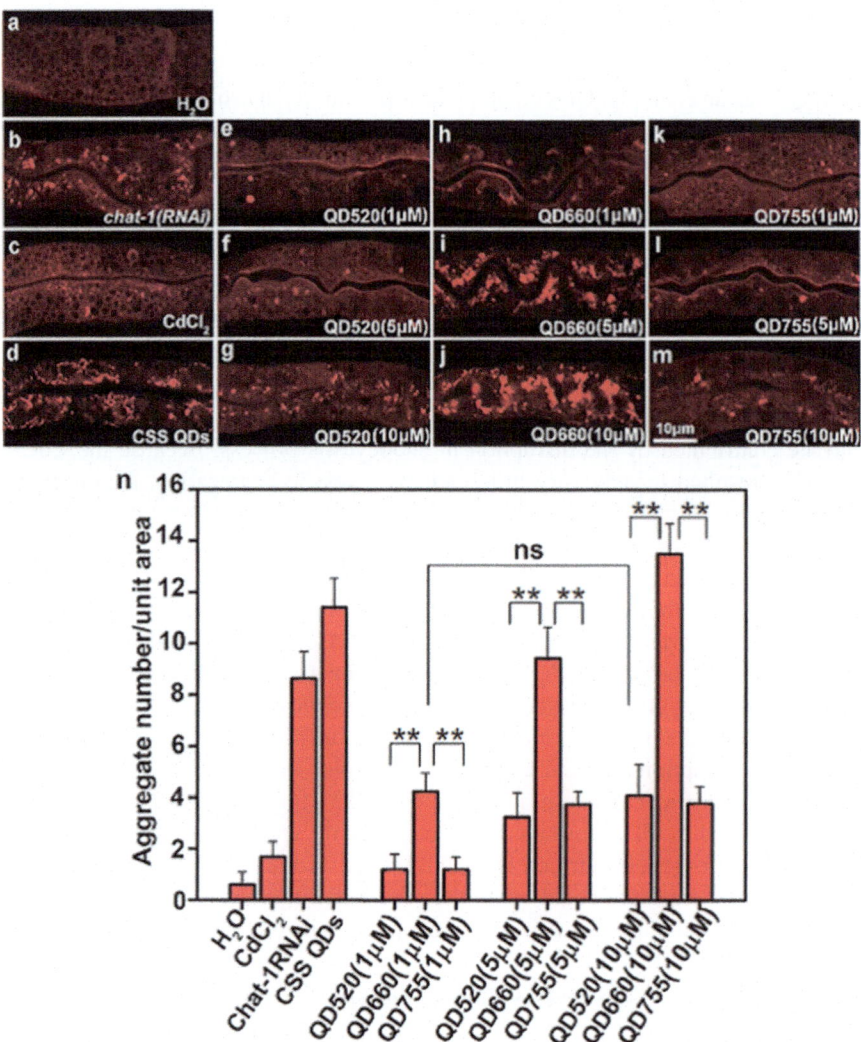

Fig. 6.16 Defective endocytosis caused by QDs [27]
Bars represent mean ± SEM. *$P < 0.05$; **$P < 0.01$

6.3 Physiological Mechanisms for Nanotoxicity Formation

6.3.1 Environmental Factors

In the environment, the toxicity of ENMs may be potentially altered by different environmental factors, such as the sunlight. By comparing the phototoxicity of ZnO-NPs and bulk ZnO under natural sunlight (NSL) versus ambient artificial laboratory light (AALL) illumination, it has been shown that exposure to ZnO-NPs or bulk ZnO under NSL could cause the greater mortality than that under AALL [28]. Additionally, ZnO-NPs showed greater phototoxicity than bulk ZnO despite their similar size of aggregates, suggesting the important role of primary particle size in determining the phototoxicity [28].

6.3.2 Exposure

6.3.2.1 Media Compositions

During the analysis on the toxicity of ENMs exposure, the media compositions should be carefully considered. For example, the K^+ medium has a much higher ionic strength, and the calculated 24 h LC_{50} value for ZnO-NPs in the presence of K^+ medium (4.9 mg/L) was much higher than that in pure water (2.2 mg/L) [29]. The Ag-NPs toxicity in the moderately hard reconstituted water was much higher than that in a K^+ medium, and the Ag-NPs in K^+ medium would aggregate with diameter of 1–1.6 μm that rapidly settled from the suspension [30], implying that a lower ionic strength medium would result in greater toxicity of ENMs in nematodes. Besides these, it has been further noticed that including the food of OP50 in the test medium could significantly enhance the Ag-NPs toxicity on nematodes compared to that in the absence of food [31].

6.3.2.2 Exposure Duration

In nematodes, different exposure durations would result in different toxicity performances for certain ENMs. For example, after exposure from L4-larvae for 24 h, DMSA-coated Fe_2O_3-NPs at concentrations more than 50 mg/L exhibited the toxicity at different aspects on nematodes [32]. In contrast, DMSA-coated Fe_2O_3-NPs at concentrations more than 500 μg/L could cause the toxicity at different aspects on nematodes after exposure from L1-larvae to adult day-1 [32]. With GO as another example, although GO at concentrations of 0.001–1 mg/L did not affect the survival of nematodes after exposure from adult day-1 to adult day-8, GO at the concentration of 1 mg/L could significantly decreased the survival of nematodes after exposure from L1-larvae to adult day-8 (Fig. 6.17) [24]. Additionally, GO only at concentrations more than 1 mg/L could significantly decrease the locomotion behavior, induce the intestinal autofluorescence, and induce the intestinal ROS production after exposure from adult day-1 to adult day-8 in nematodes (Fig. 6.17) [24]. In contrast, GO at concentrations more than 0.01 mg/L significantly decreased the locomotion behavior, induced the intestinal autofluorescence, and induced the intestinal ROS production after exposure from L1-larvae to adult day-8 (Fig. 6.17) [24], suggesting that exposure to GO from L1-larvae to adult day-8 could potentially induce more severe toxicity on the functions of both the primary and the secondary targeted organs than exposure from adult day-1 to adult day-8 in nematodes.

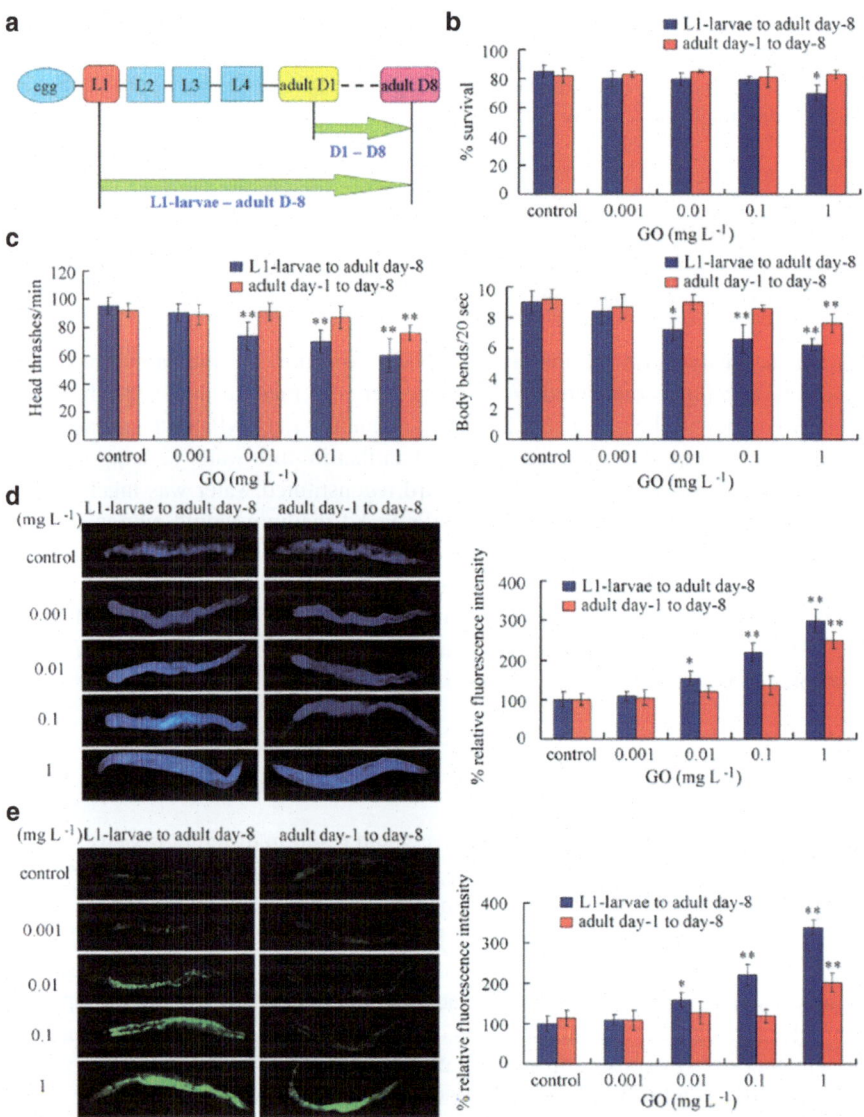

Fig. 6.17 Chronic toxicity assessment of GO using two different assay systems [24]
(**a**) Diagram of two assay systems for chronic GO exposure. (**b**) Effects of chronic GO exposure on survival of nematodes at the stage of adult day-8. (**c**) Effects of chronic GO exposure on locomotion behavior. (**d**) Effects of chronic GO exposure on intestinal autofluorescence. (**e**) Effects of chronic GO exposure on intestinal ROS production. GO exposure was performed from L1-larvae to adult day-8 or from adult day-1 to adult day-8. Bars represent mean ± SEM. *$p < 0.05$; **$p < 0.01$

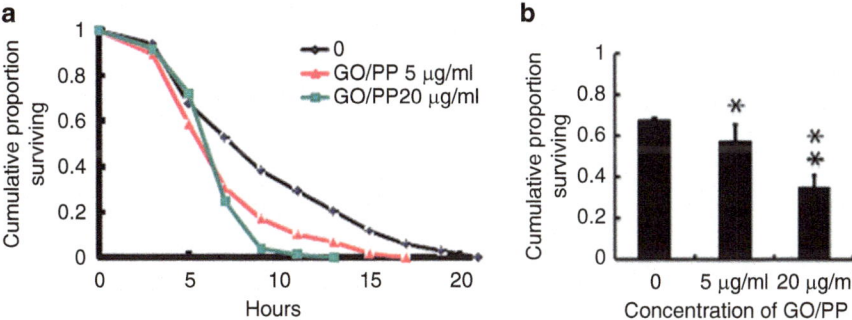

Fig. 6.18 Biological effects of GO/PP on nematodes [34]
(**a**) The survival curve of worms treated with GO/PP. (**b**) Survival curves against acute juglone exposure after a 2-day GO/PP treatment

6.3.3   Physiological State of Nematodes

Some of the studies have been performed for ENMs exposure under the oxidative state or certain stress in nematodes. For example, Zhang et al. investigated the sensitivity of CeO_2-NP-s pretreated nematodes to oxidative stress or thermal stress [33]. The CeO_2-NPs exposure not only impaired the thermotolerance but also weakened the tolerance to juglone, a ROS generator [33]. Nematodes pretreated with CeO_2-NPs could exhibit more intestinal ROS production under the oxidative state, and the intracellular accumulation of ROS induced by juglone might be negatively correlated with the survival of nematodes [33]. Moreover, it has been further observed that the mean lifespan of nematodes pretreated with PEGylated graphene oxide (GO/PP) was significantly reduced under oxidative state induced by juglone or heat stress at 35 °C, which could not be formed due to the relatively safe property of GO/PP under normal conditions (Fig. 6.18) [34].

6.3.4   Developmental Stages of Nematodes

A hermaphrodite adult stage usually can last 2–3 weeks, which is preceded by four larval stages (L1, L2, L3, and L4). Nematodes at different developmental stages exhibit different sensitivity to ENMs. Toxicity of Al_2O_3-NPs to L1-larval, L4-larval, or young adult was evaluated in nematodes. After exposure from the L1-larval stage, the significant increases of lethality, stress response, and intestinal lipofuscin autofluorescence could be detected in nematodes exposed to Al_2O_3-NPs at concentrations of 6.3–203.9 mg/L (Fig. 6.19) [35]. In contrast, after exposure from the L4-larval or the young adult stage, the significant increases of lethality, and intestinal lipofuscin autofluorescence could only be observed in nematodes exposed to Al_2O_3-NPs at concentrations of 12.7–203.9 mg/L (Fig. 6.19) [35], which

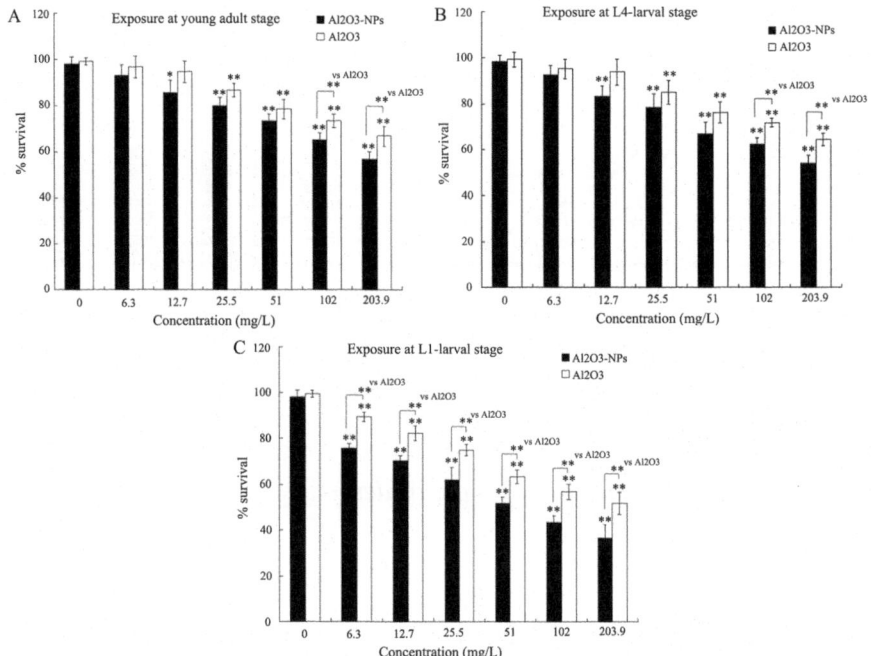

Fig. 6.19 Percentages of survival animals in nematodes exposed to different concentrations of Al₂O₃-NPs or bulk Al₂O₃ at the young adult, L4-larval, or L1-larval stage [35]
Bars represent mean ± SD. *$P < 0.05$ vs. 0 mg/L (if not specially indicated), **$P < 0.01$ vs. 0 mg/L (if not specially indicated)

demonstrates that the ENMs exposure in L1-larvae may induce more severe toxicity on nematodes than in L4-larvae or in young adults in nematodes.

6.3.5 Hormesis

Hormesis is one of the important forms of adaptive responses, and the adaptive response can be affected by different factors, including developmental stage, length of pretreatment duration, pretreated doses of the agents, length of the subsequent treatment duration, and dose of the subsequent treated agents in nematodes [36]. With Ag-NPs as an example, it has been found that the EC₅₀ of Ag-NPs for effects on the endpoint of reproduction during the life cycle was substantially lower than that in the short-term test [37], which suggests that the degree of Ag-NPs toxicity on reproduction of nematodes may be dependent on the life stage exposed and the duration of the exposure. In contrast, after the long-term exposure, the observed hormesis effects on the endpoint of reproduction during the short-term exposure could not be obviously generated [37]. To determine the hormesis effects for Ag-NPs, full life cycle toxicity tests were performed in nematodes [37].

6.4 Perspectives

In this chapter, with the concern on the cellular mechanisms of nanotoxicity formation, we systematically introduce the important roles of release of metal ion, oxidative stress, intestinal permeability, defecation behavior, bioavailability to targeted organs, acceleration in aging process, innate immune response, mitochondrial damage and DNA damage, developmental fate, and deficit in cellular endocytosis in intestinal cells in the toxicity formation of ENMs in nematodes. Nevertheless, there are still many other aspects on the underlying cellular mechanisms of nanotoxicity formation need to be further elucidated in nematodes. More importantly, besides the oxidative stress, the underlying chemical and molecular basis for these raised cellular mechanisms of nanotoxicity formation is still largely unclear.

Moreover, with the concern on the physiological mechanisms of nanotoxicity formation, we also introduce the important roles of environmental factors, exposure, physiological state, and hormesis of nematodes, and developmental stages of nematodes in the toxicity formation of ENMs in nematodes. These discussed physiological mechanisms will be helpful for our selecting and designing the suitable exposure method or route with the certain aim to assess the toxicity of certain ENMs or to elucidate the toxicological mechanisms of certain ENMs in nematodes.

References

1. Brenner S (1974) The genetics of *Caenorhabditis elegans*. Genetics 77:71–94
2. Leung MC, Williams PL, Benedetto A, Au C, Helmcke KJ, Aschner M, Meyer JN (2008) *Caenorhabditis elegans*: an emerging model in biomedical and environmental toxicology. Toxicol Sci 106:5–28
3. Zhao Y-L, Wu Q-L, Li Y-P, Wang D-Y (2013) Translocation, transfer, and *in vivo* safety evaluation of engineered nanomaterials in the non-mammalian alternative toxicity assay model of nematode *Caenorhabditis elegans*. RSC Adv 3:5741–5757
4. Wang D-Y (2016) Biological effects, translocation, and metabolism of quantum dots in nematode *Caenorhabditis elegans*. Toxicol Res 5:1003–1011
5. Wang H, Wick RL, Xing B (2009) Toxicity of nanoparticulate and bulk ZnO, Al_2O_3 and TiO_2 to the nematode *Caenorhabditis elegans*. Environ Pollut 157:1171–1177
6. Roh J, Sim SJ, Yi J, Park K, Chung KH, Ryu D, Choi J (2009) Ecotoxicity of silver nanoparticles on the soil nematode *Caenorhabditis elegans* using functional ecotoxicogenomics. Environ Sci Technol 43:3933–3940
7. Yang X, Gondikas AP, Marinakos SM, Auffan M, Liu J, Hsu-Kim H, Meyer JN (2012) Mechanism of silver nanoparticle toxicity is dependent on dissolved silver and surface coating in *Caenorhabditis elegans*. Environ Sci Technol 46:1119–1127
8. Franklin NM, Rogers NJ, Apte SC, Batley GE, Gadd GE, Casey PS (2007) Comparative toxicity of nanoparticulate ZnO, bulk ZnO, and $ZnCl_2$ to a freshwater microalga (*Pseudokirchneriella subcapitata*): the importance of particle solubility. Environ Sci Technol 41:8484–8490
9. Hsu PL, O'Callaghan M, Al-Salim N, Hurst MRH (2012) Quantum dot nanoparticles affect the reproductive system of *Caenorhabditis elegans*. Environ Toxicol Chem 31:2366–2374

10. Zhao Y-L, Wang X, Wu Q-L, Li Y-P, Wang D-Y (2015) Translocation and neurotoxicity of CdTe quantum dots in RMEs motor neurons in nematode *Caenorhabditis elegans*. J Hazard Mater 283:480–489

11. Yu S-H, Rui Q, Cai T, Wu Q-L, Li Y-X, Wang D-Y (2011) Close association of intestinal autofluorescence with the formation of severe oxidative damage in intestine of nematodes chronically exposed to Al_2O_3-nanoparticle. Environ Toxicol Pharmacol 32:233–241

12. Lim D, Roh J, Eom H, Choi J, Hyun J, Choi J (2012) Oxidative stress-related PMK-1 P38 MAPK activation as a mechanism for toxicity of silver nanoparticles to reproduction in the nematode *Caenorhabditis elegans*. Environ Toxicol Chem 31:585–592

13. Wu Q-L, Zhou X-F, Han X-X, Zhuo Y-Z, Zhu S-T, Zhao Y-L, Wang D-Y (2016) Genome-wide identification and functional analysis of long noncoding RNAs involved in the response to graphene oxide. Biomaterials 102:277–291

14. Wu Q-L, Yin L, Li X, Tang M, Zhang T, Wang D-Y (2013) Contributions of altered permeability of intestinal barrier and defecation behavior to toxicity formation from graphene oxide in nematode *Caenorhabditis elegans*. Nanoscale 5(20):9934–9943

15. Li Y-X, Wang W, Wu Q-L, Li Y-P, Tang M, Ye B-P, Wang D-Y (2012) Molecular control of TiO_2-NPs toxicity formation at predicted environmental relevant concentrations by Mn-SODs proteins. PLoS One 7(9):e44688

16. Li Y-X, Yu S-H, Wu Q-L, Tang M, Pu Y-P, Wang D-Y (2012) Chronic Al_2O_3-nanoparticle exposure causes neurotoxic effects on locomotion behaviors by inducing severe ROS production and disruption of ROS defense mechanisms in nematode *Caenorhabditis elegans*. J Hazard Mater 219-220:221–230

17. Ding X-C, Wang J, Rui Q, Wang D-Y (2018) Long-term exposure to thiolated graphene oxide in the range of μg/L induces toxicity in nematode *Caenorhabditis elegans*. Sci Total Environ 616-617:29–37

18. Wu Q-L, Rui Q, He K-W, Shen L-L, Wang D-Y (2010) UNC-64 and RIC-4, the plasma membrane associated SNAREs syntaxin and SNAP-25, regulate fat storage in nematode *Caenorhabditis elegans*. Neurosci Bull 26(2):104–116

19. Branicky R, Hekimi S (2006) What keeps *C. elegans* regular: the genetics of defecation. Trends Genet 22:571–579

20. Zhao L, Qu M, Wong G, Wang D-Y (2017) Transgenerational toxicity of nanopolystyrene particles in the range of μg/L in nematode *Caenorhabditis elegans*. Environ Sci Nano 4:2356–2366

21. Wu Q-L, Zhao Y-L, Zhao G, Wang D-Y (2014) microRNAs control of *in vivo* toxicity from graphene oxide in *Caenorhabditis elegans*. Nanomed Nanotechnol Biol Med 10:1401–1410

22. Pluskota A, Horzowski E, Bossinger O, von Mikecz A (2009) In *Caenorhabditis elegans* nanoparticle-bio-interactions become transparent: silica-nanoparticles induce reproductive senescence. PLoS One 4(8):e6622

23. Scharf A, Piechulek A, von Mikecz A (2013) Effect of nanoparticles on the biochemical and behavioral aging phenotype of the nematode *Caenorhabditis elegans*. ACS Nano 7(12):10695–10703

24. Wu Q-L, Zhao Y-L, Fang J-P, Wang D-Y (2014) Immune response is required for the control of *in vivo* translocation and chronic toxicity of graphene oxide. Nanoscale 6:5894–5906

25. Ahn J, Eom H, Yang X, Meyer JN, Choi J (2014) Comparative toxicity of silver nanoparticles on oxidative stress and DNA damage in the nematode, *Caenorhabditis elegans*. Chemosphere 108:343–352

26. Zhao Y-L, Wang X, Wu Q-L, Li Y-P, Tang M, Wang D-Y (2015) Quantum dots exposure alters both development and function of D-type GABAergic motor neurons in nematode *Caenorhabditis elegans*. Toxicol Res 4:399–408

27. Wang Q, Zhou Y, Song B, Zhong Y, Wu S, Cui R, Cong H, Su Y, Zhang H, He Y (2016) Linking subcellular disturbance to physiological behavior and toxicity induced by quantum dots in *Caenorhabditis elegans*. Small 23:3143–3154

28. Ma H, Kabengi NJ, Bertsch PM, Unrine JM, Glenn TC, Williams PL (2011) Comparative phototoxicity of nanoparticulate and bulk ZnO to a free-living nematode *Caenorhabditis elegans*: the importance of illumination mode and primary particle size. Environ Pollut 159:1473–1480

29. Ma H, Bertsch PM, Glenn TC, Kabengi NJ, Williams PL (2009) Toxicity of manufactured zinc oxide nanoparticles in the nematode *Caenorhabditis elegans*. Environ Toxicol Chem 28(6):1324–1330

30. Meyer JN, Lord CA, Yang XY, Turner EA, Badireddy AR, Marinakos SM, Chilkoti A, Wiesner MR, Auffan M (2010) Intracellular uptake and associated toxicity of silver nanoparticles in *Caenorhabditis elegans*. Aquat Toxicol 100(2):140–150

31. Ellegaard-Jensen L, Jensen KA, Johansen A (2012) Nano-silver induces dose-response effects on the nematode *Caenorhabditis elegans*. Ecotoxicol Environ Saf 80:216–223

32. Wu Q-L, Li Y-P, Tang M, Wang D-Y (2012) Evaluation of environmental safety concentrations of DMSA coated Fe_2O_3-NPs using different assay systems in nematode *Caenorhabditis elegans*. PLoS One 7(8):e43729

33. Zhang H, He X, Zhang Z, Zhang P, Li Y, Ma Y, Kuang Y, Zhao Y, Chai Z (2011) Nano-CeO_2 exhibits adverse effects at environmental relevant concentrations. Environ Sci Technol 45(8):3725–3730

34. Zhang W, Wang C, Li Z, Lu Z, Li Y, Yin J, Zhou Y, Gao X, Fang Y, Nie G, Zhao Y (2012) Unraveling stress-induced toxicity properties of graphene oxide and the underlying mechanism. Adv Mater 24:5391–5397

35. Wu S, Lu J-H, Rui Q, Yu S-H, Cai T, Wang D-Y (2011) Aluminum nanoparticle exposure in L1 larvae results in more severe lethality toxicity than in L4 larvae or young adults by strengthening the formation of stress response and intestinal lipofuscin accumulation in nematodes. Environ Toxicol Pharmacol 31:179–188

36. Zhao Y-L, Wang D-Y (2012) Formation and regulation of adaptive response in nematode *Caenorhabditis elegans*. Oxid Med Cell Longev 2012:564093

37. Tyne W, Little S, Spurgeon DJ, Svendsen C (2015) Hormesis depends upon the life-stage and duration of exposure: examples for a pesticide and a nanomaterial. Ecotoxicol Environ Saf 120:117–123

Chapter 7
Molecular Mechanisms of Nanotoxicity Formation

Abstract Based on the conserved property of molecular events, signal transduction pathways, epigenetic marks, and the homology of approximately 45% genes in *Caenorhabditis elegans* to human genome, *C. elegans* has the important potentials for the elucidation of underlying molecular mechanisms of toxicity induced by engineered nanomaterials (ENMs). We here introduced the functions and the underlying molecular mechanisms of some important signaling pathways in the regulation of nanotoxicity formation, and these signaling pathways mainly include apoptosis signaling pathway, DNA damage signaling pathway, MAPK signaling pathways, insulin signaling pathway, innate immune response signaling pathway, Wnt signaling pathway, TGF-beta signaling pathway, developmental timing control-related signals, and neurotransmission-related signals. We also systematically introduced the functions and the underlying molecular basis for microRNAs and long noncoding RNAs in the regulation of nanotoxicity formation based on the clues from the omics study performed in nematodes.

Keywords Molecular mechanism · Nanomaterials · Nanotoxicity · *Caenorhabditis elegans*

7.1 Introduction

The important potentials for *Caenorhabditis elegans* for the elucidation of underlying molecular mechanisms for the observed toxicity of engineered nanomaterials (ENMs) are at least based on the following reasons. One of those is that many molecular events and regulation, signal transduction pathways, and epigenetic marks in *C. elegans* are conserved compared with those in mammals or humans. The second reason is that the completion of *C. elegans* genome has approximately 45% of the genes having the corresponding human homologues, including numerous disease-related genes.

In this chapter, we first introduce the molecular basis for induction of oxidative stress in ENM-exposed nematodes. Again, we will introduce some important signaling pathways involved in the regulation of nanotoxicity formation in nematodes, and these signaling pathways include apoptosis signaling pathway, DNA damage signaling pathway, MAPK signaling pathways, insulin signaling pathway, innate immune response signaling pathway, Wnt signaling pathway, TGF-beta signaling pathway, developmental timing control-related signals, and neurotransmission-related signals. Moreover, we systematically introduce the molecular basis for nanotoxicity formation based on the omics study, especially on the microRNAs and the long noncoding RNAs, in nematodes.

7.2   Molecular Basis for Induction of Oxidative Stress in ENM-Exposed Nematodes

7.2.1   Alteration in Primary Molecular Mechanism for the Control of Oxidative Stress

In *C. elegans*, MEV-1, GAS-1, ISP-1, and CLK-1 may play a crucial role in the primary molecular mechanism for the control of oxidative stress [1–4]. *mev-1* encodes an ortholog of succinate dehydrogenase cytochrome b560 subunit of mitochondrial respiratory chain complex II that is required for the oxidative phosphorylation, *gas-1* encodes a subunit of mitochondrial complex I, *isp-1* encodes a "Rieske" iron-sulfur protein, and *clk-1* encodes a ubiquinone biosynthesis protein COQ7. With the graphene oxide (GO) as an example, prolonged exposure (from L1-larvae to adult day-1) to GO (100 mg/L) could induce the significant intestinal reactive oxygen species (ROS) production in nematodes [5]. Meanwhile, prolonged exposure to GO (100 mg/L) resulted in a significant decrease in the expression level of *gas-1* and a significant increase in the expression levels of *isp-1* and *clk-1* (Fig. 7.1) [5], which implies that the primary molecular machinery may be activated to lead to the induction of oxidative stress in GO-exposed nematodes.

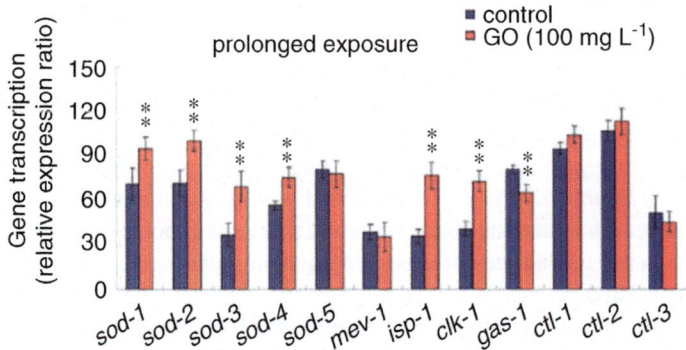

Fig. 7.1 Comparison of expression pattern for genes required for oxidative stress in GO (100 mg/L)-exposed nematodes [5]
Prolonged exposure was performed from L1-larvae to adult day-1. Bars represent mean ± SEM. $^{**}p < 0.01$

7.2.2   Induction of Expression for Proteins with the Functions to Defend Against the Oxidative Stress

In *C. elegans*, *sod-1-5* encode different SODs, and *ctl-1-3* encode catalases, which function in being against the induction of oxidative stress for animals. Further with GO as an example, prolonged exposure (from L1-larvae to adult day-1) to GO (100 mg/L) could significantly increase the expression levels of *sod-1*, *sod-2*, *sod-3*, and *sod-4* (Fig. 7.1) [5]. Besides this, prolonged exposure (from L1-larvae to young adults) to TiO_2-nanoparticles (TiO_2-NPs) with different nanosizes could also significantly increase the expression levels of *sod-2* and *sod-3* among the examined genes required for the control of oxidative stress in nematodes [6]. These data imply that the toxicity caused by certain ENM exposure can further induce a protection response in nematodes, which also reflects the formation of oxidative stress in ENM-exposed nematodes. Nevertheless, the formed protection response may be not enough to counteract the ENM-induced oxidative damage on nematodes.

7.2.3   Suppression in Expressions of Genes Mediating the Protection Response Defending Against Oxidative Stress in Nematodes After Chronic Exposure to ENMs

Usually, a protection response will be induced in nematodes after acute or prolonged exposure to certain ENMs. However, once the very severe toxicity would be induced for ENMs after chronic exposure, the expressions of genes mediating the protection response defending against the oxidative stress would be further suppressed in nematodes. With the Al_2O_3-NPs as an example, the noticeable decrease in expressions of *sod-2* and *sod-3* was detected in nematodes after chronic exposure (from young adults for 10 days) to Al_2O_3-NPs (8.1–23.1 mg/L). Meanwhile, the obvious decrease in expressions of *sod-2* and *sod-3* was also be detected in nematodes after chronic exposure to bulk Al_2O_3 (Fig. 7.2).

7.2.4   Molecular Signals Involved in the Regulation of Induction of Oxidative Stress in ENM-Exposed Nematodes

With the GO as an example, the potential molecular signals involved in the regulation of induction of oxidative stress in exposed nematodes were screened from 20 strains with mutations of genes required for stress response or oxidative stress [8]. Among these 20 mutants, the results demonstrated that mutation of *gas-1*, *sod-2*, *sod-3*, or *aak-2* could induce the greater GO translocation into the body and the

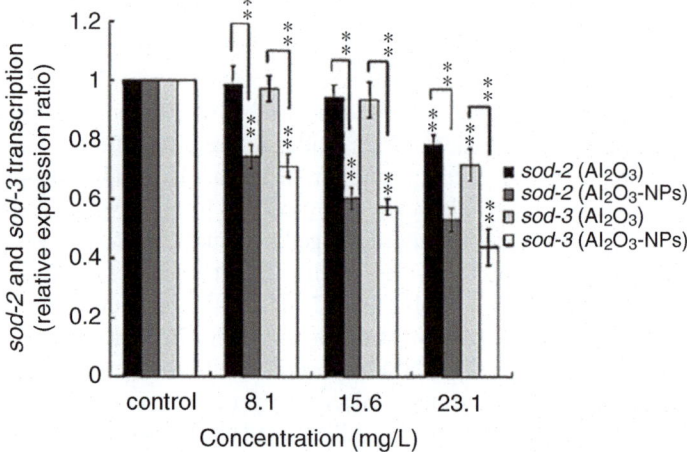

Fig. 7.2 Expressions of *sod-2* and *sod-3* in nematodes after chromic exposure to Al₂O₃-NPs or bulk Al₂O₃ from young adults for 10 days [7]

Bars represent mean ± SEM. $^{**}P < 0.01$ *vs* control (if not otherwise indicated)

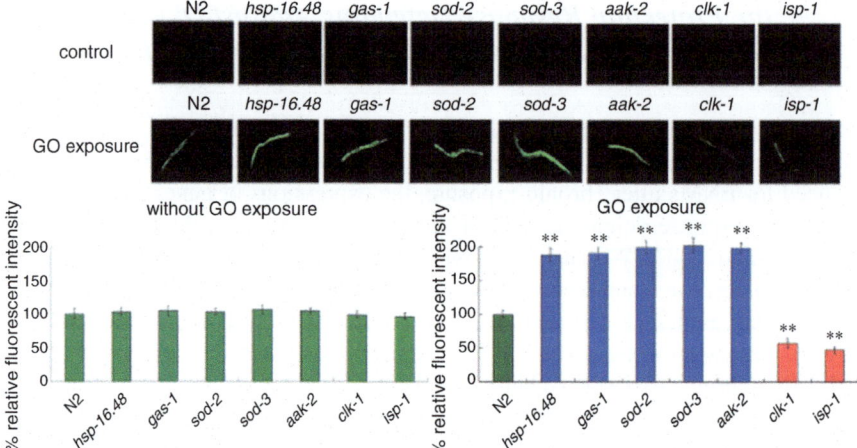

Fig. 7.3 Comparison of induction of intestinal ROS production between wild-type and mutant nematodes exposed to GO (100 mg/L) [8]

GO exposure was performed from L1-larvae to young adult. Bars represent mean ± SEM. $^{**}P < 0.01$ *vs* wild-type

greater induction of intestinal ROS production compared with those in wild-type nematodes (Fig. 7.3) [8]. In contrast, mutation of *isp-1* or *clk-1* could cause the significant decrease in both the GO translocation into the body and the induction of intestinal ROS production compared with those in wild-type nematodes (Fig. 7.3) [8]. In nematodes, *aak-2* encodes a catalytic alpha subunit of AMP-activated protein kinase.

7.3   Important Signaling Pathways Involved in the Regulation of Nanotoxicity Formation in Nematodes

Here, we mainly selected the GO as an example to induce the involvement of important signaling pathways in the regulation of nanotoxicity formation in nematodes.

7.3.1   Apoptosis Signaling Pathway

In *C. elegans*, the core signaling pathway of apoptosis is constituted by CED-3, CED-4, and CED-9 [9]. The CED-3, a cysteine aspartate protease, is required for the execution of apoptosis and acts in a conserved genetic pathway with the CED-4 [47]. The CED-9 is the sole homolog of mammalian cell-death inhibitor Bcl-2 and negatively regulates the CED-4 activity to prevent the cells from undergoing apoptosis [9]. With the GO as an example, prolonged exposure (from L1-larvae to young adults) to GO (10 mg/L) could significantly increase the expressions of *ced-3* and *ced-4* and decrease the expression of *ced-9* (Fig. 7.4) [10]. Meanwhile, mutation of *ced-3* or *ced-4* inhibited the germline apoptosis induced by GO exposure; however, mutation of *ced-9* enhanced the germline apoptosis induced by GO exposure (Fig. 7.4) [10]. These results imply the potential involvement of core signaling pathway of apoptosis constituted by CED-3, CED-4, and CED-9 in the regulation of reproductive toxicity of GO. That is, prolonged exposure to GO may potentially induce the germline apoptosis by activating the CED-4-CED-3 apoptosis signaling pathway and inhibiting the activity of CED-9 (Fig. 7.4) [10].

7.3.2   DNA Damage Signaling Pathway

In *C. elegans*, EGL-1 is a protein containing a region similar to BH3 domain of mammalian cell death activators with the functions as an upstream activator in the core apoptosis signaling pathway and as a DNA damage checkpoint [9, 11]. CEP-1 is an ortholog of human tumor suppressor p53 with the function in promoting the DNA damage-induced apoptosis by activating EGL-1 [12]. CLK-2 is an ortholog of telomere length-regulating protein Tel2p, and HUS-1 is required for the function of CEP-1 in activating DNA damage-induced apoptosis [11]. Further with GO as an example, prolonged exposure (from L1-larvae to young adults) to GO (10 mg/L) significantly increased the expressions of *hus-1*, *clk-2*, *cep-1*, and *egl-1* (Fig. 7.5) [10]. Meanwhile, mutation of *hus-1*, *clk-2*, *cep-1*, or *egl-1* could noticeably suppress the germline apoptosis induced by GO exposure (Fig. 7.5) [10]. These results imply the further involvement of signaling pathway for DNA damage checkpoints in the control of reproductive toxicity of GO in nematodes. That is, prolonged exposure to

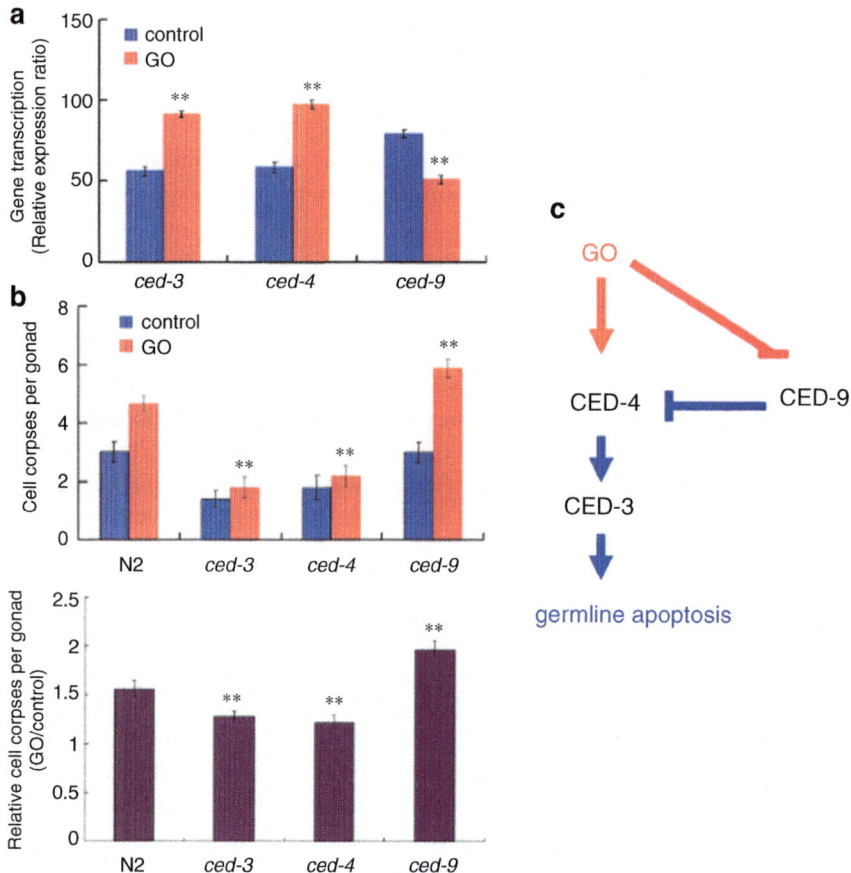

Fig. 7.4 Role of the core apoptosis signaling pathway in the control of GO toxicity in inducing germline apoptosis [10]

(**a**) GO exposure altered expression patterns of genes encoding the core apoptosis signaling pathway. Bars represent mean ± SEM. $^{**}P < 0.01$ *vs* control. (**b**) Mutations of genes encoding the core apoptosis signaling pathway affected the germline apoptosis in nematodes exposed to GO. The used strains were wild-type N2, *ced-3(n717)*, *ced-4(n1162)*, and *ced-9(n1950)*. Bars represent mean ± SEM. $^{**}P < 0.01$. (**c**) A model for the core apoptosis signaling pathway in the control of GO toxicity in inducing germline apoptosis. GO exposure concentration was 10 mg/L. Prolonged exposure to GO was performed from L1-larvae to young adults

GO may induce the DNA damage by activating the signaling pathway for DNA damage checkpoints in germ cells (Fig. 7.5) [10].

7.3.3   MAPK Signaling Pathways

MAPK signals act as the central signaling hubs to regulate various important cellular processes by transducing extracellular cues into the cells.

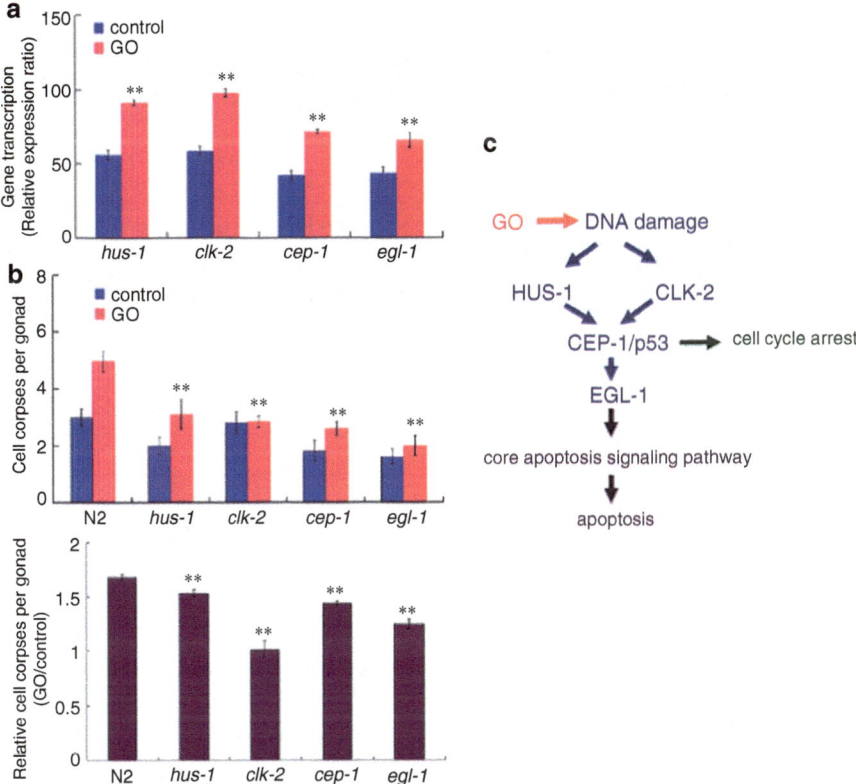

Fig. 7.5 Role of the signaling pathway for DNA damage checkpoints in the control of GO toxicity in inducing germline apoptosis [10]

(**a**) GO exposure altered expression patterns of genes encoding the signaling pathway for DNA damage checkpoints. Bars represent mean ± SEM. $^{**}P < 0.01$ vs control. (**b**) Mutations of genes encoding the signaling pathway for DNA damage checkpoints affected the germline apoptosis in nematodes exposed to GO. The used strains were wild-type N2, *egl-1(n1084n3082)*, *hus-1(op241)*, *cep-1(gk138)*, and *clk-2(mn159)*. Bars represent mean ± SEM. $^{**}P < 0.01$. (**c**) A model for the signaling pathway for DNA damage checkpoints in the control of GO toxicity in inducing germline apoptosis. GO exposure concentration was 10 mg/L. Prolonged exposure to GO was performed from L1-larvae to young adults

7.3.3.1 JNK Signaling Pathway

In nematodes, the c-Jun N-terminal kinase (JNK) signaling pathway mainly contains members of MEK-1, JKK-1, and JNK-1, which has been shown to be at least involved in the control of stress response [13]. *mek-1* and *jkk-1* encode MAP kinase kinases, homolog of human MKK-7a, and act as an activator of JNK. *jnk-1* encodes a serine/threonine kinase, homolog of human JNK, and acts as the sole member of the JNK subgroup of MAP kinase. In GO-exposed nematodes, the transcriptional expressions of *mek-1*, *jkk-1*, and *jnk-1* were significantly decreased [14]. Moreover,

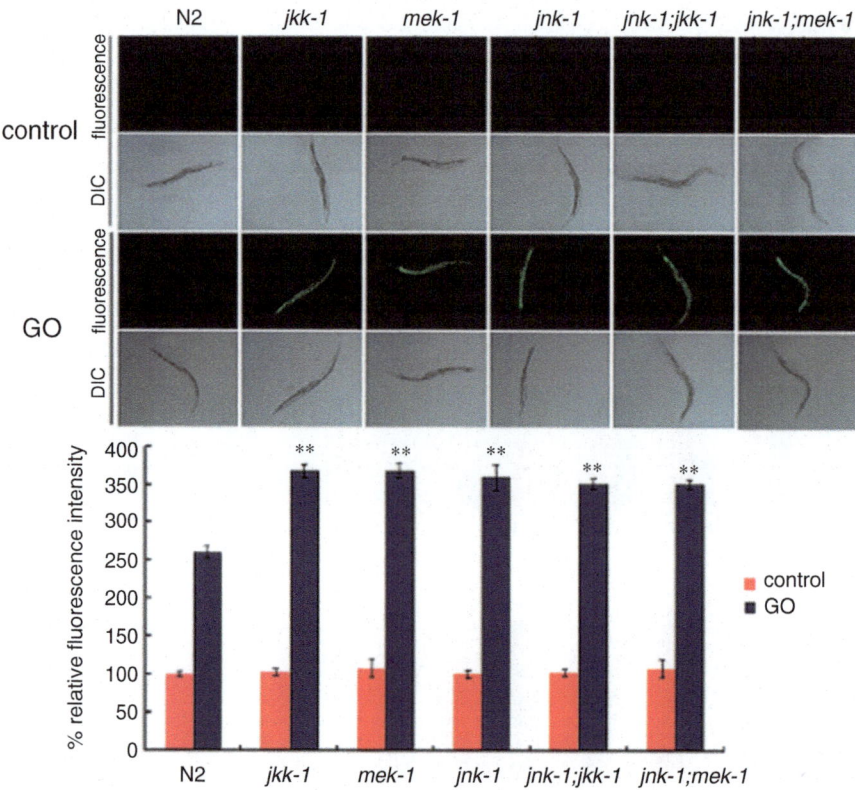

Fig. 7.6 Effects of mutations in the gene encoding JNK signaling pathways on intestinal ROS production in GO-exposed nematodes [14]
Bars represent the mean ± SEM. ${}^{**}P < 0.01$ vs wild-type N2. GO (10 mg/L) was exposed from L1-larvae to adult day-1

mutation of *mek-1*, *jkk-1*, and *jnk-1* could induce a susceptibility to GO toxicity using brood size, locomotion behavior, and intestinal ROS production as the endpoints (Fig. 7.6) [14]. The genetic interaction analysis further indicated that the *jnk-1(gk7);mek-1(ks54)* double mutant exhibited the similar GO toxicity to that in the *mek-1(ks54)* or the *jnk-1(gk7)* single mutant using brood size, locomotion behavior, and intestinal ROS production as the endpoints (Fig. 7.6) [14]. Meanwhile, the *jnk-1(gk7);jkk-1(km2)* double mutant showed the similar GO toxicity to that in the *jkk-1(km2)* or the *jnk-1(gk7)* single mutant using brood size, locomotion behavior, and intestinal ROS production as the endpoints (Fig. 7.6) [14]. These results demonstrated that the MEK-1 and the JKK-1 may act genetically in the same pathway with JNK-1 in the regulation of GO toxicity in nematodes.

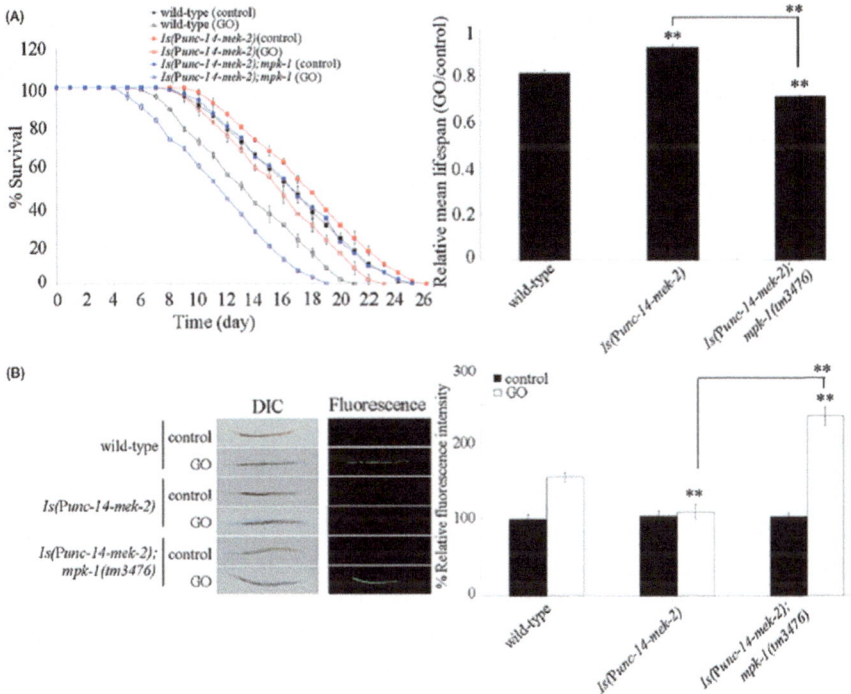

Fig. 7.7 Genetic interaction between MEK-2 and MPK-1 in the regulation of GO toxicity [15] (**a**) Genetic interaction between MEK-2 and MPK-1 in the regulation of GO toxicity in reducing lifespan. (**b**) Genetic interaction between MEK-2 and MPK-1 in the regulation of GO toxicity in inducing ROS production. Prolonged exposure was performed from L1-larvae to young adults. GO exposure concentration was 10 mg/L. Bars represent mean ± SD. **$p < 0.01$ versus wild-type (if not specially indicated)

7.3.3.2 ERK Signaling Pathway

7.3.3.2.1 Involvement of ERK Signaling Pathway in the Regulation of ENM Toxicity

In *C. elegans*, the ERK signaling pathway contains *mek-2*-encoded MAPK kinase MEK and *mpk-1*-encoded ERK. Using lifespan and intestinal ROS production as the toxicity assessment endpoints, mutation of *mpk-1* or *mek-2* resulted in the more severe reduction in lifespan and induction of intestinal ROS production, suggesting that mutation of *mpk-1* or *mek-2* may induce a susceptibility to GO toxicity [15]. Genetic interaction analysis further demonstrated that mutation of *mpk-1* could significantly suppress the resistance of nematodes overexpressing the neuronal *mek-2* to GO toxicity in reducing lifespan and in inducing intestinal ROS production (Fig. 7.7) [15], suggesting that the MPK-1 acts downstream of the neuronal MEK-2 to regulate the response of nematodes to GO exposure.

Fig. 7.8 A diagram
showing the functions and
interactions of ERK,
MAPK, and insulin
signaling pathways in the
regulation of response to
GO exposure [15]

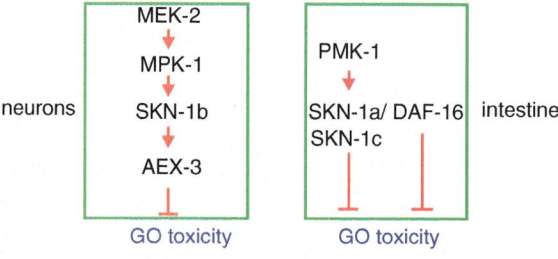

7.3.3.2.2 Neuron-Specific Activity of ERK Signaling Pathway in the Regulation of ENM Toxicity

In nematodes, MPK-1 is expressed in both the neurons and the germline. Expression of the *mpk-1* in the neurons could suppress the susceptibility of *mpk-1(tm3476)* mutant to GO toxicity in reducing lifespan and in inducing intestinal ROS production [15]. In contrast, using *rrf-1(pk1417)* as a genetic tool to perform the germline-specific RNAi knockdown of certain genes, it has been found that germline-specific RNAi knockdown of *mpk-1* could not significantly induce a susceptibility to GO toxicity like the phenotypes observed in the *mpk-1(tm3476)* mutant [15]. Moreover, expression of the *mek-2* in the neurons could also significantly suppress the susceptibility of *mek-2(n1989)* mutant to GO toxicity in reducing lifespan and in inducing intestinal ROS production [15]. These results suggest that the neuronal signaling cascade of MEK-2-MPK-1 may play a crucial role in regulating the response of nematodes to GO exposure.

7.3.3.2.3 Identification of Potential Downstream Targets for Neuronal MPK-1 in the Regulation of Response to ENM Exposure

In nematodes, the ERK signaling can act upstream of SKN-1 to regulate the longevity [16]. SKN-1 is a bZip transcriptional factor Nrf and acts downstream of the insulin receptor DAF-2 in the insulin signaling pathway to regulate the stress response [17]. Mutation of *skn-1* could significantly suppress the resistance of nematodes overexpressing neuronal *mpk-1* to GO toxicity in reducing lifespan and in inducing intestinal ROS production [15], suggesting that the SKN-1 can further act as a downstream target for neuronal MPK-1 in the regulation of GO toxicity (Fig. 7.8).

In nematodes, AEX-3 can also act downstream of MPK-1 to regulate different biological processes, such as the protein degradation [18]. Mutation of *aex-3* could significantly suppress the resistance of nematodes overexpressing neuronal *mpk-1* to GO toxicity in reducing lifespan and in inducing intestinal ROS production [15], suggesting that the AEX-3 can act as another downstream target for neuronal MPK-1 in the regulation of GO toxicity (Fig. 7.8).

Moreover, mutation of *aex-3* could further significantly suppress the resistance of nematodes overexpressing the neuronal *skn-1b* to GO toxicity in reducing lifespan and in inducing intestinal ROS production [15], suggesting the formation of neuronal signaling cascade of MEK-1-PMK-1-SKN-1b-AEX-3 in the regulation of GO toxicity in nematodes (Fig. 7.8).

7.3.3.2.4   ERK Signal Mediates a Protection Response for Nematodes to ENM Exposure

In nematodes, prolonged exposure to GO at the concentration of 1 mg/L did not influence the expressions of *mpk-1* and *mek-2* [15]. In contrast, prolonged exposure to GO at concentrations of 10 and 100 mg/L could significantly increase both the transcriptional expression of *mpk-1* and the transcriptional expression of *mek-2* [15]. Therefore, the increase in expression of ERK signal may mediate a protection mechanism for nematodes in response to GO exposure.

7.3.3.3   p38 MAPK Signaling Pathway

7.3.3.3.1   Involvement of p38 MAPK Signaling Pathway in the Regulation of ENM Toxicity

In *C. elegans*, the core p38 MAPK signaling pathway contains *pmk-1*-encoded MAPK, *sek-1*-encoded MAPK kinase (MAPKK), and *nsy-1*-encoded MAPK kinase kinase (MAPKKK). With the GO as an example, it has been found that prolonged exposure to GO (100 mg/L) induced the formation of more severe reduction in lifespan, decreased locomotion behavior, and induced intestinal ROS production in *pmk-1(km25)*, *sek-1(ag1)*, or *nsy-1(ag3)* mutant than those in wild-type nematodes (Fig. 7.9) [19]. Under the normal conditions, *pmk-1(km25)*, *sek-1(ag1)*, or *nsy-1(ag3)* mutant has the normal lifespan and locomotion behavior and exhibits no obvious induction of intestinal ROS production (Fig. 7.9) [19]. Therefore, mutations of genes encoding the core p38 MAPK signaling pathway may induce a susceptibility to GO toxicity in nematodes.

In contrast, overexpression of *pmk-1*, *sek-1*, or *nsy-1* could significantly decrease the induction of intestinal ROS production [19]. Under normal conditions, the nematodes with the overexpression of *pmk-1*, *sek-1*, or *nsy-1* do not exhibit the obvious induction of intestinal ROS production [19].

In wild-type nematodes, prolonged exposure to GO (100 mg/L) significantly increased the expression of *isp-1* and decreased the expression of *gas-1* [5]. Moreover, mutation of *pmk-1*, *sek-1*, or *nsy-1* induced the more severely increase in the expression of *isp-1* and decrease in the expression of *gas-1* [19], implying that the mutation of *pmk-1*, *sek-1*, or *nsy-1* may further affect the molecular basis for the control of oxidative stress in GO-exposed nematodes.

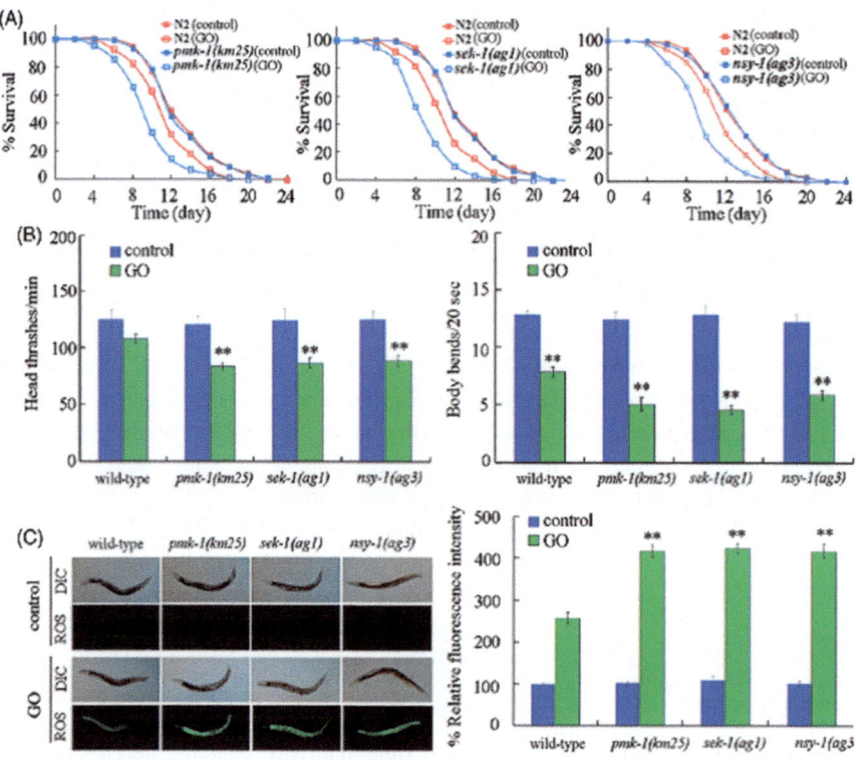

Fig. 7.9 Toxicity assessment of GO exposure on mutants of genes encoding p38 MAPK signaling pathway [19]

(**a**) Toxicity assessment of GO exposure on lifespan of mutants of genes encoding p38 MAPK signaling pathway. (**b**) Toxicity assessment of GO exposure on locomotion behavior of mutants of genes encoding p38 MAPK signaling pathway. (**c**) Toxicity assessment of GO exposure in inducing ROS production in mutants of genes encoding p38 MAPK signaling pathway. Prolonged exposure was performed from L1-larvae to young adults. GO exposure concentration was 100 mg/L. Bars represent mean ± SEM. $^{**}p < 0.01$ vs wild-type

7.3.3.3.2 Intestinal Signaling Cascade of p38 MAPK Signaling Pathway Involved in the Regulation of ENM Toxicity

In *C. elegans*, *pmk-1* is broadly expressed in multiple tissues including the intestine; *sek-1* is expressed in the excretory canal, the intestine, and the neurons; and *nsy-1* is expressed in the intestine, the hypodermis, and the neurons. Among these tissues, intestinal RNAi knockdown of *pmk-1*, *sek-1*, or *nsy-1* induced a susceptibility to GO toxicity on lifespan in nematodes using the transgenic VP303 as a genetic tool [19], suggesting that the signaling cascade of NSY-1-SEK-1-PMK-1 can act in the intestine to regulate the GO toxicity (Fig. 7.10).

In *C. elegans*, SKN-1 usually functions as a downstream target of the p38 MAPK signaling pathway to regulate biological processes, such as the activation of

Fig. 7.10 A diagram showing the intestinal p38 MAPK-SKN-1/Nrf signaling cascade involved in the control of GO toxicity in nematodes [19]

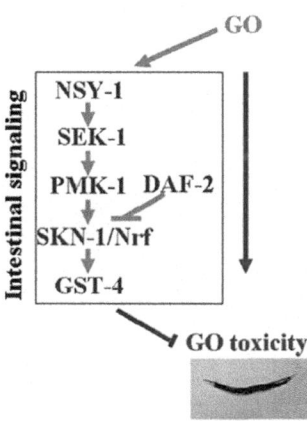

oxidative stress [20]. Intestinal RNAi knockdown of *skn-1* also caused a susceptibility of nematodes to GO toxicity in reducing lifespan and in inducing intestinal ROS production [19], suggesting that the SKN-1/Nrf can also act in the intestine to regulate the GO toxicity in nematodes. Moreover, intestinal RNAi knockdown of *skn-1* could suppress the resistance of nematodes overexpressing intestinal *pmk-1* to GO toxicity in reducing lifespan [19], implying that the intestinal core p38 MAPK signaling cascade may regulate the GO toxicity through the function of SKN-1/Nrf in nematodes (Fig. 7.10).

In nematodes, one of the key molecular mechanisms for SKN-1/Nrf in regulating stress response is that SKN-1 can function through Phase II detoxification genes [20]. One of those important Phase II detoxification proteins is the *gst-4*-encoded glutathione-requiring prostaglandin D synthase [20]. GST-4 is expressed in the intestine, the pharynx, and the hypodermis. Intestinal RNAi knockdown of *gst-4* could also induce a susceptibility to GO toxicity in reducing lifespan and in inducing intestinal ROS production [19]. Meanwhile, the genetic interaction analysis further indicated that the SKN-1 and the GST-4 could act in the same genetic pathway to regulate the GO toxicity, since exposure to GO (100 mg/L) caused the similar toxicity on lifespan in *skn-1(RNAi); gst-4(ok2108)* to that in *skn-1(RNAi)* strain or in *gst-4(ok2108)* mutant nematodes [19].

7.3.3.3.3   p38 MAPK Signaling Cascade May Mediate a Protection Response for Nematodes to ENM Exposure

In nematodes, prolonged exposure to GO (100 mg/L) could significantly increase the transcriptional expressions of *pmk-1*, *sek-1*, and *nsy-1* (Fig. 7.11) [19]. Meanwhile prolonged exposure to GO (100 mg/L) could significantly increase the expression of intestinal PMK-1::GFP (Fig. 7.11) [19]. Prolonged exposure to GO (100 mg/L) could significantly increase the percentage of PMK-1::GFP nucleus localization in the intestinal cells (Fig. 7.11) [19]. Additionally, the Western blotting

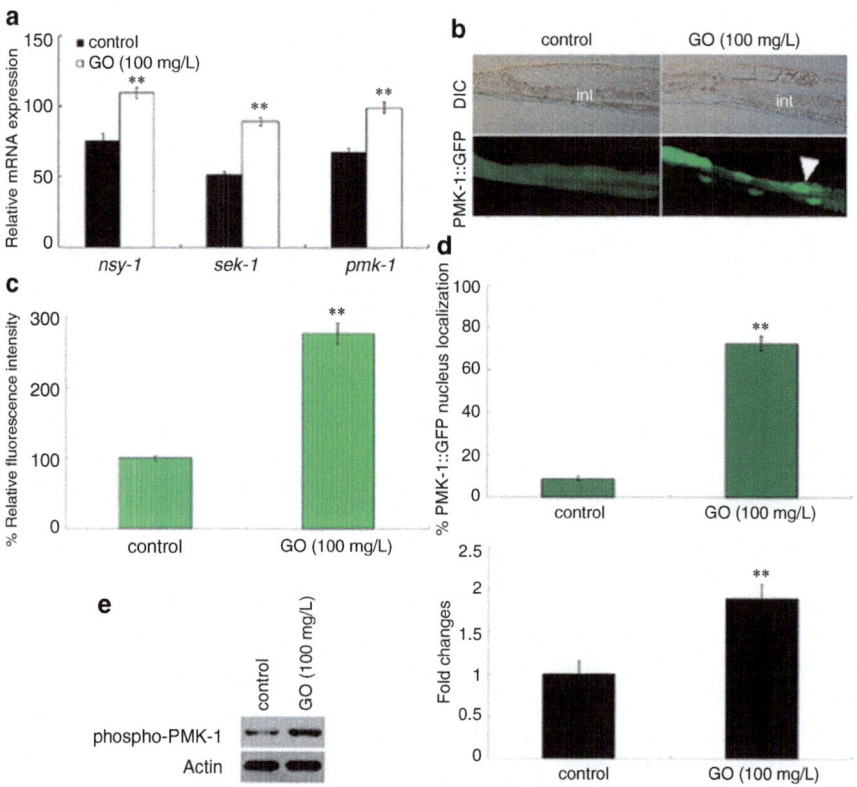

Fig. 7.11 Effects of GO exposure on expression pattern of genes encoding p38 MAPK signaling pathway in nematodes [19]
(**a**) GO exposure altered the transcriptional expression of genes encoding p38 MAPK signaling pathway. (**b**) GO exposure affected PMK-1::GFP expression in the intestine. *int*, intestine. Arrowhead indicates the localization of PMK-1::GFP in nucleus of intestinal cells. (**c**) Comparison of relative fluorescence intensity of PMK-1::GFP in the intestine. (**d**) GO exposure influenced nucleus translocation of PMK-1::GFP. (**e**) Western blotting analysis of the effect of GO exposure on expression level of phosphorylated PMK-1. Actin protein was used as the loading control. Prolonged exposure was performed from L1-larvae to young adults. Bars represent mean ± SEM. $^{**}p < 0.01$ vs control

analysis further demonstrated that prolonged exposure to GO (100 mg/L) could also obviously increase the expression of phosphorylated PMK-1 (Fig. 7.11) [19]. Besides these, it has been further found that prolonged exposure to GO (100 mg/L) significantly enhanced the expression of intestinal SKN-1::GFP in nematodes [19]. Additionally, prolonged exposure to GO (100 mg/L) significantly increased the percentage of SKN-1::GFP nucleus localization in intestinal cells and the expression of intestinal GST-4::GFP [19]. These observations suggest that the p38 MAPK signaling cascade may mediate a protection response for nematodes to GO exposure. The moderate but significant increase in the expression of SKN-1::GFP in ASI sensory neurons and the expression of GST-4::GFP in the pharynx and in the hypodermis were also observed in GO (100 mg/L)-exposed nematodes [19]; however, the exact functions for these observations are still unclear. In nematodes, the protective function of p38 MAPK signaling cascade was also observed to be activated in Ag-NPs nematodes [21, 22].

Moreover, the decreased expression in PMK-1::GFP was observed in *sek-1(ag1)* or *nsy-1(ag3)* mutant nematodes after prolonged exposure to GO (100 mg/L) [19]. Meanwhile, the decreased phosphorylation of p38 MAPK/PMK-1 was further detected in *sek-1(ag1)* or *nsy-1(ag3)* mutant nematodes after prolonged exposure to GO (100 mg/L) [19]. Furthermore, the decreased expression of SKN-1::GFP in both the ASI sensory neurons and the intestine was observed in nematodes with intestinal RNAi knockdown of *pmk-1* after GO (100 mg/L) exposure [19]. The decreased expression in GST-4::GFP was further found in *pmk-1(km25)* or *skn-1(zj15)* mutant nematodes [19]. These results further support the formation of signaling cascade of NSY-1-SEK-1-PMK-1-SKN-1-GST-4 in the regulation of GO toxicity (Fig. 7.10).

7.3.4 Insulin Signaling Pathway

7.3.4.1 Involvement of Insulin Signaling Pathway in the Regulation of ENM Toxicity

The insulin/insulin-like growth factor (IGF) signaling pathway has been implicated as a key molecular mechanism for various biological processes, such as the longevity, in nematodes [23, 24]. In *C. elegans*, insulin ligands bind to DAF-2/IGF-1 receptor (InR), activate tyrosine kinase activity, and then initiate the cascade of several kinases (AGE-1/phosphatidylinositol 3-kinase (PI3K), PDK-1/3-phosphoinositide-dependent kinase 1, AKT-1/2/serine/threonine kinase Akt/PKB, and SGK-1/serine or threonine-protein kinase). The AKT and the SGK-1 further phosphorylate and inactivate the transcription factor DAF-16/FOXO, which thereby blocks the transcription of its target genes, such as SOD-3 [23, 24].

With GO as an example, prolonged exposure (from L1-larvae to young adults) to GO (100 mg/L) could dysregulate the expressions of *daf-2*, *age-1*, *akt-1*, *akt-2*, *daf-18*, and *daf-16* among the genes in the insulin signaling pathway [25]. Moreover, using the corresponding mutants, it has been shown that mutation of *daf-16* or *daf-18* induced a susceptibility to GO toxicity in decreasing locomotion behavior and in reducing lifespan, whereas mutation of *daf-2*, *age-1*, *akt-1*, or *akt-2* induced a resistance to GO toxicity in decreasing locomotion behavior and in reducing lifespan (Fig. 7.12) [25]. Furthermore, genetic interaction analysis demonstrated that DAF-16 may act downstream of DAF-2, AGE-1, AKT-1, and AKT-2 to regulate the GO toxicity on longevity, because mutation of *daf-16* could effectively reduce the lifespan in *daf-2(e1370)*, *age-1(hx546)*, *akt-1(ok525)*, or *akt-2(ok393)* mutant nematodes exposed to GO [25]. Additionally, mutation of *daf-18* could further effectively

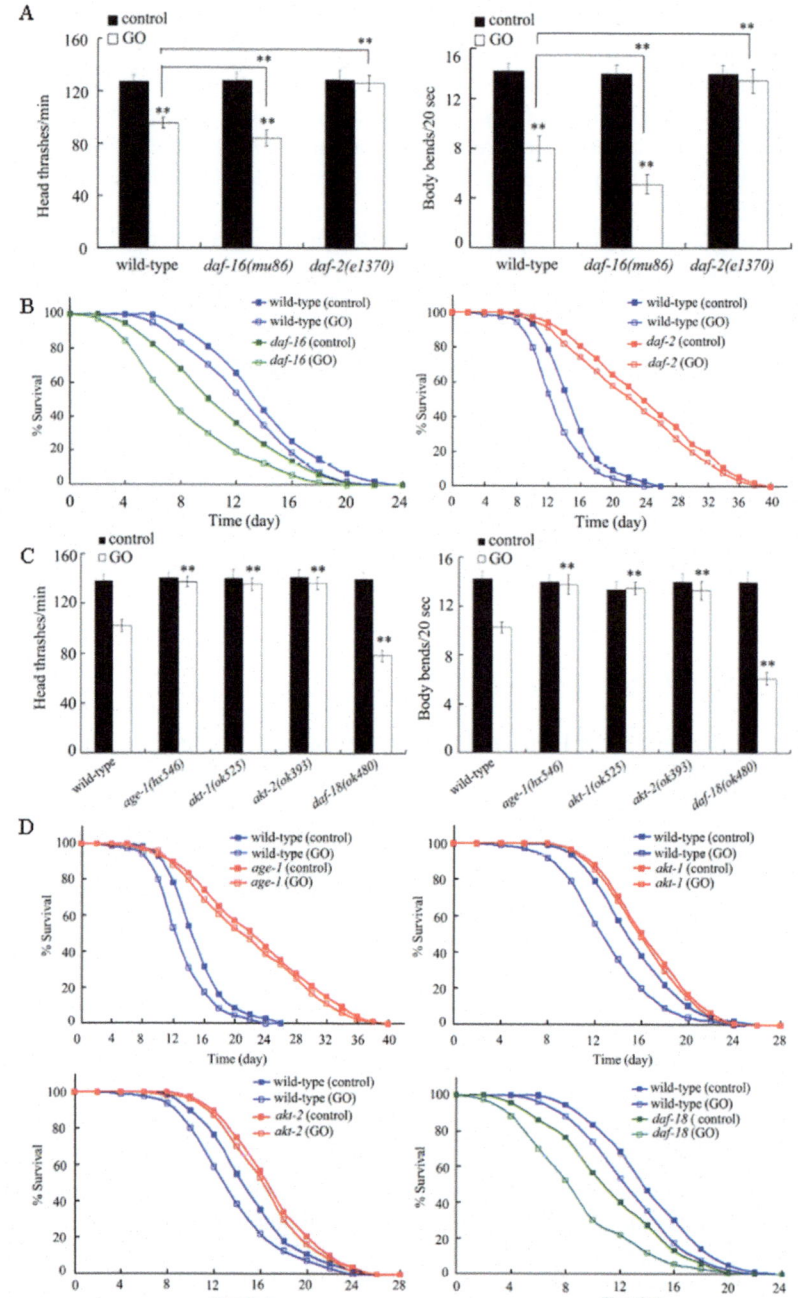

Fig. 7.12 Effects of *daf-16*, *daf-2*, *age-1*, *akt-1*, *akt-2*, or *daf-18* mutation on nematodes exposed to GO [25]

(**a**) Effects of *daf-16* or *daf-2* mutation on locomotion behavior in nematodes exposed to GO. (**b**) Effects of *daf-16* or *daf-2* mutation on lifespan in nematodes exposed to GO. (**c**) Mutations of *age-1*, *akt-1*, *akt-2*, or *daf-18* affected GO toxicity on locomotion behavior in nematodes. (**d**) Mutations of *age-1*, *akt-1*, *akt-2*, or *daf-18* affected GO toxicity on lifespan in nematodes. GO exposure concentration was 100 mg/L. Prolonged exposure was performed from L1-larvae to young adults. Bars represent mean ± SEM. **P < 0.01 *vs* control (if not specially indicated)

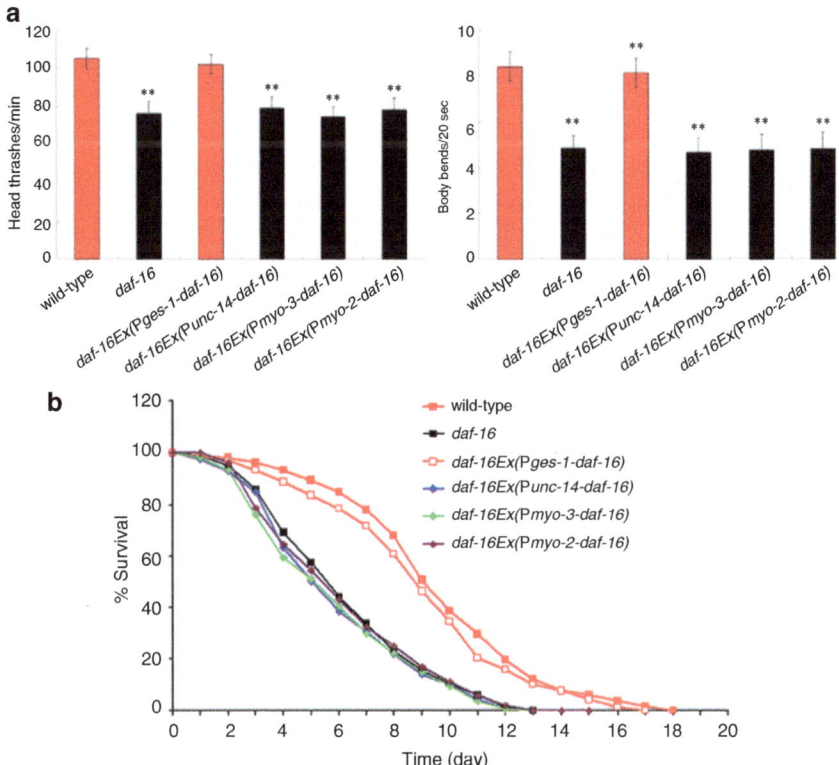

Fig. 7.13 Tissue-specific activity of DAF-16 in regulating the GO toxicity in nematodes [25] (**a**) Tissue-specific activity of DAF-16 in regulating GO toxicity on locomotion behavior in nematodes. (**b**) Tissue-specific activity of DAF-16 in regulating GO toxicity on lifespan in nematodes. GO exposure concentration was 100 mg/L. Prolonged exposure was performed from L1-larvae to young adults. Bars represent mean ± SEM. $^{**}P < 0.01$ *vs* wild-type

reduce the lifespan in *age*(*hx546*) mutant exposed to GO [25], which suggests the suppressor role of DAF-18 on the function of AGE-1 in the regulation of GO toxicity in nematodes.

7.3.4.2   Intestine-Specific Activity of Insulin Signaling Pathway in the Regulation of ENM Toxicity

In *C. elegans*, *daf-16* is expressed in almost all tissues, including the intestine, the neurons, the muscle, and the pharynx. In nematodes, expression of *daf-16* in the neurons, the muscle, or the pharynx could not significantly affect the GO toxicity in decreasing locomotion behavior and in reducing lifespan in *daf-16*(*mu86*) mutant nematodes (Fig. 7.13) [25]. In contrast, expression of *daf-16* in the intestine could effectively augment the decreased locomotion behavior or reduced lifespan in

GO-exposed *daf-16(mu96)* mutant nematodes (Fig. 7.13) [25], demonstrating that the DAF-16 acts primarily in the intestine to regulate the GO toxicity. Moreover, intestine-specific RNAi knockdown of *daf-2*, *age-1*, *akt-1*, or *akt-2* could result in the resistance to GO toxicity in reducing lifespan, whereas intestine-specific RNAi knockdown of *daf-16* or *daf-18* could cause the susceptibility to GO toxicity in reducing lifespan in nematodes [25]. Therefore, the insulin signaling pathway can act in the intestine to regulate the GO toxicity in nematodes.

7.3.4.3   Identification of Targets for DAF-16 in the Regulation of ENM Toxicity in Nematodes

During the control of longevity, SOD-3, a mitochondrial iron/manganese superoxide dismutase, is a normally considered target for DAF-16 in nematodes [23, 24]. SOD-3 is expressed in the pharynx in the head, the intestine, the muscle, the vulva, and the tail. Intestinal RNAi knockdown of *sod-3* could induce a susceptibility to GO toxicity in reducing lifespan in nematodes [25]. Genetic interaction analysis further suggested that DAF-16 may act upstream of SOD-3 to regulate the GO toxicity, because the resistant transgenic strain of *Ex(Pges-1-daf-16)* overexpressing intestinal DAF-16 to GO toxicity in reducing lifespan and in inducing intestinal ROS production could be obviously inhibited by *sod-3* mutation [25].

In nematodes, DAF-16 regulates biological processes by modulating the activities of multiple targeted genes [26–29]. Among the candidate targeted genes for DAF-16, some genes (*lys-1*, *lys-7*, *lys-8*, *dod-6*, *F55G11.4*, *spp-1*, *spp-12*, and *dod-22*) encoding potential antimicrobial proteins have also been identified [26–29]. Among these genes encoding potential antimicrobial proteins, RNAi knockdown of *lys-1*, *dod-6*, *F55G11.4*, *lys-8*, or *spp-1* could significantly suppress the resistance to GO toxicity in inducing intestinal ROS production and in decreasing locomotion behavior in nematodes overexpressing intestinal *daf-16* [30], suggesting that LYS-1, DOD-6, F55G11.4, LYS-8, and SPP-1 act as downstream targets for intestinal DAF-16 in the regulation of GO toxicity. Among LYS-1, DOD-6, F55G11.4, LYS-8, and SPP-1, it has been further found that F55G11.4 and SPP-1 acted downstream of SOD-3 in the regulation of GO toxicity, since only RNAi knockdown of *F55G11.4* or *spp-1* could significantly suppress the resistance of nematodes overexpressing intestinal *sod-3* to GO toxicity in inducing intestinal ROS production and in decreasing locomotion behavior (Fig. 7.14) [30]. The antimicrobial proteins of F55G11.4 and SPP-1 affected the expression of *gas-1* encoding a subunit of mitochondrial complex I that is required for the oxidative phosphorylation in GO-exposed nematodes [30], implying the effect of F55G11.4 and SPP-1 on GAS-1-mediated molecular basis for oxidative stress in GO-exposed nematodes.

Moreover, it has been observed that RNAi knockdown of *sod-3* could suppress the resistance of nematodes overexpressing intestinal *dod-6* to GO toxicity in inducing intestinal ROS production and in decreasing locomotion behavior in transgenic strain; however, RNAi knockdown of *sod-3* could not affect the resistance of nematodes overexpressing intestinal *lys-1* or *lys-8* to GO toxicity in inducing intestinal

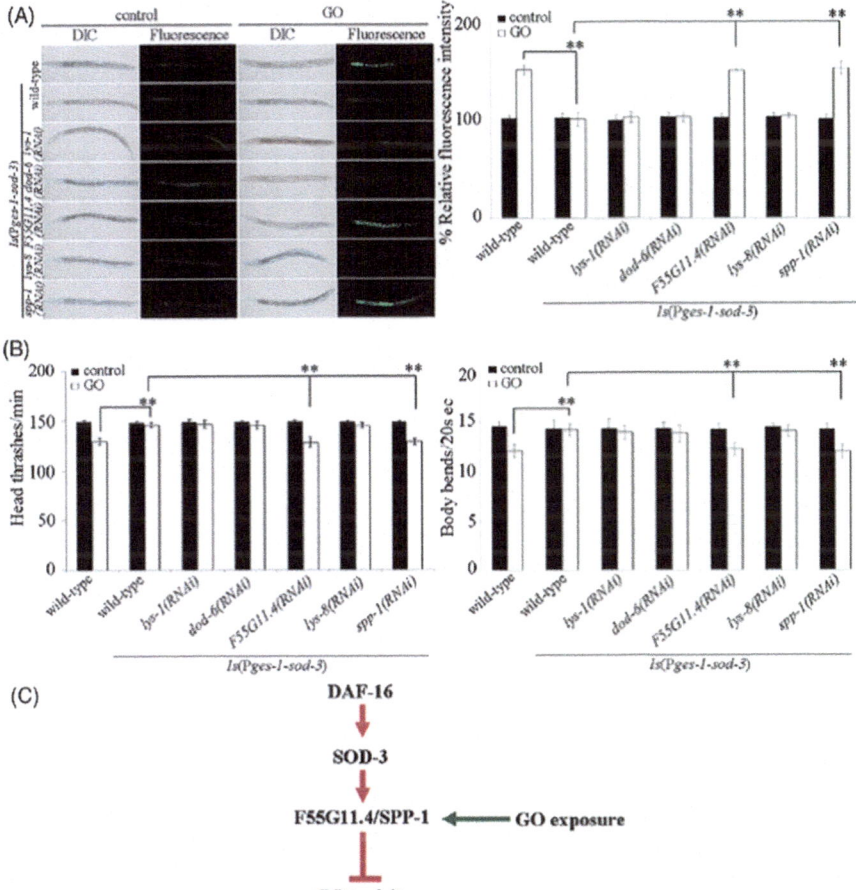

Fig. 7.14 Antimicrobial genes of F55G11.4 and SPP-1 acted downstream of SOD-3 to regulate the GO toxicity [30]

(**a**) Antimicrobial genes of F55G11.4 and SPP-1 acted downstream of SOD-3 to regulate the GO toxicity in inducing intestinal ROS production. (**b**) Antimicrobial genes of F55G11.4 and SPP-1 acted downstream of SOD-3 to regulate the GO toxicity in decreasing locomotion behavior. (**c**) A diagram showing the interaction between SOD-3 and antimicrobial proteins of F55G11.4 or SPP-1 in the regulation of GO toxicity. Prolonged exposure was performed from L1-larvae to young adults. GO exposure concentration was 10 mg/L. Bars represent mean ± SD. $^{**}p < 0.01$

ROS production and in decreasing locomotion behavior (Fig. 7.15) [30]. These results imply that LYS-1 and LYS-8 may act in a different genetic pathway from the DAF-16-SOD-3 signaling cascade, and DOD-6 may act upstream of SOD-3 in the regulation of GO toxicity. LYS-1 and LYS-8 are two members of the lysozyme family. The genetic interaction analysis further demonstrated that LYS-1 and LYS-8 may function redundantly in the regulation of GO toxicity, because the GO-exposed double mutant of *lys-8(ok3504);lys-1(ok2445)* exhibited the more severe induction

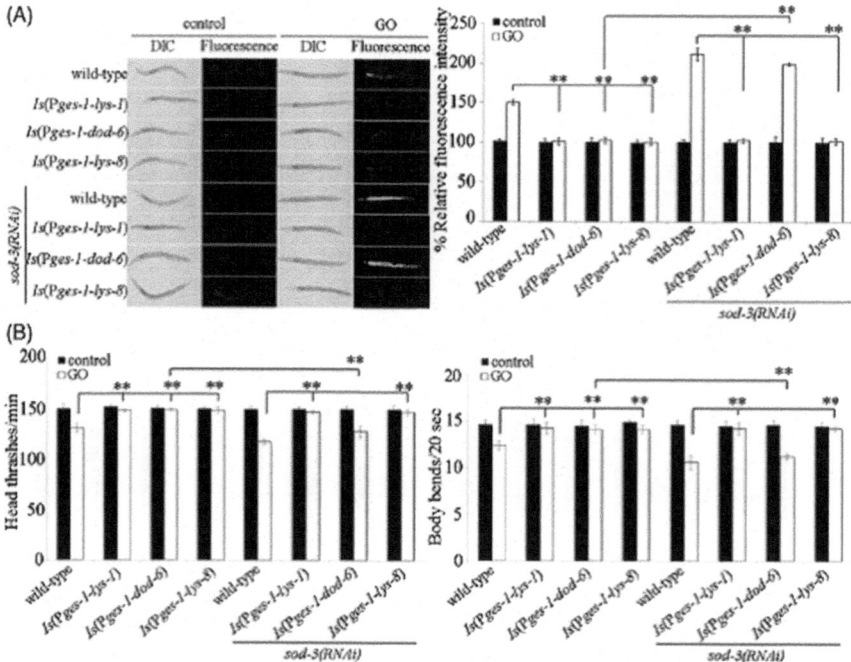

Fig. 7.15 Genetic interaction between SOD-3 and LYS-1, DOD-6, or LYS-8 in the regulation of GO toxicity [30]

(**a**) Genetic interaction between SOD-3 and LYS-1, DOD-6, and LYS-8 in the regulation of GO toxicity in inducing intestinal ROS production. (**b**) Genetic interaction between SOD-3 and LYS-1, DOD-6, or LYS-8 in the regulation of GO toxicity in decreasing locomotion behavior. Prolonged exposure was performed from L1-larvae to young adults. GO exposure concentration was 10 mg/L. Bars represent mean ± SD. $^{**}p < 0.01$

of intestinal ROS production and decrease in locomotion behavior than that in GO-exposed *lys-1(ok2445)* mutant or in GO-exposed *lys-8(ok3504)* mutant [30]. In nematodes, after GO (10 mg/L) exposure, mutation of *lys-1* or *lys-8* did not influence the expressions of *clk-1*, *gas-1*, and *isp-1*, which are required for the control of oxidative stress [30].

In nematodes, mutation of *lys-1* significantly decreased the transcriptional expression of *tub-2* in GO (10 mg/L)-exposed nematodes, and intestine-specific RNAi of *tub-2* significantly suppressed the resistance of nematodes overexpressing intestinal LYS-1 to GO toxicity in inducing ROS production and in decreasing locomotion behavior [30]. *tub-2* encodes an ortholog of human tubby-like protein 4, and intestine-specific RNAi of *tub-2* caused a susceptibility to GO toxicity [30]. These results suggest the formation of signaling cascade of DAF-16-LYS-1-TUB-2 in the regulation of GO toxicity in nematodes.

7.3.4.4   Genetic Interaction Between SKN-1 and DAF-16 or DAF-2 in the Regulation of Response to ENM Exposure

In nematodes, both the FOXO transcriptional factor DAF-16 and the FOXO transcriptional factor SKN-1/Nrf can act downstream of the insulin receptor DAF-2 in the insulin signaling pathway to regulate various biological processes, such as the stress response [17]. Using intestinal ROS production as the toxicity assessment endpoint, it has been shown that the GO toxicity in inducing intestinal ROS production in *daf-16(mu86);skn-1(RNAi)* was more severe than that in *daf-16(mu86)* or in *skn-1(RNAi)* [15], suggesting that the SKN-1 and the DAF-16 may act in parallel signaling pathways to regulate the GO toxicity in nematodes.

Besides this, the genetic interaction between DAF-2 in insulin signaling pathway and SKN-1 in p38 MAPK signaling pathway in regulating GO toxicity was also examined in nematodes. It has been shown that prolonged exposure to GO (100 mg/L) could cause the similar toxicity on lifespan in double mutant of *daf-2(e1370);skn-1(RNAi)* to that in *skn-1(RNAi)* nematodes [19], suggesting that RNAi knockdown of *skn-1* may potentially suppress the resistance of *daf-2* mutant to GO toxicity. Therefore, both the core signaling cascade of p38 MAPK signaling pathway and the insulin receptor DAF-2 in the insulin signaling pathway can act upstream of SKN-1 to regulate the GO toxicity in nematodes.

7.3.5   Innate Immune Response Signaling Pathway

7.3.5.1   Induction of Antimicrobial Proteins After Acute Exposure to ENMs

With GO as the example, it has been observed that acute exposure to GO (10 mg/L) could induce the significant increase in expressions of *lys-1*, *dod-6*, *F55G11.4*, *lys-8*, and *spp-1* in nematodes (Fig. 7.16) [30]. *lys-1* and *lys-8* encode lysozymes, *dod-6* encodes a protein downstream of DAF-16, *F55G11.4* encodes a protein containing a CUB-like domain, and *spp-1* encodes a caenopore. LYS-1 is expressed in the neurons; and the intestine, F55G11.4, is expressed in the intestine; and LYS-8 is expressed in the neurons and the intestine. Acute exposure to GO (10 mg/L) could further significantly increase the expression of LYS-1::GFP in both the neurons and the intestine, the intestinal expression of F55G11.4::GFP, and the expression of LYS-8::GFP in both the neurons and the intestine (Fig. 7.16) [30]. These results suggest that the expression of some antimicrobial protein-mediated innate immune response may be activated for nematodes against the toxicity from short-term exposure to GO.

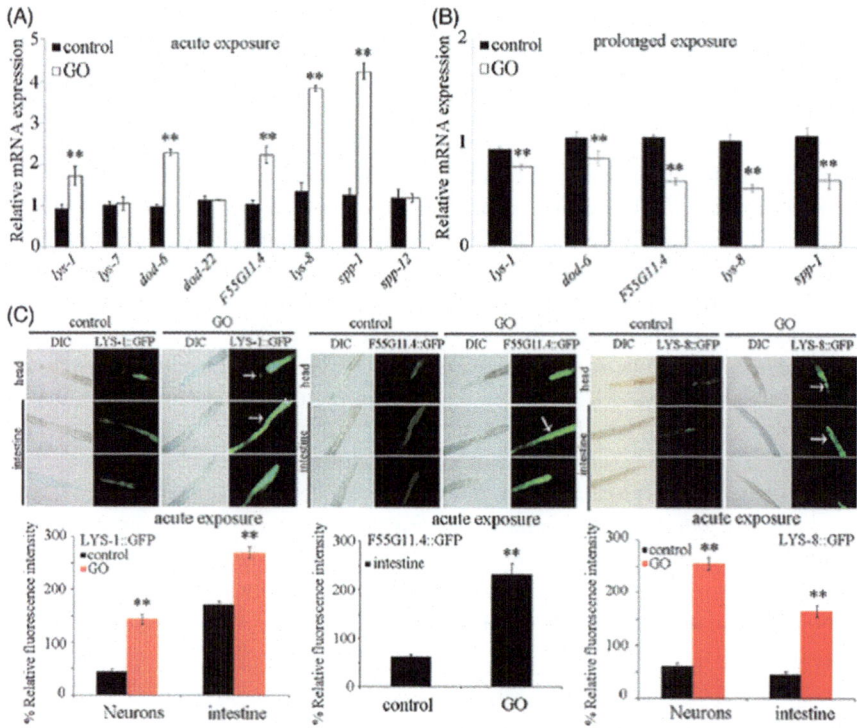

Fig. 7.16 Response of antimicrobial proteins to GO exposure [30]
(**a**) Effect of acute exposure to GO on the expressions of antimicrobial genes. (**b**) Effect of prolonged exposure to GO on the expressions of antimicrobial genes. (**c**) Effect of acute exposure to GO on the expressions of LYS-1::GFP, F55G11.4::GFP, and LYS-8::GFP. Arrowheads indicate the neurons in the head and the intestine, respectively. Acute exposure to GO was performed from young adult for 24 h. Prolonged exposure was performed from L1-larvae to young adults. GO exposure concentration was 10 mg/L. Bars represent mean ± SD. $^{**}p < 0.01$ *vs* control

7.3.5.2 Decrease in the Expression of Antimicrobial Proteins After Prolonged or Chronic Exposure to ENMs

Different from the effect of acute exposure to GO on expression pattern of antimicrobial genes, prolonged exposure (from L1-larave to young adults) to GO (10 mg/L) significantly decreased the expressions of *lys-1*, *dod-6*, *F55G11.4*, *lys-8*, and *spp-1* (Fig. 7.16) [30]. Additionally, prolonged exposure to GO (10 mg/L) also significantly decreased the expressions of LYS-1::GFP in both the neurons and the intestine, the intestinal expression of F55G11.4::GFP, and the expression of LYS-8::GFP in both the neurons and the intestine (Fig. 7.16) [30]. Moreover, chronic GO exposure from L1-larvae to adult day-8 induced a significant decrease in the expressions of genes encoding antimicrobial proteins (*F08G5.6*, *pqm-1*, *K11D12.5*, *prx-11*, *spp-1*, *lys-7*, *lys-2*, *abf-2*, *acdh-1*, *lys-8*, *nlp-29*) together with the severe accumulation of OP50 in the intestine [31]. p38 MAPK signaling pathway is one of the key signaling pathways controlling the innate immune response to bacterial infection, and chronic exposure to GO could further significantly decrease the expressions of *nsy-1*, *sek-1*, and *pmk-1* [31]. Therefore, the expression of antimicrobial proteins or certain signaling pathway-mediated innate immune response may be suppressed after long-term exposure to GO, which will in turn accelerate the GO toxicity in nematodes.

7.3.5.3 Molecular Control of ENM Toxicity by Innate Immune Response-Related Signals

In nematodes, it has been observed that mutation of *lys-1*, *lys-8*, or *spp-1* induced a susceptibility to GO toxicity in inducing intestinal ROS production and in decreasing locomotion behavior (Fig. 7.17) [30]. Additionally, RNAi knockdown of *dod-6* or *F55G11.4* also induced a susceptibility to GO toxicity in inducing intestinal ROS production and in decreasing locomotion behavior (Fig. 7.17) [30]. These results suggest that these antimicrobial proteins may negatively regulate the GO toxicity in nematodes.

Meanwhile it has been found that intestine-specific RNAi knockdown of the antimicrobial gene of *lys-1*, *dod-6*, *F55G11.4*, *lys-8*, or *spp-1* also caused a susceptibility to GO toxicity in nematodes using the transgenic strain of VP303 as an intestine-specific RNAi knockdown tool [30], which implies that these antimicrobial proteins can further act in the intestine in response to GO exposure.

7.3.6 Wnt Signaling Pathway

7.3.6.1 Involvement of Certain Wnt Ligands in the Regulation of ENM Toxicity

In nematodes, five Wnt ligands (LIN-44, EGL-20, MOM-2, CWN-1, and CWN-2) have been identified in the Wnt signaling pathway. Among these Wnt ligands, it has been found that mutation of *mom-2* or *egl-20* could not affect the formation of GO toxicity in inducing intestinal ROS production and in decreasing locomotion behavior in nematodes (Fig. 7.18) [32]. In contrast, mutation of *cwn-1* or *lin-44* induced a resistance to GO toxicity in inducing intestinal ROS production and in decreasing locomotion behavior, whereas mutation of *cwn-2* induced a susceptibility to GO toxicity in inducing intestinal ROS production and in decreasing locomotion behavior (Fig. 7.18) [32], which implies that CWN-1, CWN-2, and LIN-44 play an important role in the regulation of GO toxicity in nematodes.

Meanwhile, mutation of *cwn-2* enhanced the intestinal permeability in GO-exposed nematodes, and mutation of *cwn-1* inhibited the intestinal permeability in GO-exposed nematodes as indicated by the analysis on Nile red signal [32].

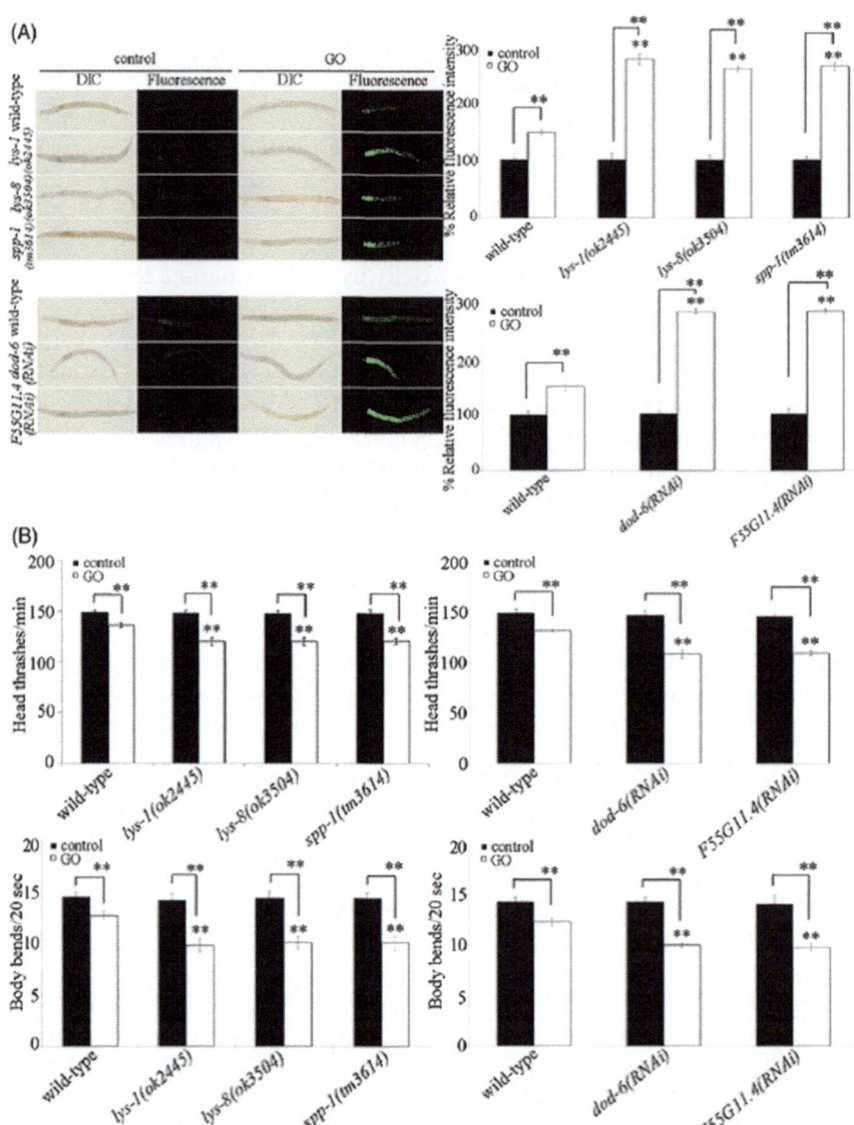

Fig. 7.17 Mutation or RNAi knockdown of antimicrobial genes induced an enhanced susceptibility to GO toxicity [30]

(**a**) Mutation or RNAi knockdown of antimicrobial genes induced an enhanced susceptibility to GO toxicity in inducing intestinal ROS production. (**b**) Mutation or RNAi knockdown of antimicrobial genes induced an enhanced susceptibility to GO toxicity in decreasing locomotion behavior. Prolonged exposure was performed from L1-larvae to young adults. GO exposure concentration was 10 mg/L. Bars represent mean ± SD. $^{**}p < 0.01$ vs wild-type (if not specially indicated)

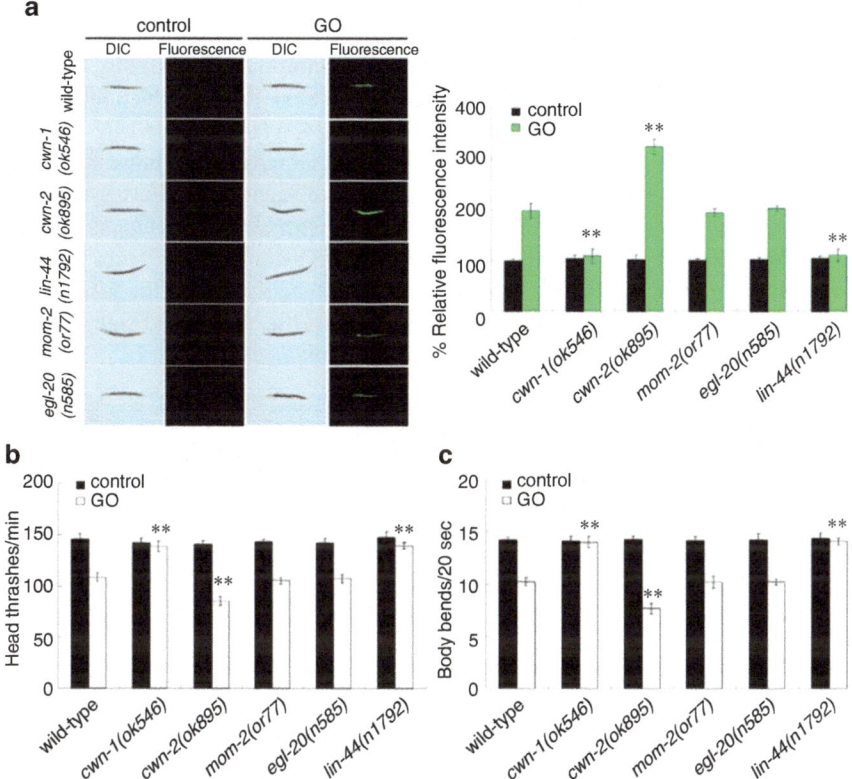

Fig. 7.18 Effects of *cwn-1*, *cwn-2*, *mom-2*, *egl-20*, or *lin-44* mutation on GO toxicity in nematodes [32]
(**a**) Effects of *cwn-1*, *cwn-2*, *mom-2*, *egl-20*, or *lin-44* mutation on GO toxicity in inducing intestinal ROS production. (**b**) Effects of *cwn-1*, *cwn-2*, *mom-2*, *egl-20*, or *lin-44* mutation on GO toxicity in decreasing locomotion behavior in nematodes. GO exposure concentration was 100 mg/L. Prolonged exposure was performed from L1-larvae to young adults. Bars represent mean ± SD. $^{**}P < 0.01$ *vs* wild-type

Mutation of *lin-44* only moderately suppressed the intestinal permeability in GO-exposed nematodes [32]. Moreover, mutation of *cwn-2* did not obviously alter the mean defecation cycle length in GO-exposed nematodes; however, mutation of *lin-44* could significantly decrease the mean defecation cycle length in GO-exposed nematodes [32]. Mutation of *cwn-1* could only moderately decrease the mean defecation cycle length in GO-exposed nematodes [32]. These results imply that these three Wnt ligands may regulate the GO toxicity through different underlying cellular mechanisms in nematodes.

7.3.6.2 Genetic Interactions of Wnt Ligands in Regulating ENM Toxicity

In nematodes, mutation of *cwn-1* or *lin-44* could inhibit the susceptibility of *cwn-2(ok895)* mutant to GO toxicity in inducing intestinal ROS production and in decreasing locomotion behavior [32]. Moreover, the CWN-1 and the LIN-44 may function in the same genetic pathway to regulate the GO toxicity in inducing intestinal ROS production and in decreasing locomotion behavior [32]. Furthermore, it has found that the *lin-44(n1792); cwn-1(ok546);cwn-2(ok895)* triple mutant showed the similar induction of intestinal ROS production and locomotion behavior to those in *cwn-1(ok546); cwn-2(ok895)*, *lin-44(n1792); cwn-1(ok546)*, or *lin-44(n1792); cwn-2(ok895)* double mutant [32], which implies that the Wnt ligand signals may, as a whole, positively regulate the GO toxicity in nematodes.

7.3.6.3 Involvement of Canonical Wnt/β-Catenin Signaling Pathway in the Regulation of ENM Toxicity

In *C. elegans*, the major effector in canonical Wnt signaling pathway is the transcriptional factor β-catenin BAR-1 [33]. After the binding of Wnt ligand(s) to a Frizzled receptor (LIN-17, MOM-5, MIG-1, or CFZ-2), the APC complex containing APR-1/Axin, PRY-1/CK1a, KIN-19, and GSK-3 can be inhibited by Dishevelled proteins (MIG-5, DSH-1, and/or DSH-2) so as to lead to the stabilization of β-catenin BAR-1 and its translocation into the nucleus [33]. Further with GO as an example, it has been found that mutation of *bar-1* induced a susceptibility to GO toxicity in inducing intestinal ROS production and in decreasing locomotion behavior in nematodes [34].

Among the genes encoding the APC complex, prolonged exposure (from L1-larvae to young adults) to GO (1 mg/L) could significantly increase the expressions of *apr-1* and *gsk-3* [34]. Meanwhile, mutation of *apr-1* or *gsk-3* induced a resistance to GO toxicity in inducing intestinal ROS production and in decreasing locomotion behavior in nematodes [34]. Moreover, RNAi knockdown of *bar-1* could significantly suppress the resistance of *apr-1* or *gsk-3* mutant to GO toxicity in inducing intestinal ROS production and in decreasing locomotion behavior [34], suggesting that APR-1 and GSK-3 act upstream of BAR-1 in the regulation of GO toxicity in nematodes.

Among the genes encoding the Dishevelled proteins, prolonged exposure to GO (1 mg/L) could significantly decrease the expressions of *dsh-1* and *dsh-2* [34]. Meanwhile, mutation of *dsh-1* or *dsh-2* induced a susceptibility to GO toxicity in inducing intestinal ROS production and in decreasing locomotion behavior [34]. Moreover, RNAi knockdown of *apr-1* or *gsk-3* could significantly suppress the susceptibility of *dsh-1* or *dsh-2* mutant to GO toxicity in inducing intestinal ROS production and in decreasing locomotion behavior [34], which suggests that DSH-1 and DSH-2 can further act upstream of APR-1 and GSK-3 in the regulation of GO toxicity in nematodes.

Among the genes encoding the Frizzled receptor, prolonged exposure to GO (1 mg/L) could significantly decrease the expressions of *mom-5* and *cfz-2* [34]. Meanwhile, mutation of *mom-5* or *cfz-2* induced a susceptibility to GO toxicity in inducing intestinal ROS production and in decreasing locomotion behavior [34]. Additionally, the *mom-5(gk812);dsh-1(RNAi)* exhibited the similar induction of intestinal ROS production and locomotion behavior to those in *mom-5(gk812)* or *dsh-1(RNAi)* after GO exposure, the *dsh-1(RNAi);cfz-2(ok1201)* exhibited the similar induction of intestinal ROS production and locomotion behavior to those in *cfz-2(ok1201)* and *dsh-1(RNAi)* after GO exposure, the *mom-5(gk812);dsh-2(RNAi)* exhibited the similar induction of intestinal ROS production and locomotion behavior to those in *mom-5(gk812)* and *dsh-2(RNAi)* after GO exposure, and the *dsh-2(RNAi);cfz-2(ok1201)* exhibited the similar induction of intestinal ROS production and locomotion behavior to those in *cfz-2(ok1201)* and *dsh-2(RNAi)* after GO exposure [34].

7.3.6.4   Identification of Downstream Targets for β-Catenin BAR-1 in the Regulation of ENM Toxicity

The homeobox proteins of LIN-39, MAB-5, and EGL-5 are normally considered downstream targets for β-catenin BAR-1-depemdent signaling [35, 36]. Among these potential targets, prolonged exposure to GO (1 mg/L) could only significantly decrease the expression of *egl-5* (Fig. 7.19) [34]. Mutation of *egl-5* induced a susceptibility to GO toxicity in inducing intestinal ROS production and in decreasing locomotion behavior (Fig. 7.19) [34]. The GO toxicity in inducing intestinal ROS production in *egl-5(n945);bar-1(ga80)* was similar to that in *egl-5(n945)* or in *bar-1(ga80)* mutant (Fig. 7.19) [34], suggesting that BAR-1 and EGL-5 may act in the same genetic pathway in the regulation of GO toxicity in nematodes. In *C. elegans*, *egl-5* encodes a homeodomain transcription factor, orthologous to *Drosophila* Abd-B and the vertebrate Hox9-13 proteins. Additionally, mutation of *egl-5* could significantly increase the expression of *clk-1*, *sod-1*, *sod-3*, and *sod-4* and decrease the expression of *gas-1* in GO-exposed nematodes (Fig. 7.19) [34]. Thus, during the control of GO toxicity, a signaling cascade of MOM-5/CFZ-2-DSH-1/DSH-2-APR-1/GSK-3-BAR-1-EGL-5 in the canonical Wnt/β-catenin signaling pathway has been raised (Fig. 7.19) [34].

7.3.6.5   Genetic Interactions Between β-Catenin BAR-1 and Other Signaling Pathways in the Regulation of ENM Toxicity

The genetic interaction analysis demonstrated that a more severe GO toxicity in inducing intestinal ROS production and in decreasing locomotion behavior was formed in *daf-16(mu86);bar-1(RNAi)* double mutant compared with that in *daf-16(mu86)* or in *bar-1(RNAi)* nematodes (Fig. 7.20) [34], suggesting the formation of a synergistic effect between BAR-1 and DAF-16 in the regulation of GO toxicity.

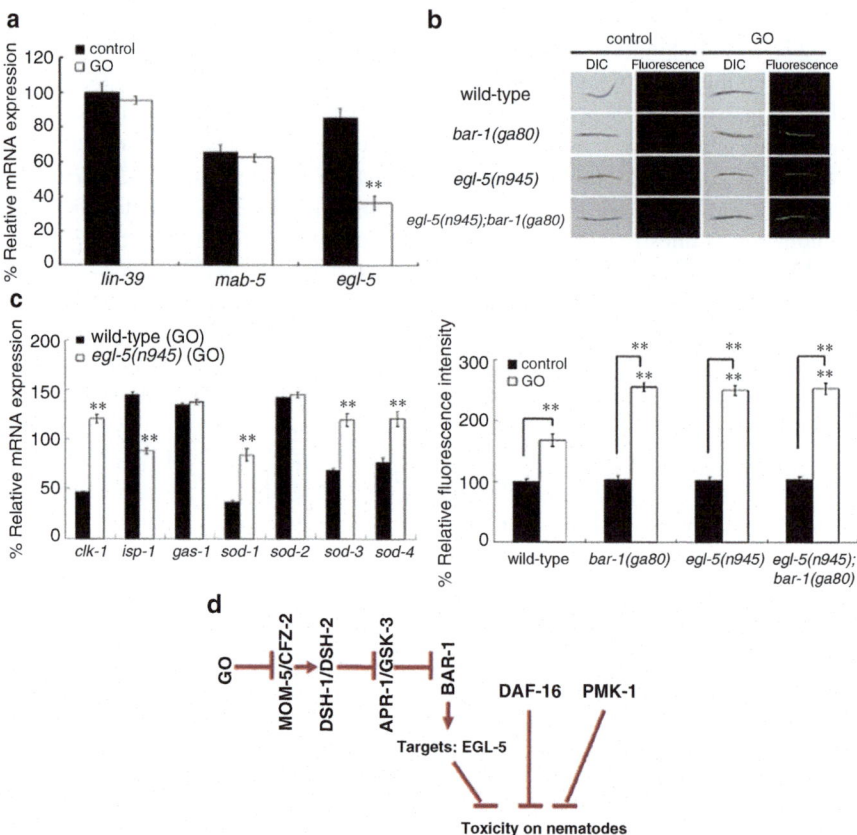

Fig. 7.19 Role of EGL-5 in the regulation of GO toxicity in nematodes [34]
(**a**) Effect of GO exposure on expression of genes encoding downstream mediators of BAR-1 in nematodes. Bars represent mean ± SD. $^{**}P < 0.01$ vs control. (**b**) Genetic interaction between BAR-1 and EGL-5 in the regulation of GO toxicity in inducing intestinal ROS production. Bars represent mean ± SD. $^{**}P < 0.01$ vs wild-type (if not specially indicated). (**c**) Effect of *egl-5* mutation on expression of genes required for the control of oxidative stress after GO exposure. Bars represent mean ± SD. $^{**}P < 0.01$ vs wild-type (GO). Prolonged exposure was performed from L1-larvae to young adults. GO exposure concentration was 1 mg/L. (**d**) A diagram showing the signaling cascade of canonical Wnt/β-catenin signaling in the regulation of GO toxicity in nematodes

Meanwhile, a more severe GO toxicity in inducing intestinal ROS production and in decreasing locomotion behavior was also observed in *pmk-1(km25);bar-1(RNAi)* double mutant compared with that in *pmk-1(km25)* or in *bar-1(RNAi)* nematodes (Fig. 7.20) [34], suggesting the formation of a synergistic effect between BAR-1 and PMK-1 in the regulation of GO toxicity. Therefore, the canonical Wnt/β-catenin signaling pathway may act in parallel with the insulin or p38 MAPK signaling pathway in the regulation of GO toxicity in nematodes.

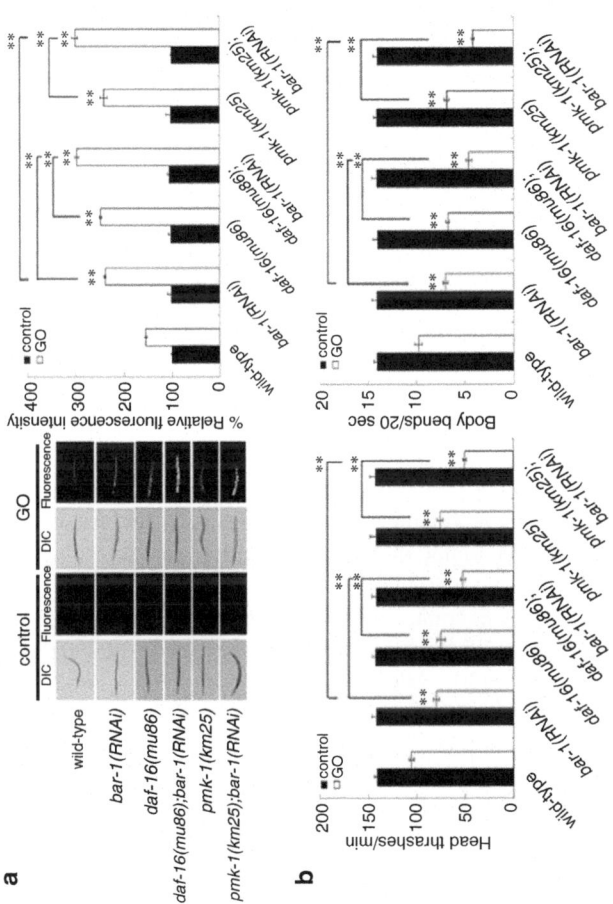

Fig. 7.20 Genetic interaction between BAR-1 and DAF-16 or p38 MAPK/PMK-1 in the regulation of GO toxicity in nematodes [34] (**a**) Genetic interaction between BAR-1 and DAF-16 or p38 MAPK/PMK-1 in the regulation of GO toxicity in inducing intestinal ROS production. Bars represent mean ± SD. $^{**}P < 0.01$ vs wild-type (if not specially indicated). (**b**) Genetic interaction between BAR-1 with DAF-16 and p38 MAPK/PMK-1 in the regulation of GO toxicity in decreasing locomotion behavior. Bars represent mean ± SD. $^{**}P < 0.01$ vs wild-type (if not specially indicated). Prolonged exposure was performed from L1-larvae to young adults. GO exposure concentration was 1 mg/L

7.3.6.6   Wnt Signaling Pathway May Mediate a Protective Response for Nematodes Against the ENM Toxicity in Nematodes

In nematodes, CWN-1 is expressed in the neurons and the intestine, CWN-2 is predominantly expressed in the pharynx and the intestine, and LIN-44 is only expressed in the posterior of the animal and the tail hypodermis. Prolonged exposure (from L1-larvae to young adults) to GO (100 mg/L) could significantly decrease the expression of CWN-1 in the intestine and the neurons, increase the expression of CWN-2 in the intestine and the pharynx, and decrease the expression of LIN-44 in the tail [32]. These results suggest that the altered Wnt ligand signaling pathway may mediate a protective response for nematodes against the GO toxicity.

7.3.7   TGF-Beta Signaling Pathway

In *C. elegans*, DAF-1, DAF-3, DAF-5, DAF-8, DAF-12, and DAF-14 are important components in the TGF-beta signaling pathway. Among these components, intestine-specific RNAi knockdown or mutation of *daf-8* induced in a susceptibility to GO toxicity in inducing intestinal ROS production, whereas intestine-specific RNAi knockdown or mutation of *daf-5* induced a resistance to GO toxicity in inducing intestinal ROS production in nematodes [30]. *daf-5* encodes a transcriptional factor, and *daf-8* encodes a R-Smad protein. Genetic interaction analysis further indicated that double mutations of *daf-8* and *daf-5* induced a resistance to GO toxicity in inducing intestinal ROS production and in decreasing locomotion behavior (Fig. 7.21) [30], suggesting that mutation of *daf-5* could suppress the susceptibility of *daf-8(e1393)* mutant to GO toxicity in nematodes.

Moreover, mutation of *lys-8* could significantly decrease the transcriptional expression of *daf-8* and increase the transcriptional expression of *daf-5* in GO (10 mg/L)-exposed nematodes [30]. Additionally, intestine-specific RNAi knockdown of *daf-8* could inhibit the resistance of nematodes overexpressing intestinal *lys-8* to GO toxicity in inducing ROS production and in decreasing locomotion behavior [30]. These results suggest the formation of a signaling cascade of intestinal DAF-16-LYS-8-DAF-8-DAF-5 in the regulation of GO toxicity in nematodes.

7.3.8   Developmental Timing Control-Related Signals

7.3.8.1   Involvement of *let-7* and Its Direct Targets in the Regulation of ENM Toxicity

In *C. elegans*, *let-7* is one of the founding members of the miRNA family firstly identified with the function in regulating the timing of larval and adult transition by suppressing the expression and the function of its direct targets of HBL-1 and

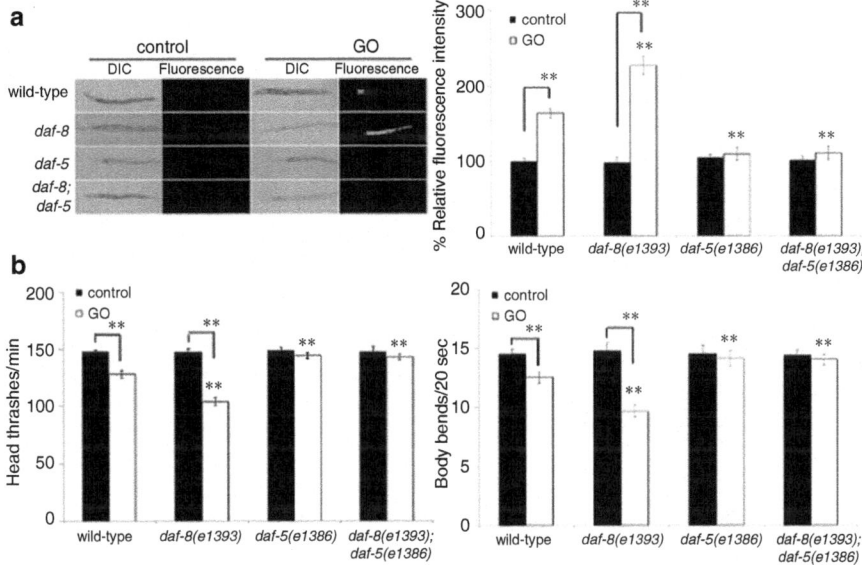

Fig. 7.21 Genetic interaction between DAF-8 and DAF-5 in the regulation of GO toxicity [30] (**a**) Genetic interaction between DAF-8 and DAF-5 in the regulation of GO toxicity in inducing intestinal ROS production. (**b**) Genetic interaction between DAF-8 and DAF-5 in the regulation of GO toxicity in decreasing locomotion behavior. Prolonged exposure was performed from L1-larvae to young adults. GO exposure concentration was 10 mg/L. Bars represent mean ± SD. $^{**}p < 0.01$ vs wild-type (if not specially indicated)

LIN-41 [37, 38]. In nematodes, mutation of *let-7* induced a resistance to multi-walled carbon nanotubes (MWCNTs) toxicity in inducing intestinal ROS production and in decreasing locomotion behavior [39]. In MWCNT-exposed nematodes, *let-7* mutation could significantly increase the expressions of both the *hbl-1* and the *lin-41* [34]. Different from the phenotype in *let-7* mutant, mutation of *hbl-1* or *lin-41* induced a susceptibility to MWCNT toxicity in inducing intestinal ROS production and in decreasing locomotion behavior [34]. Genetic interaction analysis further demonstrated that mutation of *hbl-1* or *lin-41* could suppress the resistance of *let-7(mg279)* mutant to MWCNT toxicity in inducing intestinal ROS production and in decreasing locomotion behavior [39], which suggests that HBL-1 and LIN-41 act as the targets for *let-7* in the regulation of MWCNT toxicity in nematodes.

7.3.8.2 Identification of Downstream Targets for HBL-1 in the Regulation of ENM Toxicity

In nematodes, totally 13 genes (*T01D3.6*, *F13B12.4*, *ugt-18*, *cpt-4*, *clec-60*, *F28H1.1*, *W04G3.3*, *K12B6.3*, *nurf-1*, *sym-1*, *tir-1*, *nhx-3*, and *zig-4*) could be significantly increased (more than 2.5-fold changes) by *hbl-1* overexpression under the normal conditions [40]. Among these genes, after prolonged exposure to MWCNTs (10 µg/L), mutation of *hbl-1* could significantly decrease the expressions of *tir-1*, *sym-1*, and *lpr-4* [39]. Mutation of *tir-1* or *sym-1* induced a susceptibility to MWCNT toxicity in inducing intestinal ROS production and in decreasing locomotion behavior, whereas mutation of *lpr-4* could not obviously affect the MWCNT toxicity [39]. In *C. elegans*, *tir-1* encodes a Toll-interleukin 1 receptor (TIR) domain adaptor protein, and *sym-1* encodes a protein containing 15 contiguous leucine-rich repeats (LRRs).

Genetic interaction analysis further demonstrated that mutation of *tir-1* or *sym-1* could significantly suppress the resistance of nematodes overexpressing HBL-1 to MWCNT toxicity in inducing intestinal ROS production and in decreasing locomotion behavior [39], which suggest the formation of a signaling cascade of *let-7*-HBL-1-TIR-1/SYM-1 in the regulation of MWCNT toxicity in nematodes.

7.3.8.3   Genetic Interaction Between LIN-41 and ALG-1 or ALG-2 in the Regulation of ENM Toxicity

alg-1 and *alg-2* encode two members of RDE-1 proteins. In *C. elegans*, mutation of *lin-41* could inhibit the retarded heterochronic phenotypes caused by RNAi knockdown of *alg-1/alg-2* [41]. The genetic interaction analysis suggested that mutation of *lin-41* could significantly suppress the resistance of *alg-1(gk214)* or *alg-2(ok304)* mutant to MWCNT toxicity in inducing intestinal ROS production and in decreasing locomotion behavior [39], suggesting that the ALG-1 or the ALG-2 acts upstream of the LIN-41 in the regulation of MWCNT toxicity in nematodes.

7.3.8.4   A Feedback Loop Between *let-7* and HBL-1 or LIN-41 in the Regulation of ENM Toxicity

In nematodes, it has been further observed that mutation of *hbl-1* or *lin-41* could also significantly increase the expression of *let-7::GFP* in MWCNT-exposed nematodes (Fig. 7.22) [39]. This observation implies the existence of a feedback loop between *let-7* and HBL-1 or LIN-41 in the regulation of ENM toxicity in nematodes.

7.3.8.5   *let-7* and Its Direct Targets May Mediate a Protective Response to ENMs in Nematodes

In nematodes, prolonged exposure to MWCNTs (10 µg/L) could cause the significant decrease in the expression of *let-7::GFP* [39]. Meanwhile, prolonged exposure to MWCNTs (10 µg/L) could further significantly increase the expressions of both *hbl-1* and *lin-41* [39]. Therefore, the altered *let-7* and its direct targets of HBL-1 and LIN-41 may mediate a protective response for nematodes against the MWCNT toxicity.

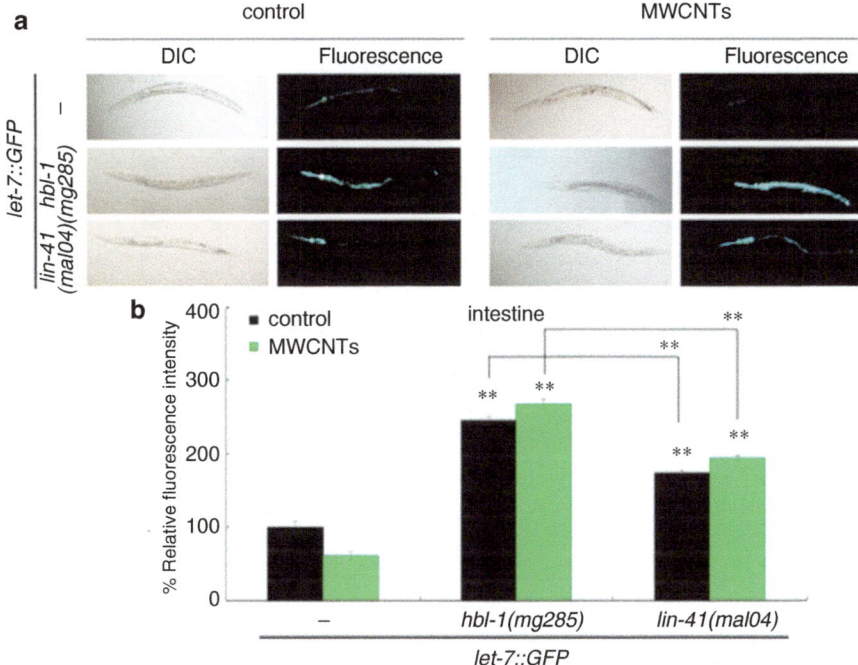

Fig. 7.22 Effect of mutation of *hbl-1* or *lin-41* on *let-7::GFP* expression in MWCNT-exposed nematodes [39]
(**a**) Comparison of *let-7::GFP* expression. (**b**) Comparison of intestinal *let-7::GFP* expression. "-" nematodes without mutation of *hbl-1* or *lin-41*. Prolonged exposure was performed from L1-larvae to young adults. Exposure concentration of MWCNTs was 10 μg/L. Bars represent mean ± SD. **P < 0.01 vs nematodes without mutation of *hbl-1* or *lin-41* (if not specially indicated)

7.3.9 Neurotransmission-Related Signals

7.3.9.1 Neurotransmitter Signals Required for the Regulation of ENM Toxicity

The nematodes contain some classic neurotransmitters found in the vertebrates, such as acetylcholine, glutamate, gamma-aminobutyric acid (GABA), serotonin, and dopamine. *bas-1* encodes an aromatic acid decarboxylase required for serotonin and dopamine synthesis, *eat-4* encodes a glutamate transporter, and *tdc-1* encodes an aromatic L-amino-acid/L-histidine decarboxylase required for making tyramine and octopamine. With the Al_2O_3-NPs (60 nm) as an example, after acute exposure to Al_2O_3-NPs (25 mg/L) for 6 or 12 h, mutations of *bas-1* or *eat-4* could significantly suppress the Al_2O_3-NP toxicity in decreasing locomotion behavior; however, mutation of *tdc-1* could not affect the Al_2O_3-NP toxicity in decreasing locomotion behavior (Fig. 7.23) [42], which implies that the glutamate, the serotonin, and/or the dopamine may regulate the formation of Al_2O_3-NP toxicity in nematodes.

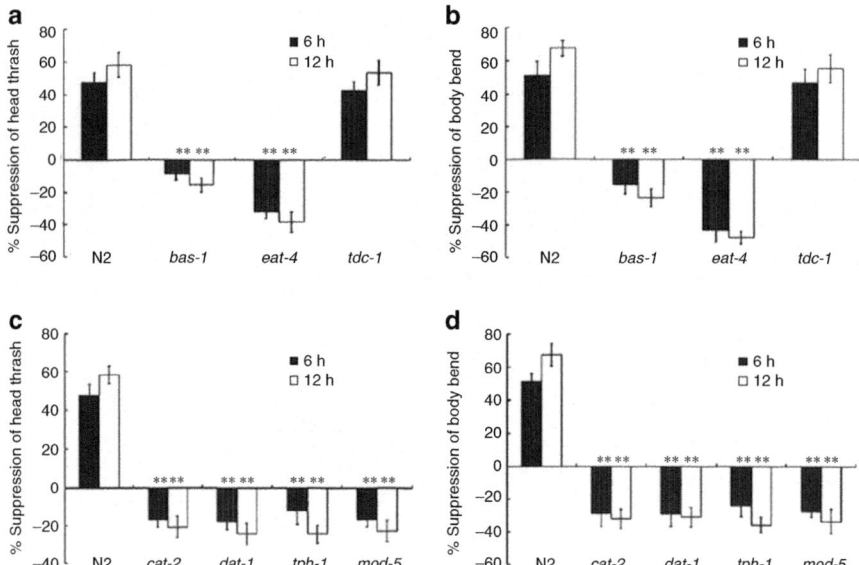

Fig. 7.23 Formation of adverse effects on locomotion behavior in Al$_2$O$_3$-NP-exposed nematodes requires glutamate, dopamine, and serotonin [42]

(**a**) Head thrashes in Al$_2$O$_3$-NP-exposed wild-type N2, *bas-1*, *eat-4*, and *tdc-1* mutant nematodes. (**b**) Body bends in Al$_2$O$_3$-NP-exposed wild-type N2, *bas-1*, *eat-4*, and *tdc-1* mutant nematodes. (**c**) Head thrashes in Al$_2$O$_3$-NP-exposed wild-type N2, *cat-2*, *dat-1*, *tph-1*, and *mod-5* mutant nematodes. (**d**) Body bends in Al$_2$O$_3$-NP-exposed wild-type N2, *cat-2*, *dat-1*, *tph-1*, and *mod-5* mutant nematodes. Al$_2$O$_3$-NP exposure at the concentration of 25 mg/L was performed at the L4 larval stage. Data are quantified as percentage suppression of locomotion behavior between naive and Al$_2$O$_3$-NP-exposed conditions for each genotype. % suppression = (locomotion behavior under the naive condition − locomotion behavior under the Al$_2$O$_3$-NP-exposed condition)/locomotion behavior under the naïve condition. Bars represent mean ± SEM. $^{**}p < 0.01$

In nematodes, *cat-2* encodes a tyrosine hydroxylase essential for dopamine synthesis, *dat-1* encodes a dopamine transporter, *tph-1* encodes a tryptophan hydroxylase required for serotonin synthesis, and *mod-5* encodes a serotonin transporter. Moreover, after acute exposure to Al$_2$O$_3$-NPs (25 mg/L) for 6 or 12 h, mutation of *cat-2* or *tph-1* could significantly suppress the Al$_2$O$_3$-NPs toxicity in decreasing locomotion behavior (Fig. 7.23) [42], suggesting that both the dopamine and the serotonin may be required for the formation of Al$_2$O$_3$-NPs toxicity in nematodes. Meanwhile, after acute exposure to Al$_2$O$_3$-NPs (25 mg/L) for 6 or 12 h, mutation of *dat-1* or *mod-5* could also significantly suppress the Al$_2$O$_3$-NPs toxicity in decreasing locomotion behavior (Fig. 7.23) [42], suggesting the possible critical functions of dopamine and serotonin transporters in the formation of Al$_2$O$_3$-NPs toxicity in nematodes.

Genetic interaction analysis further demonstrated that the more severe Al$_2$O$_3$-NPs toxicity in decreasing locomotion behavior could be formed in the *bas-1eat-4* double mutant compared that in *bas-1* or in *eat-4* single mutant, implying that the glutamate signal may function in the parallel pathways with the dopamine signal or serotonin signal in the regulation of Al_2O_3-NPs toxicity [42]. Additionally, the more severe Al_2O_3-NPs toxicity in decreasing locomotion behavior could be detected in the *tph-1cat-2* double mutant compared with that in *tph-1* or in *cat-2* single mutant, and the more severe Al_2O_3-NPs toxicity in decreasing locomotion behavior could be observed in the *mod-5;dat-1* double mutant compared with that in *mod-5* or in *dat-1* single mutant, suggesting that the dopamine signal and the serotonin signal may further function in the parallel pathways in the regulation of Al_2O_3-NPs toxicity [42].

The neurotransmitters exert their functions by acting on certain corresponding receptors. In nematodes, after acute exposure to Al_2O_3-NPs (25 mg/L) for 6 or 12 h, mutation of *dop-1* could significantly suppress the Al_2O_3-NPs toxicity in decreasing locomotion behavior; however, mutation of *dop-2*, *dop-3*, or *dop-4* could not obviously affect the Al_2O_3-NPs toxicity in decreasing locomotion behavior (Fig. 7.24) [42], suggesting that the D1-like dopamine receptor DOP-1 may be involved in the regulation of Al_2O_3-NPs toxicity in nematodes. Moreover, after acute exposure to Al_2O_3-NPs (25 mg/L) for 6 or 12 h, mutation of *mod-1* could significantly suppress the Al_2O_3-NPs toxicity in decreasing locomotion behavior; however, mutation of *ser-1*, *ser-4*, or *ser-7* could not obviously influence the Al_2O_3-NPs toxicity in decreasing locomotion behavior (Fig. 7.24) [42], suggesting that the ionotropic serotonin receptor MOD-1 may be required for the regulation of Al_2O_3-NPs toxicity in nematodes. Furthermore, after acute exposure to Al_2O_3-NPs (25 mg/L) for 6 or 12 h, mutation of *glr-2* or *glr-6* could significantly suppress the Al_2O_3-NPs toxicity in decreasing locomotion behavior; however, mutation of *glr-1*, *glr-3*, *glr-4*, *glr-5*, *glr-7*, *glr-8*, *nmr-1*, or *nmr-2* could not obviously affect the Al_2O_3-NPs toxicity in decreasing locomotion behavior (Fig. 7.24) [42]. Therefore, the formation of Al_2O_3-NPs neurotoxicity may require the functions of D1-like dopamine receptor DOP-1, ionotropic serotonin receptor MOD-1, and non-NMDA glutamate receptors GLR-2 and GLR-6 in nematodes.

7.3.9.2   Function of AEX-3 in the Regulation of GO Toxicity

In nematodes, AEX-3, a guanine exchange factor for GTPase, is required for neuronal vesicle trafficking and synaptic vesicle release [43, 44]. Mutation of *aex-3* enhanced both the induction of intestinal ROS production and the intestinal permeability in GO-exposed nematodes (Fig. 7.25) [15]. Prolonged exposure to GO (10 mg/L) could further significantly increase the transcriptional expressions of *sod-1*, *sod-2*, *sod-3*, and *sod-4* in *aex-3* mutant compared with those in wild-type nematodes (Fig. 7.25) [15]. Meanwhile, prolonged exposure to GO (10 mg/L) could also significantly increase the transcriptional expressions of *clk-1* and *isp-1* in *aex-3* mutant compared with those in wild-type nematodes (Fig. 7.25) [15]. Thus, the molecular basis for neurotransmission may be further involved in the regulation of ENM toxicity. Nevertheless, the underlying molecular mechanisms are still unclear.

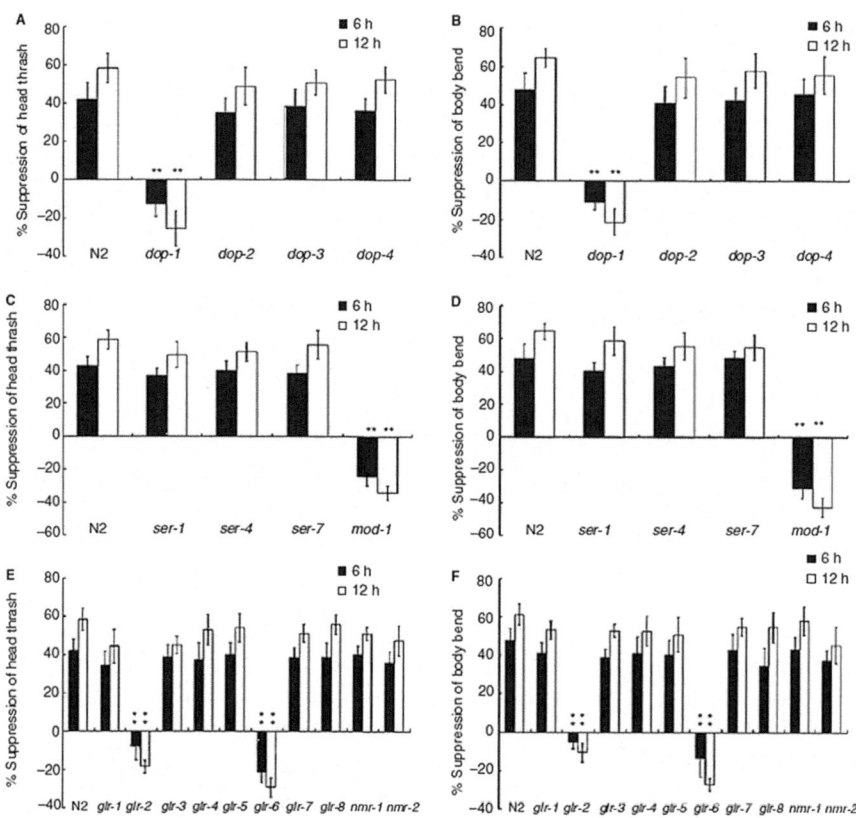

Fig. 7.24 Formation of adverse effects on locomotion behavior in Al$_2$O$_3$-NP-exposed nematodes requires dopamine receptor DOP-1, serotonin receptor MOD-1, and glutamate receptors GLR-2 and GLR-6 [42]

(**a**) Head thrashes in Al$_2$O$_3$-NP-exposed wild-type N2, *dop-1*, *dop-2*, *dop-3*, and *dop-4* mutant nematodes. (**b**) Body bends in Al$_2$O$_3$-NP-exposed wild-type N2, *dop-1*, *dop-2*, *dop-3*, and *dop-4* mutant nematodes. (**c**) Head thrashes in Al$_2$O$_3$-NP-exposed wild-type N2, *ser-1*, *ser-4*, *ser-7*, and *mod-1* mutant nematodes. (**d**) Body bends in Al$_2$O$_3$-NP-exposed wild-type N2, *ser-1*, *ser-4*, *ser-7*, and *mod-1* mutant nematodes. (**e**) Head thrashes in Al$_2$O$_3$-NP-exposed wild-type N2, *glr-1*, *glr-2*, *glr-3*, *glr-4*, *glr-5*, *glr-6*, *glr-7*, *glr-8*, *nmr-1*, and *nmr-2* mutant nematodes. (**f**) Body bends in Al$_2$O$_3$-NP-cxposed wild-type N2, *glr-1*, *glr-2*, *glr-3*, *glr-4*, *glr-5*, *glr-6*, *glr-7*, *glr-8*, *nmr-1*, and *nmr-2* mutant nematodes. Al$_2$O$_3$-NPs exposure at the concentration of 25 mg/L was performed at the L4 larval stage. Data are quantified as percentage suppression of locomotion behavior between naïve and Al$_2$O$_3$-NP-exposed conditions for each genotypes. % suppression = (locomotion behavior under the naive condition − locomotion behavior under the Al$_2$O$_3$-NPs exposed condition)/ locomotion behavior under the naive condition. Bars represent mean ± SEM. $^{**}p < 0.01$

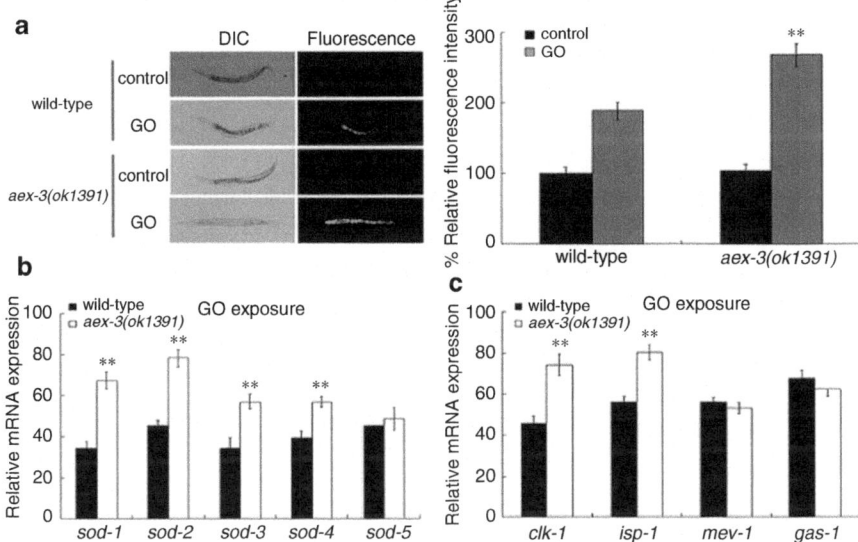

Fig. 7.25 Effect of *aex-3* mutation on the induction of oxidative stress in nematodes [15]
(**a**) Effect of *aex-3* mutation on the induction of oxidative stress. (**b**) Effect of *aex-3* mutation on the expressions of *sod* genes. (**c**) Effect of *aex-3* mutation on the expression of *clk-1*, *isp-1*, *mev-1*, and *gas-1*. Prolonged exposure was performed from L1-larvae to young adults. GO exposure concentration was 10 mg/L. Bars represent mean ± SD. **$p < 0.01$ versus wild-type

7.4   Molecular Basis for Nanotoxicity Formation Based on Omics Study in Nematodes

7.4.1   Dysregulated mRNAs by ENMs in Nematodes

7.4.1.1   Dysregulated mRNAs by Certain ENMs in Nematodes

After exposure with the selected exposure route, different ENMs can potentially induce alteration in transcriptional expressions of many genes in nematodes. With GO as the example, the transcriptomic analysis demonstrated that 1965 genes were dysregulated by prolonged exposure (from L1-larvae to adult day-1) to GO (10 mg/L) among the detected 13752 genes (Fig. 7.26) [14]. Among these dysregulated genes, 970 mRNAs were upregulated, and 995 mRNAs were downregulated based on the Illumina HiSeq™ 2000 sequencing analysis (Fig. 7.26) [14]. These dysregulated genes included those associated with the control of oxidative stress (*sod-1*, *sod-2*, *sod-3*, *sod-4*, *isp-1*, *gas-1*, and *clk-1*), the control of intestinal development (*nhx-2*, *par-6*, and *pkc-3*), and the control of defecation behaviors (*hlh-8*, *unc-93*, *fat-2*, *fat-3*, *unc-44*, *smp-1*, *itr-1*, *cab-1*, *mig-2*, *ced-10*, *unc-101*, and *egl-36*) [14].

For these dysregulated genes induced by prolonged exposure (from L1-larvae to adult day-1) to GO (10 mg/L), the gene ontology analysis was further performed for the aim of providing the ontology in terms of the associated biological processes, cellular components, and molecular functions. Based on the gene ontology analysis, the

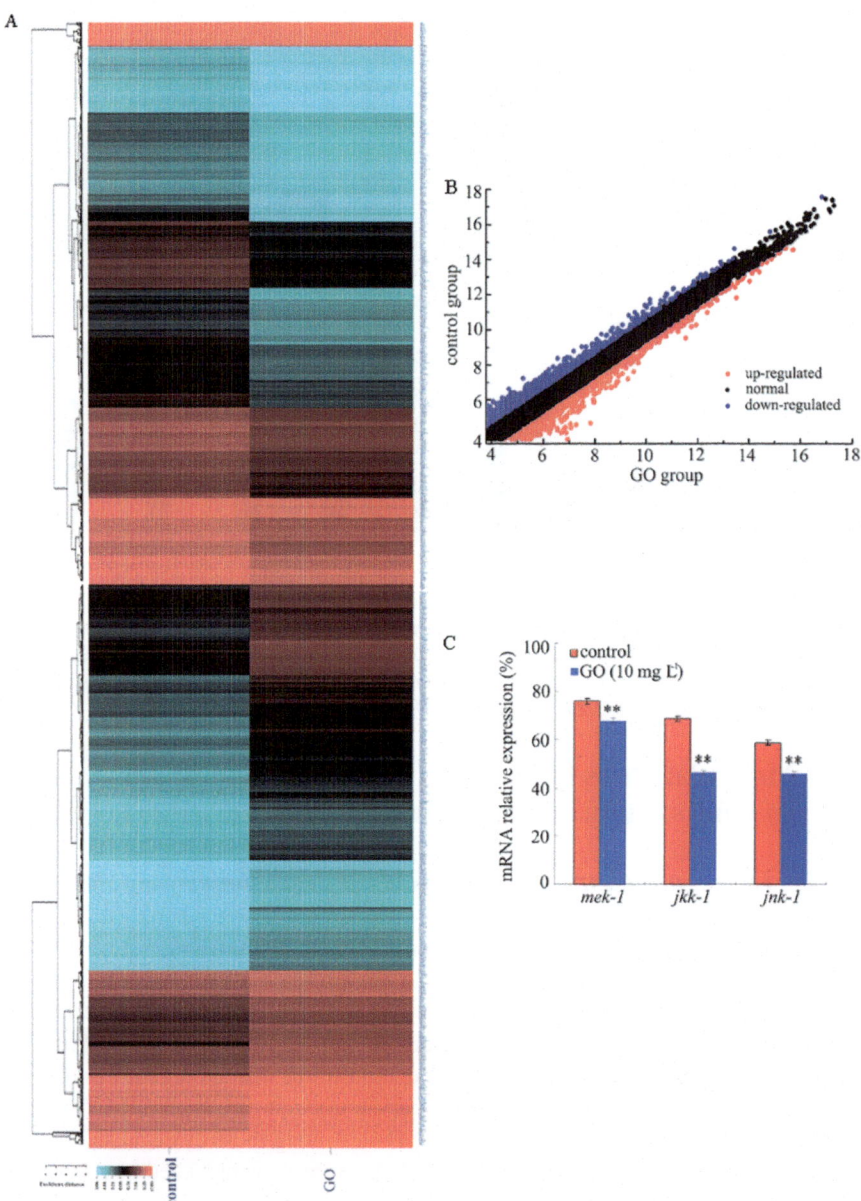

Fig. 7.26 Dysregulated mRNAs induced by GO [14]

(**a**) Heat map showing the result of the expression of mRNAs obtained from the control and GO-treated nematodes. Relatively low expression levels are represented in blue, and high levels are represented in red. (**b**) A scatter diagram of the relationship between mRNA coverage in the control group and the GO exposure group. (**c**) qRT-PCR analysis of the expression of *mek-1*, *jkk-1*, and *jnk-1* in nematodes exposed to GO. Bars represent the mean ± SEM. $^{**}P < 0.01$ vs control. GO (10 mg/L) was exposed from L1-larvae to adult day-1

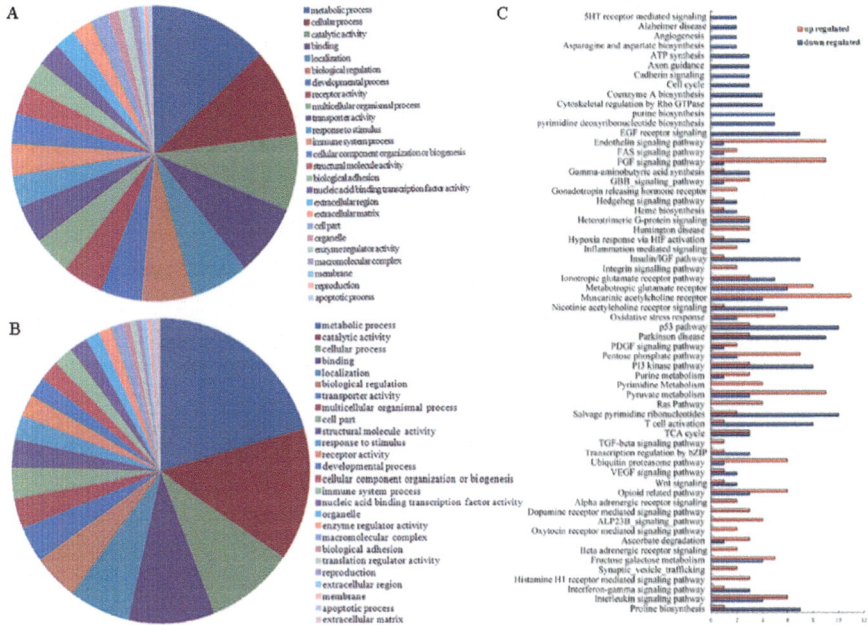

Fig. 7.27 Assessments of gene ontology terms and signaling pathways [14]
(a) Gene ontology terms with gene counts based on downregulated mRNAs in GO-exposed nematodes. (b) Gene ontology terms with gene counts based on upregulated mRNAs in GO-exposed nematodes. (c) The predicted KEGG signaling pathways based on the dysregulated mRNAs in GO-exposed nematodes

significantly affected gene ontology terms for these dysregulated genes could be classified into several categories, which were at least involved in biological processes of development, reproduction, cell adhesion, apoptosis, enzyme activity, cellular component, cellular localization and transportation, response to stimulus, immune response, and cell metabolism (Fig. 7.27) [14]. The signaling pathways mediated by these dysregulated genes induced by prolonged exposure (from L1-larvae to adult day-1) to GO (10 mg/L) were further determined by the KEGG pathway mapping, a bioinformatics resource to map the molecular data sets in genomics. Totally 50 signaling pathways for upregulated genes and 45 signaling pathways for downregulated genes were identified, and the affected signaling pathways include those related to development, cell cycle, cell death, neuronal degeneration, transcription regulation, stress response, cellular component, vesicle transportation, immune response, neuronal development, and cell metabolism (Fig. 7.27) [14].

7.4.1.2 Regulation of mRNAs Expression Profiling in GO-Exposed Nematodes by PEG Modification or FBS Coating

In nematodes, PEG surface modification or FBS surface coating can effectively reduce the GO toxicity [45]. Moreover, compared with the mRNAs expression profiling in GO-exposed nematodes, PEG surface modification could further cause the formation of 178 downregulated mRNAs and 156 upregulated mRNAs [45]. Meanwhile, compared with the mRNA expression profiling in GO-exposed nematodes, FBS surface coating could result in the formation of 298 downregulated mRNAs and 184 upregulated mRNAs [45]. These results imply that the molecular basis for PEG surface modification to reduce the GO toxicity may be somewhat different from that of FBS surface coating to reduce the GO toxicity in nematodes.

7.4.1.3   Distinct Transcriptomic Responses of Nematodes to Pristine and Sulfidized Ag-NPs

Based on the transcriptomic profiling induced by Ag-NPs and sulfidized sAg-NPs with the similar effect concentrations (EC_{30} for reproduction), the overlapped dysregulated mRNAs between Ag-NPs exposure and sAg-NPs were only approximately 5% [46], which implies that genomic responses of nematodes to the sAg- NPs may be very different from that to the pristine Ag-NPs.

7.4.2   Dysregulated microRNAs by ENMs in Nematodes

7.4.2.1   Dysregulated microRNAs by ENMs in Nematodes

With the GO as an example, the miRNA profiling induced by prolonged exposure (from L1-larvae to adult day-1) to GO (10 mg/L) was determined in nematodes using SOLiD sequencing technique. The detected miRNA sequences appeared as the size of 18–25 nucleotides, and most of the detected sequences were found to be distributed between 20 and 23 nucleotides [47]. Among the detected sequences, 31 dysregulated expressed miRNAs including 23 upregulated miRNAs and 8 downregulated miRNAs were identified in GO-exposed nematodes (Fig. 7.28) [47]. The upregulated miRNAs were *mir-259*, *mir-1820*, *mir-36*, *mir-82*, *mir-239*, *mir-246*, mir-247, *mir-392*, *mir-4806*, *mir-2217*, *mir-360*, *mir-4810*, *mir-4807*, *mir-1822*, *mir-4805*, *mir-800*, *mir-1830*, *mir-236*, *mir-244*, *mir-235*, *mir-4937*, *mir-4812*, and *mir-43*, and the downregulated miRNAs were *mir-1834*, *mir-800*, *mir-231*, *mir-5546*, *mir-42*, *mir-2214*, *mir-2210*, and *mir- 73* (Fig. 7.28) [47]. For these 31 dysregulated expressed miRNAs and their predicted targeted genes, the gene ontology analysis indicated that the possible biological processes altered by GO exposure may at least contain development, reproduction, cell adhesion, cell cycle, cellular localization and transportation, cell communication, response to stimulus, immune response, and cell metabolism [47]. Additionally, for these 31 dysregulated expressed miRNAs and their predicted targeted genes, the KEGG pathway analysis suggested that the possible influenced signaling pathways by GO exposure include the signaling pathways related to development, cell cycle, cell death, neuronal degeneration, transcription regulation, oxidative stress response, DNA damage and repair, vesicle transportation, immune response, and cell metabolism [47].

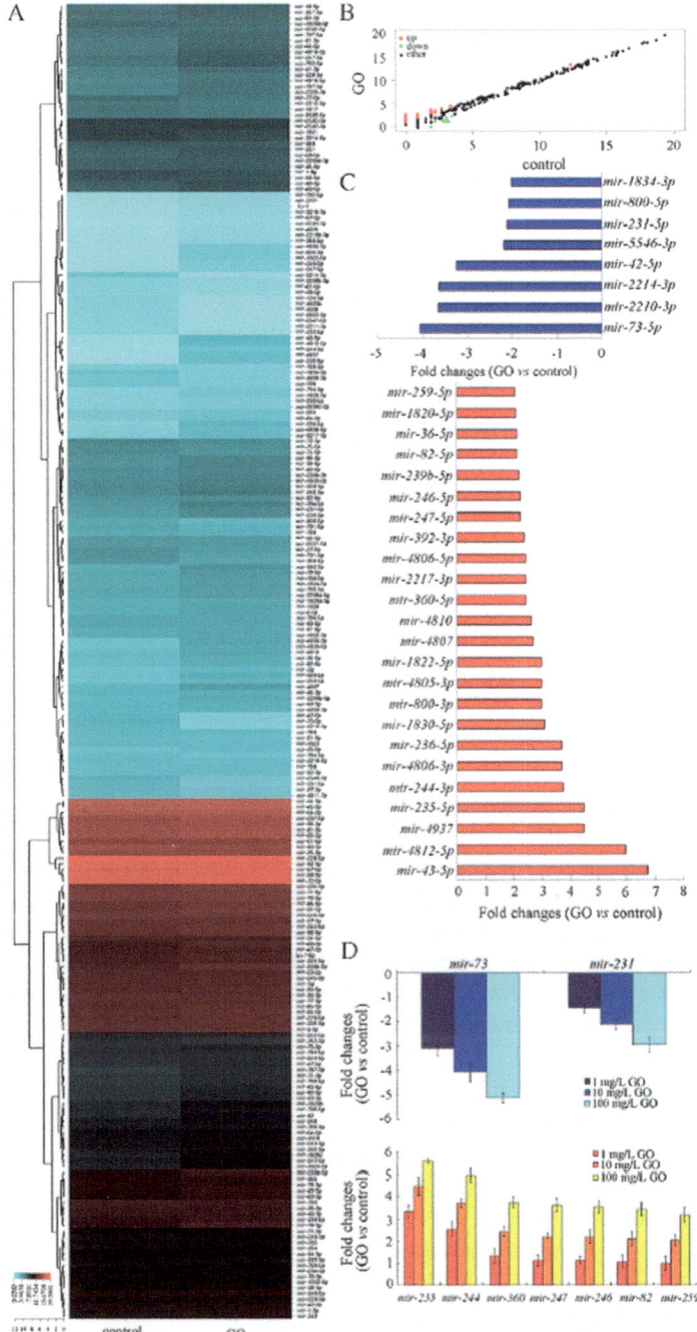

Fig. 7.28 Results of SOLiD sequencing for GO exposure [47]
(**a**) miRNA gene expression analysis by hierarchical clustering assay to reveal a characteristic molecular signature for GO (10 mg/L)-exposed nematodes. (**b**) Scatter diagram of miRNA coverage of the control group and GO treatment group. (**c**) Fold changes of up- and downregulated miRNAs in GO (10 mg/L)-exposed nematodes. (**d**) Expression pattern of mature miRNAs detected by real-time PCR after exposure to different concentrations of GO. Bars represent mean ± SEM

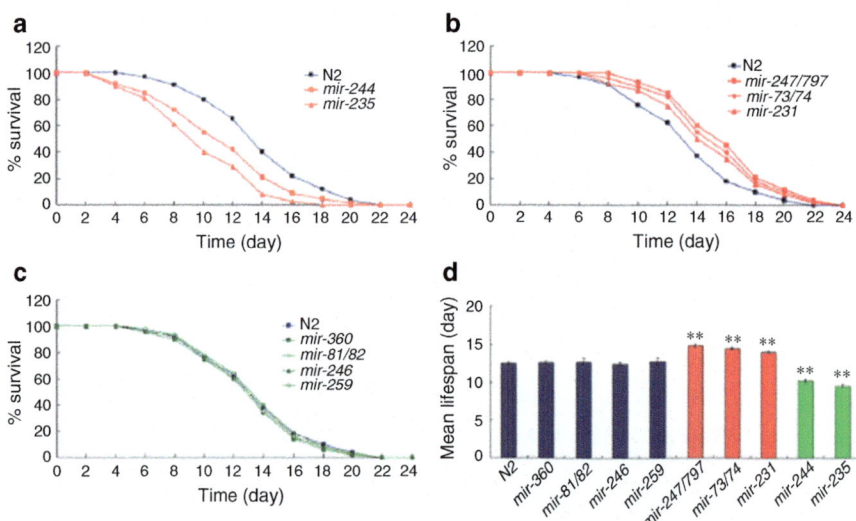

Fig. 7.29 Lifespan of mutants for some dysregulated miRNAs exposed to GO [47]
(**a**) Lifespan of wild-type, *mir-244*, and *mir-235* mutants exposed to GO. (**b**) Lifespan of wild-type, *mir-247/797*, *mir-73/74*, and *mir-231* mutants exposed to GO. (**c**) Lifespan of wild-type, *mir-360*, *mir-81/82*, *mir-246*, and *mir-259* mutants exposed to GO. (**d**) Comparison of mean lifespan. Exposure concentration was 10 mg/L. Bars represent mean ± SEM. $^{**}P < 0.01$

Using the available mutants for the identified dysregulated miRNAs, it has been observed that mutant of *mir-244* of *mir-235* induced a susceptibility to GO toxicity in reducing lifespan and in enhancing aging-related phenotypes as reflected by the endpoints of induction of intestinal autofluorescence, intestinal ROS production, and locomotion behavior, and mutation of *mir-247/797*, *mir-73/74*, or *mir-231* induces a resistance to GO toxicity in reducing lifespan and in enhancing aging-related phenotypes (Fig. 7.29) [47]. In contrast, mutation of *mir-360*, *mir-81/82*, *mir-246*, or *mir-259* did not obviously affect the GO toxicity in reducing lifespan in nematodes (Fig. 7.29) [47].

The molecular mechanism for dysregulated miRNAs in regulating the GO toxicity in reducing lifespan has been further elucidated in nematodes. Among the predicted targeted genes (*daf-16*, *daf 18*, *pdk-1*, *akt-2*, *sgk-1*, *smk-1*, *hcf-1*, *aak-2*, *unc-51*, *daf-15*, *raga-1*, *rheb-1*, *pha-4*, *daf-9*, *daf-12*, and *kri-1*) involved in the control of longevity for the dysregulated miRNAs, the expressions of *daf-16*, *daf-18*, *pdk-1*, *sgk-1*, *smk-1*, *daf-15*, and *kri-1* could be further dysregulated by prolonged exposure to GO (10 mg/L) (Fig. 7.30) [47]. In nematodes, *daf-16*, *daf-18*, *pdk-1*, and *sgk-1* encode the insulin/IGF signaling pathway, *smk-1* encodes a DAF-16 transcriptional coregulator, *daf-15* encodes a component for TOR signaling pathway, and *kri-1* encodes a component for germline signaling pathway. Therefore, the insulin/insulin-like growth factor (IGF), target of rapamycin (TOR), and germline signaling pathways may be all involved in the regulation of GO toxicity in reducing lifespan in nematodes.

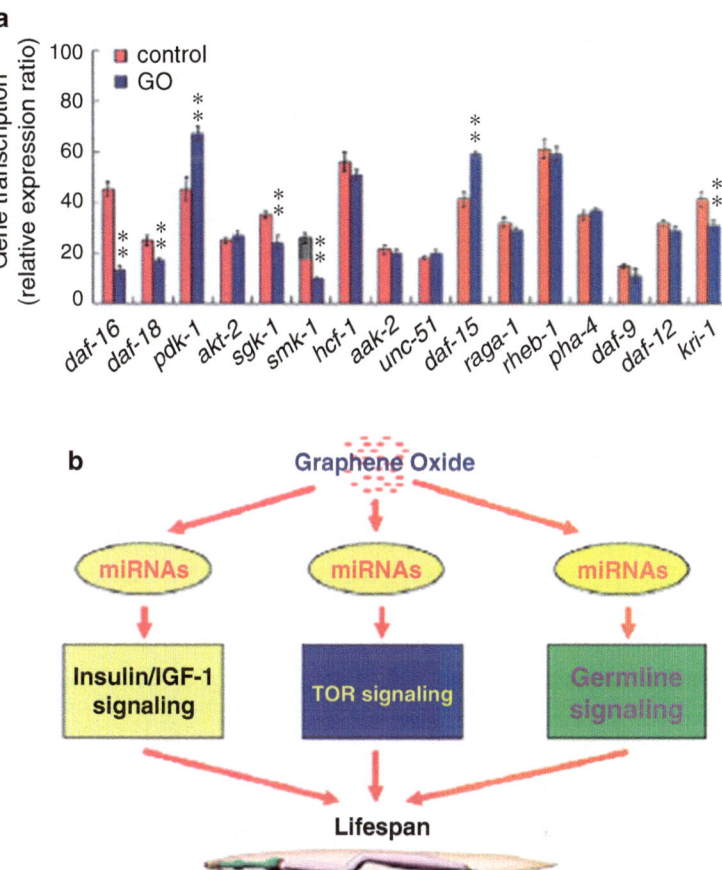

Fig. 7.30 Expression of possible targeted genes for dysregulated miRNAs in GO-exposed nematodes [47]
(**a**) Expression patterns of targeted genes for dysregulated miRNAs in GO-exposed nematodes. Exposure concentration was 10 mg/L. Bars represent mean ± SEM. $^{**}P < 0.01$. (**b**) A model for miRNAs in regulating the GO-induced lifespan reduction in nematodes

7.4.2.2   Different Functions of miRNAs in Response to ENMs and the Underlying Molecular Mechanisms

We here selected the miRNAs dysregulated by GO exposure to introduce the different functions of miRNAs in response to ENMs and the underlying molecular mechanisms.

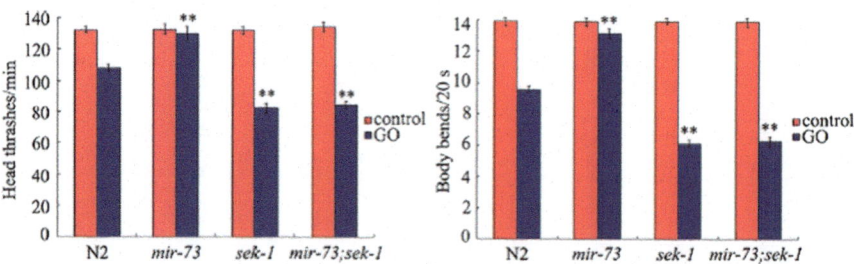

Fig. 7.31 Genetic interaction of *mir-73* and *sek-1* in regulating GO toxicity on locomotion behavior in nematodes [14]
Bars represent the mean ± SEM. $^{**}P < 0.01$ *vs* wild-type N2. GO (10 mg/L) was exposed from L1-larvae to adult day-1

7.4.2.2.1 Biological Function of *mir-73*

In *C. elegans*, SEK-1, a mitogen-activated protein kinase kinase (MAPKK), is an important member of the p38 MAPK signaling pathway. Using locomotion behavior as the toxicity assessment endpoint, mutation of *mir-73* induced a resistance to GO toxicity, and mutation of *sek-1* induced a susceptibility to GO toxicity [14]. Moreover, the GO-exposed double mutant of *mir-73;sek-1* showed the similar locomotion behavior to that in the GO-exposed *sek-1* mutant (Fig. 7.31) [14]. That is, the *sek-1* mutation suppressed the phenotype of *mir-73* mutant in GO-exposed nematodes, which suggests the role of SEK-1 as the molecular target for *mir-73* in regulating GO toxicity in nematodes.

In nematodes, GO exposure decreased the expression of *mir-73* [47]. This implies that the decreased *mir-73* may mediate a protection response or function for nematodes against the GO toxicity by increasing the function of SEK-1 in the p38 MAPK signaling pathway.

7.4.2.2.2 Biological Function of *mir-360*

In nematodes, some corresponding miRNAs for *ced-3*, *ced-4*, and *ced-9* encoding the core apoptosis signaling pathway and for *hus-1*, *clk-2*, *cep-1*, and *elg-1* encoding the signaling pathway for DNA damage checkpoints were identified in GO-exposed nematodes based on bioinformatics analysis [10]. Among the dysregulated miRNAs induced by GO, *mir-2210*, *mir-4810*, *mir-4805*, *mir-259*, *mir-82*, *mir-360*, and *mir-4807* might be further involved in the control of reproductive toxicity in inducing germline apoptosis in GO-exposed nematodes [10]. Among these miRNAs, mutation of *mir-360* significantly enhanced the reproductive toxicity of GO in inducing germline apoptosis, whereas *mir-360* overexpression significantly suppressed the reproductive toxicity of GO in inducing germline apoptosis in nematodes (Fig. 7.32) [10].

For the underlying molecular mechanism of *mir-360* in regulating the reproductive toxicity of GO in inducing germline apoptosis, CEP-1 was identified as a

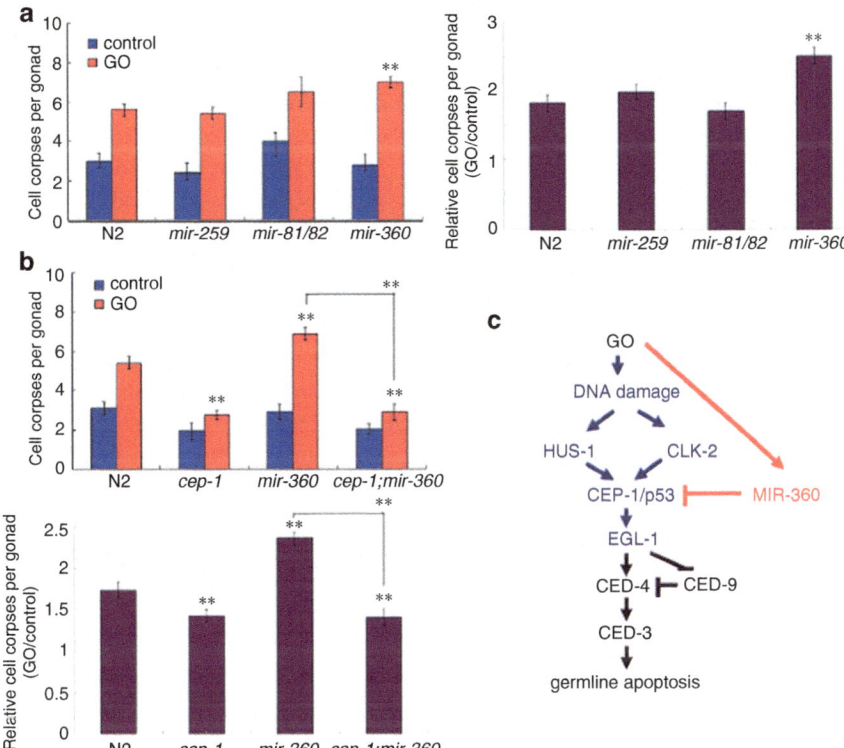

Fig. 7.32 *mir-360* negatively regulated the functions of signaling pathways of DNA damage checkpoints and apoptosis in the control of reproductive toxicity in GO-exposed nematodes [10] (**a**) Germline apoptosis in *mir-259*, *mir-81/82*, or *mir-360* mutant exposed to GO. The used mutants were wild-type N2, *mir-360(n4635)*, *mir-259(n4106)*, and *mir-81&82(nDf54)*. (**b**) Genetic interaction of *mir-360* with *cep-1* in regulating germline apoptosis in nematodes exposed to GO. The used strains were wild-type N2, *mir-360(n4635)*, *cep-1(RNAi)*, and *cep-1(RNAi);mir-360(n4635)*. (**c**) A summary model for the molecular control of reproductive toxicity of GO in inducing germline apoptosis in nematodes. GO exposure concentration was 10 mg/L. Prolonged exposure to GO was performed from L1-larvae to young adults. Bars represent mean ± SEM. $^{**}P < 0.01$

molecular target for *mir-360* [10]. The double mutant of *cep-1(RNAi);mir-360(n4635)* exposed to GO (10 mg/L) showed the similar phenotype of germline apoptosis to that in *cep-1(RNAi)* nematodes exposed to GO (10 mg/L) (Fig. 7.32) [10], implying the role of CEP-1 as the potential target for *mir-360* in regulating the reproductive toxicity of GO in inducing germline apoptosis in nematodes.

In nematodes, GO exposure increased the expression of *mir-360* [47]. This implies that the increased *mir-360* may mediate a protection response or function for nematodes against the reproductive toxicity of GO by suppressing the function of CEP-1 in the DNA damage checkpoint signaling pathway.

7.4.2.2.3 Biological Function of *mir-231*

In nematodes, the tissue-specific activity analysis suggested that *mir-231* acted in the intestine to regulate the GO toxicity, since expression of *mir-231* in the intestine could significantly suppress the resistance to GO toxicity in reducing lifespan and in decreasing locomotion behavior in *mir-231(n4571)* mutant nematodes [48]. In contrast, *mir-231* had no such function in the neurons, the pharynx, and the hypodermis [48]. Meanwhile, intestinal overexpression of *mir-231* induced a susceptibility to GO toxicity in reducing lifespan, in decreasing locomotion behavior, and in inducing intestinal ROS production in nematodes [47].

Using the TargetScan tool, SMK-1 is a predict target for *mir-231*. In nematodes, *smk-1* encodes a protein that is orthologous to the mammalian SMEK (suppressor of MEK null) proteins. The expression of *smk-1* was significantly enhanced by *mir-231* mutation in GO-exposed nematodes, and mutation of *mir-231* induced a susceptibility to GO toxicity in reducing lifespan, in decreasing locomotion behavior, and in inducing intestinal ROS production [47]. Genetic interaction analysis further demonstrated that the GO toxicity in reducing lifespan, in decreasing locomotion behavior, and in inducing intestinal ROS production in double mutant of *mir-231(n4571);smk-1(mn156)* was similar to that in single mutant of *smk-1(mn156)* (Fig. 7.33) [48], suggesting that the resistance of *mir-231(n4571)* mutant to GO toxicity can be suppressed by *smk-1* mutation in nematodes.

In nematodes, *smk-1* is expressed in the intestine, the pharynx, the neurons, the muscle, and the hypodermis. The tissue-specific activity analysis suggested that SMK-1 could also act in the intestine to regulate the GO toxicity in reducing lifespan and in decreasing locomotion behavior in nematodes, because expression of *smk-1* in the intestine could significantly increase both the lifespan and the locomotion behavior in GO-exposed *smk-1(mn156)* mutant nematodes [48]. In nematodes, intestinal overexpression of *smk-1* induced a resistance to GO toxicity in reducing lifespan, in decreasing locomotion behavior, and in inducing intestinal ROS production [48]. In contrast, SMK-1 had no such functions in the pharynx, the neurons, the muscle, and the hypodermis [48].

During the control of GO toxicity, DAF-16 was identified as a downstream target of SMK-1 [48]. After prolonged exposure to GO, both the lifespan and the locomotion behavior in double mutant of *daf-16(RNAi);smk-1(mn156)* were similar to those in single mutant of *smk-1(mn156)* or *daf-16(RNAi)* nematodes [48]. More importantly, RNAi knockdown of *daf-16* could significantly inhibit the resistance of nematodes overexpressing intestinal SMK-1 to GO toxicity in reducing lifespan and in decreasing locomotion behavior [48]. These results suggest the formation of intestinal signaling cascade of *mir-231*-SMK-1-DAF-16 in the regulation of GO toxicity in nematodes.

In nematodes, GO exposure decreased the transcriptional expression of *mir-231* [47]. Additionally, prolonged exposure to GO could further decrease the expression of *mir-231::GFP* in the pharynx, the intestine, the neurons, and the hypodermis [48]. Therefore, the altered *mir-231* may mediate a protection response for nematodes against the GO toxicity by suppressing the function of signaling cascade of SMK-1-DAF-16.

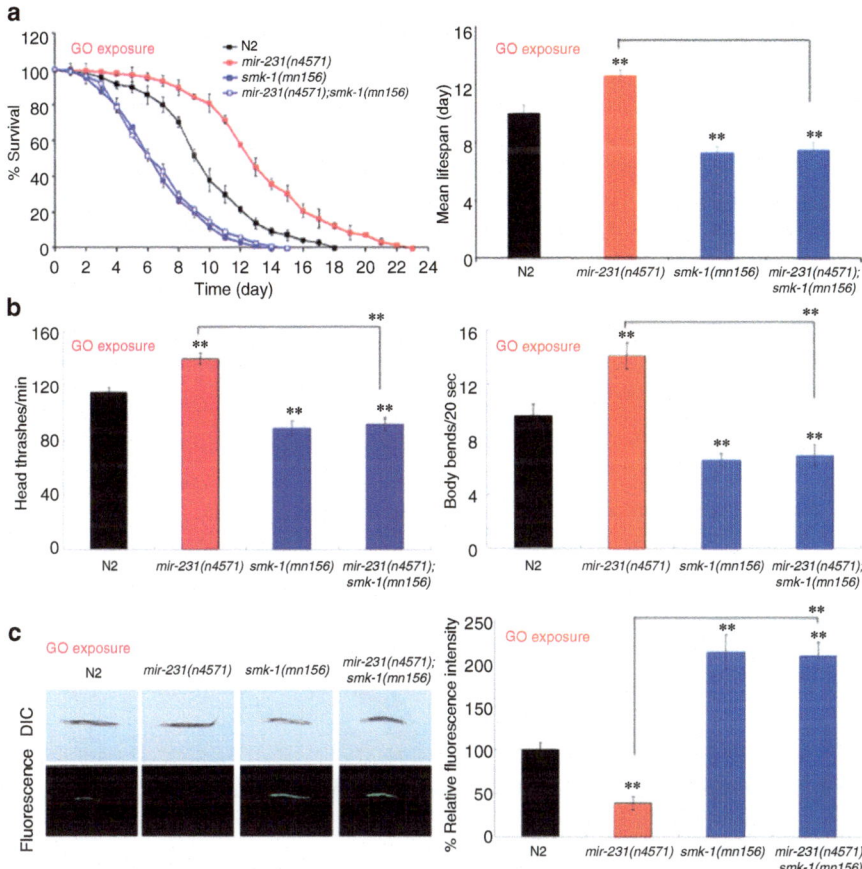

Fig. 7.33 Genetic interaction between *mir-231* and SMK-1 in regulating GO toxicity in nematodes [48]
(**a**) Genetic interaction between *mir-231* and SMK-1 in regulating GO toxicity in reducing lifespan. (**b**) Genetic interaction between *mir-231* and SMK-1 in regulating GO toxicity in decreasing locomotion behavior. (**c**) Genetic interaction between *mir-231* and SMK-1 in regulating GO toxicity in inducing intestinal ROS production. GO exposure concentration was 100 mg/L. Prolonged exposure was performed from L1-larvae to young adults. Bars represent mean ± SD. **$P < 0.01$ *vs* N2 (if not specially indicated)

7.4.2.3   mRNA-miRNA Network Involved in the Control of ENM Toxicity

With the GO as an example, a mRNA-miRNA network involved in the control of GO toxicity was raised based both the dysregulated mRNAs and the dysregulated miRNAs in GO-exposed nematodes [14, 47]. Among the dysregulated miRNAs induced by GO exposure, the further bioinformatics analysis demonstrate that *mir-259, mir-1820, mir-36, mir-82, mir-239, mir-246, mir-392, mir-2217, mir-360, mir-4810, mir-4807, mir-4805, mir-800, mir-1830, mir-236, mir-4806, mir-244, mir-235, mir-4812, mir-43, mir-231, mir-42, mir-2210,* and *mir-73* might be involved in the control of GO toxicity with some of the dysregulated genes as their targets in GO-exposed nematodes (Table 7.1) [14]. Among these potential targeted genes, *gas-1* might act as a molecular target for *mir-4810* to regulate the activation of oxidative stress in GO-exposed nematodes, and *par-6* might act as a molecular target for *mir-1820* to regulate intestinal development and function in GO-exposed nematodes (Table 7.1) [14]. Additionally, *mir-231, mir-236, mir-259, mir-36, mir-42, mir-43, mir-73, mir-82,* and *mir-4805* might act through the functions of *unc-44, mig-2, itr-1, ced-10, unc-93, fat-2, smp-1,* and/or *cab-1* to regulate the defecation behavior in GO-exposed nematodes (Table 7.1) [14]. Therefore, a certain miRNA-mRNA network may exist to regulate the GO toxicity by influencing some important biological processes, such as oxidative stress, intestinal development and function, and defecation behavior, in nematodes.

7.4.3   Dysregulated Long Noncoding RNAs (lncRNAs) by ENMs in Nematodes

Long noncoding RNAs (lncRNAs) are normally defined as the noncoding RNAs having at least 200 nucleotides and no (or weak) protein coding ability, which have been shown to potentially regulate various biological processes [49–51].

7.4.3.1   Dysregulated Long Noncoding RNAs (lncRNAs) by ENM Exposure

With the GO as an example, the Illumina HiSeq 2000 sequencing was performed for nematodes after prolonged exposure (L1-larvae stage to young adults) to GO (100 mg/L). Totally 39 dysregulated expressed lncRNAs were identified in GO-exposed nematodes (Fig. 7.34) [45]. Among these dysregulated lncRNAs, ten known lncRNAs including nine downregulated lncRNAs (*linc-14, anr-24, linc-68, linc-103, linc-83, linc-24, anr-36, linc-5,* and *linc-37*) and one upregulated lncRNA (*linc-3*) were detected (Fig. 7.34) [45]. Among the known lncRNAs, *linc-14, linc-68, linc-103, linc-83, linc-24, linc-5, linc-37,* and *linc-3* are intergenic lncRNAs, and *anr-36* and *anr-24* are antisense lncRNAs (Fig. 7.34) [45]. Additionally, 29 novel lncRNAs including 24 downregulated lncRNAs and 5 upregulated lncRNAs were also detected (Fig. 7.34) [45]. Among these novel lncRNAs, *XLOC_002902, XLOC_013698, XLOC_002296, XLOC_001403, XLOC_008926, XLOC_012387, XLOC_015092, XLOC_000665, XLOC_004416, XLOC_015091, XLOC_003882, XLOC_003625, XLOC_012820, XLOC_016657, XLOC_015843, XLOC_013642, XLOC_007207, XLOC_005996, XLOC_001973,* and *XLOC_002997* are intergenic lncRNAs; *XLOC_010849, XLOC_007829, XLOC_016843,* and *XLOC_013549* are antisense lncRNAs; *XLOC_013655, XLOC_001936, XLOC_002933,* and *XLOC_015166* are intronic lncRNAs; and *XLOC_007959* is sense lncRNA (Fig. 7.34) [45].

Table 7.1 MicroRNA-mRNA networks involved in the control of GO toxicity [14]

miRNA	Targeted genes
mir-1820	valv-1, T16G1.2, T02E9.5, T01G1.2, rol-8, rgs-6, *par-6*, osm-10, nhr-227, nhr-125, F26F12.3, F18C5.5, ceh-18, C01H6.4, ajm-1, adt-1
mir-1830	T12G3.1, slt-1, pqn-73, C34D10.2
mir-2210	F13H6.1
mir-2217	flp-16, igcm-1
mir-231	Y43F8B.3, Y43E12A.2, *unc-44*, mig-13, col-17, C23H4.8
mir-235	ZK470.1, zag-1, unc-41, tag-97, R06C7.6, pqn-32, peb-1, osm-11, mlt-10, mlc-2, ham-2, H18N23.2, gnrr-5, fkh-7, F59B10.4, F53B1.4, F52B10.3, F45E4.3, F33D4.6, cutl-28, cut-6, cut-3, col-45, col-43, col-110, C53A5.2, C32F10.4, B0244.9, atf-2, alr-1
mir-236	ZC190.4, Y47D3B.6, Y18D10A.8, *unc-44*, tps-1, T25F10.3, T25E4.1, sym-2, pcca-1, nhl-1, ncs-3, fkh-9, dnj-13, DH11.5, D2005.6, col-168, C18E9.7, C11E4.6
mir-239	W06A7.4, unc-6, T25E4.1, R144.6, R09E10.6, pqn-73, pcca-1, nhl-1, M60.7, K01B6.3, hlh-11, H18N23.2, grl-16, grd-1, F55A4.2, egl-3, dur-1, D2005.6, clec-1, clc-5, C33A11.1, C29H12.6
mir-244	valv-1, T14B1.1, olrn-1, nlp-6, F30H5.3, C11E4.6, alr-1, adt-1
mir-246	Y11D7A.14, valv-1, tth-1, T04C12.7, M01H9.3, F45D3.3, egl-13, col-169, C38H2.2
mir-259	somi-1, *mig-2*, *mek-1*, ceh-18, C11E4.6
mir-360	ZK470.2, ZC190.4, tag-243, T22F3.11, T10E9.3, M60.7, M163.1, *jnk-1*, F13H8.11, C33A11.1, B0302.5
mir-36	Y11D7A.14, uig-1, trp-4, snf-1, rgs-6, hlh-11, F13H8.5, F13H6.1, egl-3, egl-13, egl-1, ceh-18, *cab-1*, C06E1.3, alr-1
mir-392	Y43E12A.2, tsp-11, T05A8.3, srab-4, snf-1, fkh-7, F13H8.5, adt-1
mir-42	Y11D7A.14, uig-1, trp-4, snf-1, rgs-6, hlh-11, F13H8.5, F13H6.1, egl-3, egl-13, egl-1, ceh-18, *cab-1*, C06E1.3, alr-1
mir-43	ZK470.2, ZC518.1, Y75B8A.6, Y73F8A.5, Y65A5A.1, Y60A3A.19, Y18D10A.8, uig-1, twk-13, trk-1, tag-97, tag-89, tag-218, T27F2.2, somi-1, sem-4, sek-1, rpy-1, rpl-41, R144.6, ptr-5, prl-1, pqn-73, peb-1, nhx-5, nhr-35, mnm-2, *mek-1*, lin-11, K07H8.12, K07E3.4, *itr-1*, irk-3, hlh-30, ham-2, H20J04.6, H18N23.2, fut-8, flp-27, fkh-7, fbn-1, F59F4.2, F54G2.1, F53B1.4, F10D7.2, elt-3, egl-9, egl-38, D2023.3, D2005.6, cki-1, *ced-10*, C54G4.5, C48E7.6, C44B7.7, C39D10.2, C34D10.2, C18B12.2, C01B12.4, B0507.3, alr-1
mir-4805	ZC449.7, W05H12.1, *unc-93*, unc-7, twk-13, T22F3.11, T04A8.18, R12A1.3, olrn-1, mnm-2, mlt-9, mlt-7, M01A8.1, ida-1, hlh-11, hlh-1, *fat-2*, F45E1.1, F13D12.3, F09C8.2, col-84, ceh-18, C48E7.6, C34E7.4, C18A3.10, brp-1, bam-2, adt-1, aak-2
mir-4806	bus-8, egl-3, T27F2.2
mir-4807	W06A7.4, ttx-1, ida-1, F59F4.2
mir-4810	olrn-1, K02E10.4, hlh-30, his-35, *gas-1*, F08G12.5, dnj-13, des-2, dct-1, cutl-12, clec-1
mir-4812	lec-2, hlh-11
mir-73	zag-1, Y64G10A.7, Y37F4.8, T10H10.2, T06E4.8, T04C12.7, sym-2, *smp-1*, sek-1, scd-1, pqn-31, M02B1.2, hst-3.2, F59B10.4, F45E1.1, F26F12.3, F13H8.5, elt-3, coel-1, ceh-14, C28C12.4, C13F10.1, bus-19
mir-800	col-43, col-36, ceh-14
mir-82	ZK970.1, ZK470.2, ZK180.5, ZK1248.11, zag-1, Y57G11C.42, unc-41, ugt-8, tag-97, tag-89, tag-297, T27F2.2, T27D12.1, T16G1.2, T01B10.5, srv-1, snr-3, snf-1, *smp-1*, scd-2, ram-2, pqn-35, pqn-18, pen-2, osm-11, oac-14, nlp-8, nas-7, mxl-3, mnm-2, mkk-4, M01A8.1, lys-2, lin-1, K07F5.12, K06A5.8, hrg-4, hil-3, H20J04.6, grsp-1, grl-4, flp-11, fkh-9, fat-5, F59B10.4, F58H1.7, F32H5.3, F32G8.3, F23F1.4, F18E3.11, F11D5.1, egl-1, DH11.5, col-115, cog-1, cdh-9, C48E7.6, C46H11.10, C44B7.7, C33A11.1, C27H5.4, C18B12.2, C14A4.12, B0507.3, B0495.6, B0454.6, alr-1, adt-1

Note: The blue color indicates genes associated with the control of oxidative stress, the purple color indicates genes associated with the control of intestinal development, and the red color indicates genes associated with the control of defecation behavior.

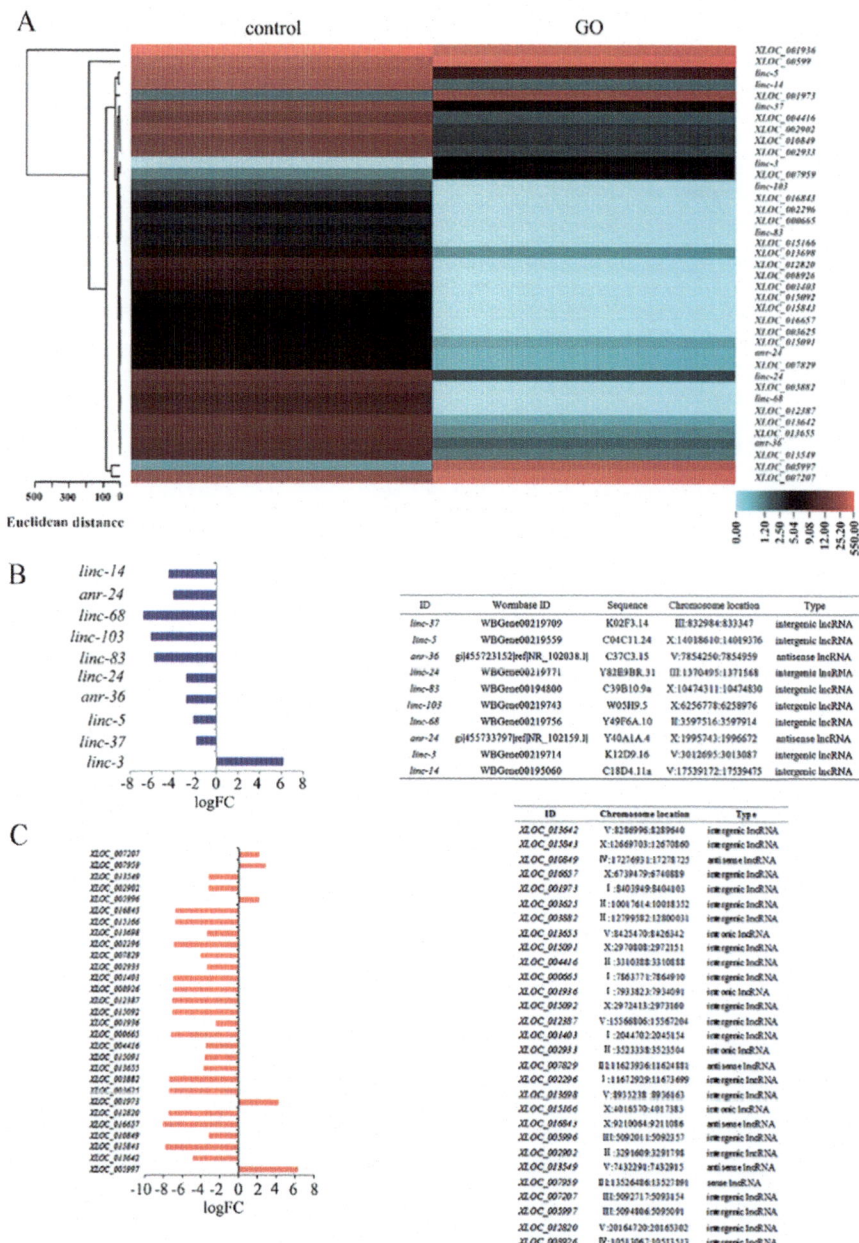

Fig. 7.34 Dysregulated lncRNAs induced by GO exposure in nematodes [45]
(**a**) Hierarchical clustering assay to show the characteristic molecular signature of 39 dysregulated lncRNAs in GO-exposed nematodes. (**b**) Dysregulated known lncRNAs induced by GO exposure in nematodes. logFC > 2 and $p < 0.05$. (**c**) Dysregulated novel lncRNAs induced by GO exposure in nematodes. logFC > 2 and $p < 0.05$

The qRT-PCR results further confirmed that 34 differentially expressed lncRNAs (30 downregulated lncRNAs and 4 upregulated lncRNAs) could be identified in GO-exposed nematodes [45]. Among these 34 dysregulated lncRNAs, 10 known lncRNAs including 9 downregulated lncRNAs and 1 upregulated lncRNA and 24 novel lncRNAs including 21 downregulated lncRNAs and 3 upregulated lncRNAs were included [45]. Among them, 24 lncRNAs are intergenic lncRNAs, 6 lncRNAs are antisense lncRNAs, 3 lncRNAs are intronic lncRNAs, and 1 lncRNA is sense lncRNA [45].

The gene ontology analysis based on the predicted targeted miRNAs implied that the possible affected biological processes by GO exposure may at least contain growth, development, reproduction, cellular localization and transportation, signal transduction, and cell metabolism [45]. The KEGG pathway mapping further suggested that the possible affected signal pathways by GO exposure may contain those related to development, cell cycle, cell death, stress response, and cell metabolism [45].

7.4.3.2   Effect of Ascorbate or Paraquat Treatment on lncRNA Profiling Induced by ENM Exposure

After pretreatment with paraquat (2 mM), a ROS generator, the expression of candidate 34 lncRNAs was more severely affected in GO-exposed nematodes [45]. In contrast, after pretreatment with ascorbate (10 mM), an antioxidant, the alteration in some of the 34 lncRNAs could be effectively reversed in GO-exposed nematodes, and the related lncRNAs were *linc-37*, *anr-36*, *linc-24*, *linc-83*, *linc-68*, *anr-24*, *linc-3*, *linc-14*, *XLOC_013642*, *XLOC_010849*, *XLOC_016657*, *XLOC_003625*, *XLOC_013655*, *XLOC_004416*, *XLOC_000665*, *XLOC_001936*, *XLOC_012387*, *XLOC_002933*, *XLOC_007829*, *XLOC_016843*, *XLOC_005996*, *XLOC_002902*, *XLOC_013549*, *XLOC_005997*, *XLOC_012820*, and *XLOC_008926* [45].

7.4.3.3   Effect of *clk-1* or *isp-1* Mutation on lncRNA Profiling Induced by ENM Exposure

In nematodes, mutation of *clk-1* or *isp-1* could induce a resistance to GO toxicity in nematodes [8]. After comparison of expression patterns of identified 34 candidate lncRNAs in GO-exposed wild-type, *clk-1* mutant, and *isp-1* mutant nematodes, the expression patterns of some dysregulated lncRNAs (*linc-37*, *linc-5*, *linc-83*, *linc-103*, *linc-68*, *anr-24*, *linc-14*, *XLOC_013642*, *XLOC_016657*, *XLOC_003625*, *XLOC_015091*, *XLOC_004416*, *XLOC_015092*, *XLOC_007829*, *XLOC_005996*, *XLOC_013549*, *XLOC_012820*, and *XLOC_008926*) induced by GO exposure were obviously reversed by mutation of *clk-1* or *isp-1* (Fig. 7.35) [45]. In details, mutation of *clk-1* significantly increased expressions of *linc-37*, *linc-5*, *linc-83*, *linc-103*, *linc-68*, *anr-24*, *linc-14*, *XLOC_013642*, *XLOC_003625*, *XLOC_015091*, *XLOC_007829*, *XLOC_013549*, and *XLOC_012820* in GO-exposed nematodes (Fig. 7.35) [45]. Meanwhile, mutation of *isp-1* significantly increased expressions

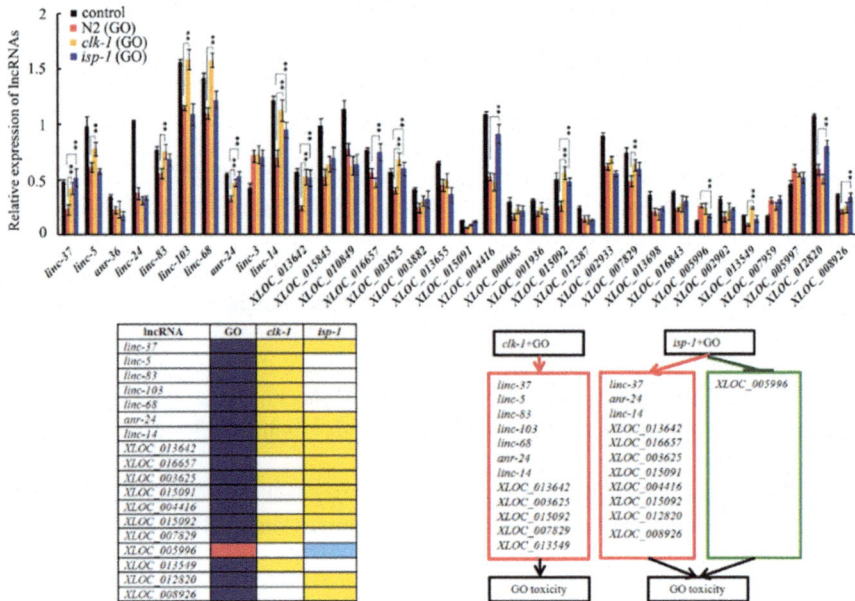

Fig. 7.35 Certain genetic mutation altered the expression patterns of some lncRNAs in GO-exposed nematodes [45]

Dark blue color indicates the decreased lncRNAs by GO, red color indicates the increased lncRNA by GO, yellow color indicates the increased lncRNAs by *clk-1* or *isp-1* mutation, and light blue color indicates the decreased lncRNA by *clk-1* or *isp-1* mutation. GO exposure concentration was 100 mg/L. Prolonged exposure to GO was performed from L1-larvae to young adults. Bars represent mean ± SEM. $^{**}P < 0.01$

of *linc-37*, *anr-24*, *linc-14*, *XLOC_013642*, *XLOC_016657*, *XLOC_003625*, *XLOC_015091*, *XLOC_004416*, *XLOC_015092*, *XLOC_012820*, and *XLOC_008926* and decreased the expression of *XLOC_005996* in GO-exposed nematodes (Fig. 7.35) [45].

More importantly, it has been found that both the *clk-1* mutation and the *isp-1* mutation could potentially decrease the expressions of *linc-37*, *anr-24*, *linc-14*, *XLOC_013642*, *XLOC_003625*, and *XLOC_015092* in GO-exposed nematodes (Fig. 7.35) [45], implying that *clk-1* or *isp-1* mutation may reduce the GO toxicity by influencing the shared functions of certain lncRNAs in nematodes.

7.4.3.4 LncRNA-miRNA Network Involved in the Control of ENM Toxicity in Nematodes

Based on the identified dysregulated miRNAs and dysregulated lncRNAs [14, 45], a lncRNA-miRNA network that responds to GO exposure was raised in nematodes using a bioinformatics analysis algorithm. In this lncRNA-miRNA network, lncRNAs may regulate the GO toxicity by affecting or altering the function of

LncRNA	Targets miRNA
linc-37	*mir-1820, mir-2214, mir-73*
linc-5	*mir-1820, mir-392, mir-4810*
anr-36	*mir-247*
linc-24	*mir-1820, mir-247, mir-42, mir-4805, mir-5546*
linc-83	*mir-800*
linc-103	*mir-2214. mir-259. mir-36. mir-360, mir-392, mir-43 mir-82, mir-5546, mir-800,*
anr-24	*mir-247, mir-4805, mir-4807, mir-4937, mir-5546*
linc-14	*mir-5546, mir-73, mir-82*
XLOC_013642	*mir-1820, mir-1834, mir-360, mir-4805, mir-4812, mir-5546, mir-82*
XLOC_015843	*mir-1820, mir-244, mir-247, mir-5546*
XLOC_010849	*mir-1820, mir-1830. mir-1834. mir-247. mir-43 mir-5546, mir-800, mir-4805, mir-4806,*
XLOC_016657	*mir-247, mir-73, mir-800, mir-82*
XLOC_003625	*mir-247, mir-4806, mir-4937*
XLOC_003882	*mir-4805*
XLOC_013655	*mir-4805, mir-5546*
XLOC_015091	*mir-1820, mir-360, mir-4812, mir-5546, mir-82*
XLOC_004416	*mir-4806*
XLOC_000665	*mir-1834, mir-231, mir-360, mir-4806, mir-4810, mir-4937*
XLOC_015092	*mir-1822*
XLOC_012387	*mir-4805*
XLOC_002933	*mir-2214, mir-4805*
XLOC_007829	*mir-1820, mir-235, mir-247, mir-259, mir-42, mir-43, mir-4805, mir-5546*
XLOC_013698	*mir-1820, mir-2214, mir-4805, mir-4806, mir-5546, mir-82*
XLOC_016843	*mir-1820, mir-246, mir-43, mir-4805, mir-5546, mir-82*
XLOC_005996	*mir-247*
XLOC_013549	*mir-1820, mir-4805*
XLOC_007959	*mir-1820, mir-247, mir-259, mir-5546*
XLOC_005997	*mir-1834, mir-259, mir-5546*
XLOC_012820	*mir-1820, mir-1822*
XLOC_008926	*mir-231, mir-4805, mir-5546*

Fig. 7.36 LncRNA-miRNA network involved in the control of GO toxicity in nematodes [45]

certain targeted miRNA(s) (Fig. 7.36) [45]. It is worthy to note the fact that these identified 34 lncRNAs might potentially only bind to a very limited number of miR-NAs to regulate the GO toxicity in nematodes (Fig. 7.36) [45].

7.4.3.5   Functional Analysis of *linc-37* and *linc-14* in the Regulation of ENM Toxicity

In nematodes, RNAi knockdown of *linc-37* or *linc-14* could induce a resistance to GO toxicity in reducing lifespan, in inducing intestinal ROS production, and in decreasing locomotion behavior [45]. In contrast, overexpression of *linc-37* or *linc-14* could induce a susceptibility to GO toxicity in inducing intestinal ROS production and in decreasing locomotion behavior [45].

The lncRNAs can regulate the transcription by serving as "ligands" for TFs [49]. After the screen for the expression patterns of candidate binding TFs of *linc-37*, the expressions of *ama-1*, *daf-16*, *nhr-28*, *elt-3*, *fos-1*, and *gmeb-1* could be significantly upregulated in GO-exposed *linc-37* RNAi knockdown nematodes (Fig. 7.37) [45]. Among these TFs, *daf-16* encodes the FOXO transcriptional factor in the insulin signaling pathway, and RNAi knockdown of *linc-37* could increase the DAF-16::GFP expression and enhance the translocation of DAF-16::GFP into the nucleus of intestinal cells (Fig. 7.37) [45]. Genetic interaction analysis further demonstrated that the GO toxicity on lifespan and locomotion behavior in double mutant of *daf-16(mu86);linc-37 RNAi* was similar to that in *daf-16(mu86)* mutant (Fig. 7.37) [45], suggesting that the *linc-37* may regulate the GO toxicity by affecting the function of DAF-16 in the insulin signaling pathway.

Besides this, it has further found that RNAi knockdown of *linc-37* could significantly affect the expressions of *sod-1*, *sod-2*, *sod-3*, *sod-4*, *isp-1*, *clk-1*, *nhx-2*, *pkc-3*, *jkk-1*, *jnk-1*, *ced-3*, *ced-4*, *cep-1*, and *daf-16* [45], implying that *linc-37* may possibly also regulate the GO toxicity by targeting proteins required for the control of oxidative stress or intestinal development and function or encoding the JNK, insulin, or apoptosis and DNA damage signaling pathways in nematodes. Additionally, RNAi knockdown of *linc-14* could also significantly influence the expressions of *sod-3*, *sod-4*, *isp-1*, *nhx-2*, *par-6*, *pkc-3*, *jnk-1*, *mek-1*, *ced-3*, *ced-4*, *clk-2*, *cep-1*, *daf-16*, *akt-1*, and *daf-18* [45], implying that *linc-14* may possibly regulate the GO toxicity by targeting proteins required for the control of oxidative stress or intestinal development and function or encoding the JNK, insulin, or apoptosis and DNA damage signaling pathways in nematodes.

7.5 Alteration in Molecular Machinery for Important Biochemical Events in ENM-exposed Nematodes

In nematodes, exposure to certain ENMs may potentially alter the molecular machinery for some important biochemical events. For example, the molecular machinery for miRNA biogenesis can be affected in certain ENM-exposed nematodes. The primary miRNA transcripts can be processed into pre-miRNAs into nucleus by a complex containing RNase III enzyme Drosha and its binding partner, dsRNA binding proteins of DRSH-1 and PASH-1. Moreover, the pre-miRNA will be exported into the cytoplasm by a homolog of exportin 5 IMB-4 and further cleaved by RNase III enzyme Dicer (DCR-1) into a miRNA and its partner strand.

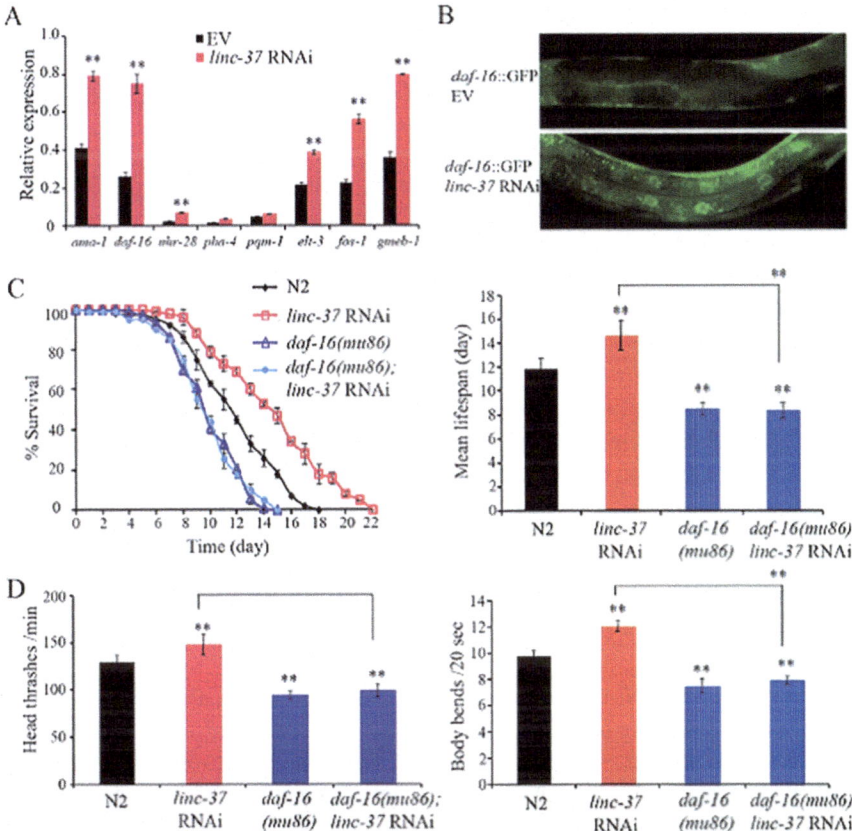

Fig. 7.37 Interaction of DAF-16/TF with *linc-37* in the regulation of GO toxicity [45]
(**a**) TF expression patterns in *linc-37* RNAi knockdown nematodes. (**b**) Subcellular localization of
DAF-16::GFP fusion protein in GO-exposed *linc-37* RNAi knockdown nematodes. (**c**) Lifespan of
wild-type N2, *daf-16(mu86)*, *linc-37* RNAi, and *daf-16(mu86);linc-37* RNAi nematodes exposed
to GO. (**d**) Locomotion behavior in wild-type N2, *daf-16(mu86)*, *linc-37* RNAi, and *daf-
16(mu86);linc-37* RNAi nematodes exposed to GO. GO exposure concentration was
100 mg/L. Prolonged exposure to GO was performed from L1-larvae to young adults. Bars represent mean ± SEM. $^{**}P < 0.01$

Argonaute protein ALG-1 is required for this Dicer cleavage, and ALG-1 and
ALG-2 belong to the same complex with DCR-1. Prolonged exposure to GO (10
mg/L) could significantly decrease the expressions of *drsh-1*, *pash-1*, and *dcr-1* in
nematodes [47], which implies that prolonged exposure to GO may adversely affect
the molecular machinery for miRNA biogenesis in nematodes.

7.6   Functional Analysis of Human Genes in Nematodes

Considering the fact that *C. elegans* genome has approximately 45% of genes having corresponding human homologues, *C. elegans* can be employed to directly predict the possible functions of human genes. Besides this, the certain human genes with or without the *C. elegans* homologues can be further expressed in nematodes to determine their potential biological functions. In humans, *SOD2* encodes the Mn-SODs, and SOD-2 and SOD-3 in nematodes are highly homologous with the SOD2 in humans. In nematodes, ectopically expression of the human *SOD2* could obviously suppress the increase in mortality, the decrease in body length, the reduction of brood size, the decrease in locomotion behavior, the increase in intestinal autofluorescence, and the induction of intestinal ROS production in nematodes exposed to TiO_2-NPs (4 nm) at the concentration of 10 mg/L (Fig. 7.38) [6], which suggests the similar function of human SOD2 to that of *C. elegans* SOD-2 and SOD-3 in the regulation of TiO_2-NPs toxicity.

7.7   Perspectives

In this chapter, we introduced the important functions of some signaling pathways in the regulation of ENM toxicity. The mainly introduced signaling pathways contain apoptosis signaling pathway, DNA damage signaling pathway, MAPK signaling pathways, insulin signaling pathway, innate immune response signaling pathway, Wnt signaling pathway, TGF-beta signaling pathway, developmental timing control-related signals, and neurotransmission-related signals. Nevertheless, the potential functions of many other important signaling pathways in the regulation of ENM toxicity, as well as in the response to ENM exposure, have not been determined yet in nematodes. Especially, so far, still no entire molecular network at the mRNA level has been raised to explain the molecular basis for the response to certain ENM exposure in different organs in nematodes. More importantly, in the future, we may need to pay more attention to the elucidation of molecular basis for the response of organism to ENM exposure at environmentally relevant concentrations.

We here also systematically introduced the molecular basis for miRNAs and lncRNAs in the regulation of RNMs toxicity, as well as the response to ENM exposure. Nevertheless, even in the identified dysregulated miRNAs and lncRNAs induced by ENM exposure, the exact functions of most of the identified dysregulated miRNAs and lncRNAs have not been examined so far. Moreover, besides the miRNAs and the lncRNAs, there are still many other types of RNAs, such as circular RNAs, in nematodes [52]. The potential involvement of those types of RNAs in the regulation of ENM toxicity and the underlying molecular mechanisms are needed to be further elucidated.

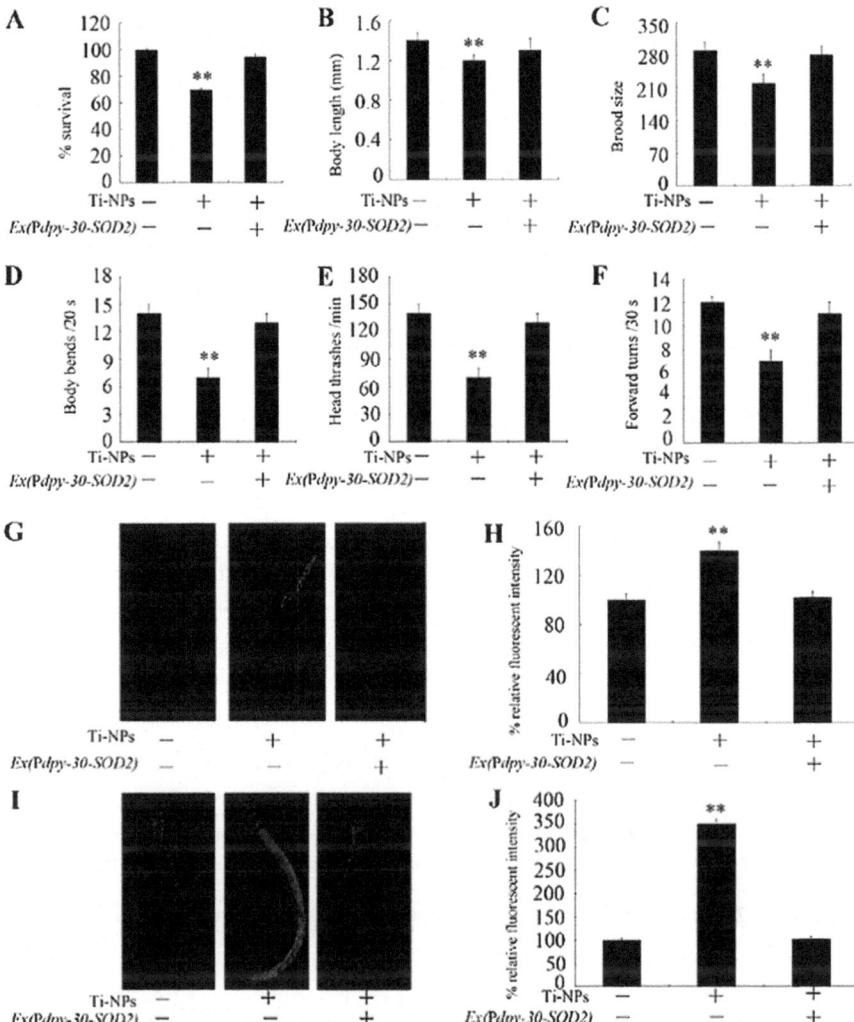

Fig. 7.38 Effects of ectopically expression of human SOD2 on the toxicity formation in nematodes exposed to TiO₂-NPs (4 nm) at the concentration of 10 mg/L [6]
(**a**) Effects of ectopically expression of human SOD2 on the survival of animals exposed to TiO₂-NPs (4 nm). (**b**) Effects of ectopically expression of human SOD2 on the growth of animals exposed to TiO₂-NPs (4 nm). (**c**) Effects of ectopically expression of human SOD2 on the reproduction of animals exposed to TiO₂-NPs (4 nm). (**d–f**) Effects of ectopically expression of human SOD2 on the locomotion behavior of animals exposed to TiO₂-NPs (4 nm). (**g–h**) Effects of ectopically expression of human SOD2 on the intestinal autofluorescences of animals exposed to TiO₂-NPs (4 nm). (**i–j**) Effects of ectopically expression of human SOD2 on the ROS production of animals exposed to TiO₂-NPs (4 nm). Exposure of Ti-NPs was performed from L1-larvae to adult day-1. Ti-NPs, TiO₂-NPs. Bars represent mean ± SEM. $^{**}p < 0.01$

References

1. Feng J, Bussiere F, Hekimi S (2001) Mitochondrial electron transport is a key determinant of life span in *Caenorhabditis elegans*. Dev Cell 1:633–644
2. Ishii N, Fujii M, Hartman PS, Tsuda M, Yasuda K, Senoo-Matsuda N, Yanase S, Ayusawa D, Suzuki K (1998) A mutation in succinate dehydrogenase cytochrome b causes oxidative stress and ageing in nematodes. Nature 394:694–697
3. Kayser EB, Morgan PG, Hoppel CL, Sedensky MM (2001) Mitochondrial expression and function of GAS-1 in *Caenorhabditis elegans*. J Biol Chem 276:20551–20558
4. Miyadera H, Amino H, Hiraishi A, Taka H, Murayama K, Miyoshi H, Sakamoto K, Ishii N, Hekimi S, Kita K (2001) Altered quinone biosynthesis in the long-lived *clk-1* mutants of *Caenorhabditis elegans*. J Biol Chem 276:7713–7716
5. Wu Q-L, Yin L, Li X, Tang M, Zhang T, Wang D-Y (2013) Contributions of altered permeability of intestinal barrier and defecation behavior to toxicity formation from graphene oxide in nematode *Caenorhabditis elegans*. Nanoscale 5(20):9934–9943
6. Li Y-X, Wang W, Wu Q-L, Li Y-P, Tang M, Ye B-P, Wang D-Y (2012) Molecular control of TiO_2-NPs toxicity formation at predicted environmental relevant concentrations by Mn-SODs proteins. PLoS One 7(9):e44688
7. Li Y-X, Yu S-H, Wu Q-L, Tang M, Pu Y-P, Wang D-Y (2012) Chronic Al_2O_3-nanoparticle exposure causes neurotoxic effects on locomotion behaviors by inducing severe ROS production and disruption of ROS defense mechanisms in nematode *Caenorhabditis elegans*. J Hazard Mater 219–220:221–230
8. Wu Q-L, Zhao Y-L, Li Y-P, Wang D-Y (2014) Molecular signals regulating translocation and toxicity of graphene oxide in nematode *Caenorhabditis elegans*. Nanoscale 6:11204–11212
9. Lettre G, Hengartner MO (2006) Developmental apoptosis in *C. elegans*: a complex CEDnario. Ann Rev Mol Cell Biol 7:97–108
10. Zhao Y-L, Wu Q-L, Wang D-Y (2016) An epigenetic signal encoded protection mechanism is activated by graphene oxide to inhibit its induced reproductive toxicity in *Caenorhabditis elegans*. Biomaterials 79:15–24
11. O'Neil N, Rose A (2006) DNA repair. WormBook. https://doi.org/10.1895/wormbook.1.54.1.
12. Kamath RK, Martinez-Campos M, Zipperlen P, Fraser AG, Ahringer J (2001) Effectiveness of specific RNA-mediated interference through ingested double stranded RNA in *C. elegans*. Genome Biol 2:1–10
13. Koga M, Zwaal R, Guan KL, Avery L, Ohshima Y (2000) A *Caenorhabditis elegans* MAP kinase kinase, MEK-1, is involved in stress responses. EMBO J 19:5148–5156
14. Zhao Y-L, Wu Q-L, Wang D-Y (2015) A microRNAs-mRNAs network involved in the control of graphene oxide toxicity in *Caenorhabditis elegans*. RSC Adv 5:92394–92405
15. Qu M, Li Y-H, Wu Q-L, Xia Y-K, Wang D-Y (2017) Neuronal ERK signaling in response to graphene oxide in nematode *Caenorhabditis elegans*. Nanotoxicology 11(4):520–533
16. Okuyama T, Inoue H, Ookuma S, Satoh T, Kano K, Honjoh S, Hisamoto N, Matsumoto K, Nishida E (2010) The ERK-MAPK pathway regulates longevity through SKN-1 and insulin-like signaling in *Caenorhabditis elegans*. J Biol Chem 285:30274–30281
17. Tullet JM, Hertweck M, An JH, Baker J, Hwang JY, Liu S, Oliveira RP, Baumeister R, Blackwell TK (2008) Direct inhibition of the longevity-promoting factor SKN-1 by insulin-like signaling in *C. elegans*. Cell 132:1025–1038
18. Shephard F, Adenle AA, Jacobson LA, Szewczyk NJ (2011) Identification and functional clustering of genes regulating muscle protein degradation from amongst the known *C. elegans* muscle mutants. PLoS One 6:e24686
19. Zhao Y-L, Zhi L-T, Wu Q-L, Yu Y-L, Sun Q-Q, Wang D-Y (2016) p38 MAPK-SKN-1/ Nrf signaling cascade is required for intestinal barrier against graphene oxide toxicity in *Caenorhabditis elegans*. Nanotoxicology 10(10):1469–1479
20. Blackwell TK, Steinbaugh MJ, Hourihan JM, Ewald CY (2015) SKN-1/Nrf, stress responses, and aging in *Caenorhabditis elegans*. Free Radic Biol Med 88:290–301

21. Lim D, Roh JY, Eom HJ, Choi JY, Hyun J, Choi J (2012) Oxidative stress-related PMK-1 P38 MAPK activation as a mechanism for toxicity of silver nanoparticles to reproduction in the nematode *Caenorhabditis elegans*. Environ Toxicol Chem 31(3):585–592

22. Chatterjee N, Eom HJ, Choi J (2014) Effects of silver nanoparticles on oxidative DNA damage-repair as a function of p38 MAPK status: a comparative approach using human Jurkat T cells and the nematode *Caenorhabditis elegans*. Environ Mol Mutagen 55(2):122–133

23. Kenyon C (2010) The genetics of ageing. Nature 464:504–512

24. Lapierre LR, Hansen M (2012) Lessons from *C. elegans*: signaling pathways for longevity. Trend Endocrinol Metab 23:637–644

25. Zhao Y-L, Yang R-L, Rui Q, Wang D-Y (2016) Intestinal insulin signaling encodes two different molecular mechanisms for the shortened longevity induced by graphene oxide in *Caenorhabditis elegans*. Sci Rep 6:24024

26. Jensen VL, Simonsen KT, Lee Y-H, Park D, Riddle DL (2010) RNAi screen of DAF-16/FOXO target genes in *C. elegans* links pathogenesis and dauer formation. PLoS One 5:e15902

27. McElwee J, Bubb K, Thomas JH (2003) Transcriptional outputs of the *Caenorhabditis elegans* forkhead protein DAF-16. Aging Cell 2:111–121

28. Murphy CT, McGarroll SA, Bargmann CI, Fraser A, Kamath RS, Ahringer J, Kenyon C (2003) Genes that act downstream of DAF-16 to influence the lifespan of *Caenorhabditis elegans*. Nature 424:277–284

29. Tepper RG, Ashraf J, Kaletsky R, Kleemann G, Murphy CT, Bussemaker HJ (2013) PQM-1 complements DAF-16 as a key transcriptional regulator of DAF-2-mediated development and longevity. Cell 154:676–690

30. Ren M-X, Zhao L, Lv X, Wang D-Y (2017) Antimicrobial proteins in the response to graphene oxide in *Caenorhabditis elegans*. Nanotoxicology 11(4):578–590

31. Wu Q-L, Zhao Y-L, Fang J-P, Wang D-Y (2014) Immune response is required for the control of *in vivo* translocation and chronic toxicity of graphene oxide. Nanoscale 6:5894–5906

32. Zhi L-T, Ren M-X, Qu M, Zhang H-Y, Wang D-Y (2016) Wnt ligands differentially regulate toxicity and translocation of graphene oxide through different mechanisms in *Caenorhabditis elegans*. Sci Rep 6:39261

33. Eisenmann DM (2005) Wnt signaling. WormBook. https://doi.org/10.1895/wormbook.1.7.1.

34. Zhi L-T, Qu M, Ren M-X, Zhao L, Li Y-H, Wang D-Y (2017) Graphene oxide induces canonical Wnt/β-catenin signaling-dependent toxicity in *Caenorhabditis elegans*. Carbon 113:122–131

35. Eisenmann DM, Maloof JN, Simske JS, Kenyon CJ, Kim SK (1998) The beta-catenin homolog BAR-1 and LET-60 Ras coordinately regulate the Hox gene *lin-39* during *Caenorhabditis elegans* vulval development. Development 125:3667–3680

36. Maloof JN, Whangbo J, Harris JM, Jongeward GD, Kenyon C (1999) A Wnt signaling pathway controls hox gene expression and neuroblast migration in *C. elegans*. Development 126:37–49

37. Sokol NS (2012) Small temporal RNAs in animal development. Curr Opin Genet Dev 22:368–373

38. Reinhart B, Slack F, Basson M, Pasquinelli AE, Bettinger JC, Rougvie AE, Horvitz HR, Ruvkun G (2000) The 21 nucleotide *let-7* RNA regulates *C. elegans* developmental timing. Nature 403:901–906

39. Zhao L, Wan H-X, Liu Q-Z, Wang D-Y (2017) Multi-walled carbon nanotubes-induced alterations in microRNA *let-7* and its targets activate a protection mechanism by conferring a developmental timing control. Part Fibre Toxicol 14:27

40. Niwa R, Hada K, Moliyama K, Ohniwa RL, Tan YM, Olsson-Carter K, Chi W, Reinke V, Slack FJ (2009) *C. elegans sym-1* is a downstream target of the hunchback-like-1 developmental timing transcription factor. Cell Cycle 8:4147–4154

41. Grishok A, Pasquinelli AE, Conte D, Li N, Parrish S, Ha I, Baillie DL, Fire A, Ruvkun G, Mello CC (2001) Genes and mechanisms related to RNA interference regulate expression of the small temporal RNAs that control *C. elegans* developmental timing. Cell 106:23–34

42. Li Y-X, Yu S-H, Wu Q-L, Tang M, Wang D-Y (2013) Transmissions of serotonin, dopamine and glutamate are required for the formation of neurotoxicity from Al$_2$O$_3$-NPs in nematode *Caenorhabditis elegans*. Nanotoxicology 7(5):1004–1013
43. Iwasaki K, Staunton J, Saifee O, Nonet ML, Thomas JH (1997) *aex-3* encodes a novel regulator of presynaptic activity in *C. elegans*. Neuron 18:613–622
44. Iwasaki K, Toyonaga R (2000) The Rab3 GDP/GTP exchange factor homolog AEX-3 has a dual function in synaptic transmission. EMBO J 19:4806–4816
45. Wu Q-L, Zhou X-F, Han X-X, Zhuo Y-Z, Zhu S-T, Zhao Y-L, Wang D-Y (2016) Genome-wide identification and functional analysis of long noncoding RNAs involved in the response to graphene oxide. Biomaterials 102:277–291
46. Starnes DL, Lichtenberg SS, Unrine JM, Starnes CP, Oostveen EK, Lowry GV, Bertsch PM, Tsyusko OV (2016) Distinct transcriptomic responses of *Caenorhabditis elegans* to pristine and sulfidized silver nanoparticles. Environ Pollut 213:314–321
47. Wu Q-L, Zhao Y-L, Zhao G, Wang D-Y (2014) microRNAs control of *in vivo* toxicity from graphene oxide in *Caenorhabditis elegans*. Nanomedicine: Nanotechnol Biol Med 10:1401–1410
48. Yang R-L, Ren M-X, Rui Q, Wang D-Y (2016) A *mir-231*-regulated protection mechanism against the toxicity of graphene oxide in nematode *Caenorhabditis elegans*. Sci Rep 6:32214
49. Rinn JL, Chang HY (2012) Genome regulation by long noncoding RNAs. Annu Rev Biochem 81:145–166
50. Guttman M, Rinn JL (2012) Modular regulatory principles of large non-coding RNAs. Nature 482:339–346
51. Ulitsky I, Bartel DP (2013) LincRNAs: genomics, evolution, and mechanisms. Cell 154:26–46
52. Chen LL, Carmichael GG (2010) Decoding the function of nuclear long non-coding RNAs. Curr Opin Cell Biol 22:357–364

Chapter 8
Distribution and Translocation of Nanomaterials

Abstract Distribution and translocation of engineered nanomaterials (ENMs) is one of the crucial cellular contributors for the toxicity induction of ENMs. After exposure, the primary targeted organs of ENMs can be defined as the intestine and the epidermis, and the secondary targeted organs of ENMs contain at least the reproductive organs, such as the gonad and the spermatheca, and the neurons in nematodes. So far, the normally considered primary targeted organ for ENMs is the intestine in nematodes. In this chapter, we first introduce the techniques for the analysis on distribution and translocation of ENMs. Moreover, the distribution and translocation of different ENMs in the primary or the secondary targeted organs, as well as the patterns of transgenerational translocation of ENMs, and the underlying cellular, molecular, and chemical metabolisms for distribution and translocation of ENMs are systematically introduced.

Keywords Distribution · Translocation · Nanomaterials · *Caenorhabditis elegans*

8.1 Introduction

Distribution and translocation of engineered nanomaterials (ENMs) is the crucial cellular contributor for the toxicity formation of certain ENMs in nematodes [1–3]. This implies that the bioavailability of ENMs to certain targeted primary or secondary organs may be the key or direct inducer for the observed toxicity of certain ENMs in targeted primary or secondary organs. Elucidation of the patterns of distribution and translocation of different ENMs will be helpful for our understanding and further deeply determining the targeted organs toxicology of different ENMs in organisms. Moreover, the patterns of distribution and translocation of ENMs will help us judge the potential formation of transgenerational toxicity of ENMs in organisms.

In nematodes, the primary targeted organs can be defined as the intestine and the epidermis [2–4]. Both the intestine and the epidermis are normally considered defense barriers against the environmental toxicants or environmental stresses [5, 6]. Nevertheless, usually the epidermal barrier is strong enough to block the

© Springer Nature Singapore Pte Ltd. 2018
D. Wang, *Nanotoxicology in* Caenorhabditis elegans,
https://doi.org/10.1007/978-981-13-0233-6_8

translocation of environmental toxicants through this biological barrier. Thus, the normally found primary targeted organ is the intestine for environmental toxicants [2, 3]. Besides these, the secondary targeted organs contain at least the reproductive organs, such as the gonad and the spermatheca, and the neurons [2, 3]. Now, seldom evidence has been raised that the environmental toxicants can be translocated into other organs, such as muscle, in nematodes.

In this chapter, we first introduce the techniques for the analysis on distribution and translocation of ENMs. Again, we systematically introduce the progress on the distribution and translocation of different ENMs in the primary or the secondary targeted organs in nematodes, as well as the patterns of transgenerational translocation of ENMs. Moreover, we also introduce the dynamic cellular, molecular, and chemical metabolisms of ENMs in the body of nematodes.

8.2   Methods for the Analysis on Distribution and Translocation of ENMs

So far, several techniques have been employed or raised to determine the distribution and translocation of ENMs in nematodes. With the aid of these powerful techniques, the distribution and translocation patterns of some important ENMs have been well described in nematodes.

8.2.1   Direct Visualization

For some nanoparticles, we can directly observe them under the normal light microscopy if the exposure concentrations are high enough. With the carboxyl-functionalized graphene (G–COOH) as an example, the distribution and accumulation of G–COOH could be directly visualized in the pharynx and the intestinal lumen under the light microscopy after prolonged exposure at the concentration (100 mg/L) (Fig. 8.1) [7]. In contrast, no obvious distribution and accumulation of G–COOH was observed in the reproductive organ, such as the gonads (Fig. 8.1) [7].

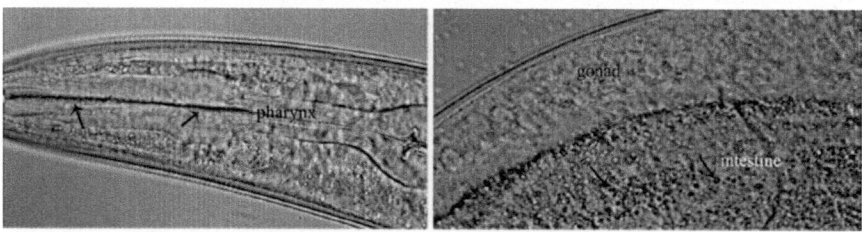

Fig. 8.1 Distribution and translocation of G–COOH in exposed nematodes [7]
Prolonged exposure was performed from L1-larvae to adult day-1. Exposure concentration for G–COOH was 100 mg/L. Arrowheads indicate the G–COOH distribution

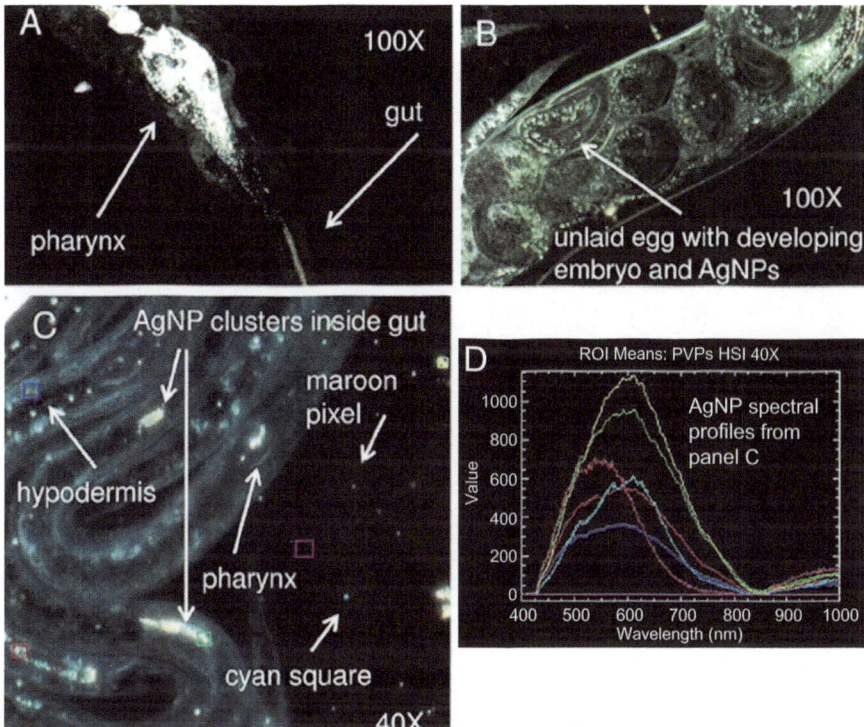

Fig. 8.2 Ag-NPs are ingested and internalized into the cells of nematodes [8]
(**a**) CIT10 Ag-NPs were taken up along with food. (**b**) Some CIT10 Ag-NPs were taken up into the cells and transferred to the offspring. (**c**) Hyperspectral image (HSI) showing the presence of PVPS Ag-NPs inside the body of nematodes. (**d**) The spectral profiles of Ag-NPs and hypodermis region of nematodes: green, red, and dark blue represent internal Ag-NPs clusters; yellow, cyan, and maroon external Ag-NP clusters; and magenta background

8.2.2 Visualization of ENMs Using Dark-Field Optical Microscopy

Some ENMs, such as the silver nanoparticles (Ag-NPs), can be directly visualized using a CytoViva nanoscale microscope employing a dark-field-based illuminator that can focus a highly collimated light at oblique angles on the examined sample in order to obtain an image with improved contrast and signal-to-noise ratio. With the aid of dark-field-based optical microscopy together with a hyperspectral imaging system, the cellular internalization of Ag-NPs was determined in nematodes. Based on the analysis using this technique, a large amount of Ag-NPs were distributed and accumulated in the pharynx and the intestine (Fig. 8.2) [8]. Additionally, the spectral features of Ag-NPs profiles in the presence and absence of *C. elegans* remained unchanged [8]. In contrast, no Ag-NPs were detected in AgNO$_3$-exposed

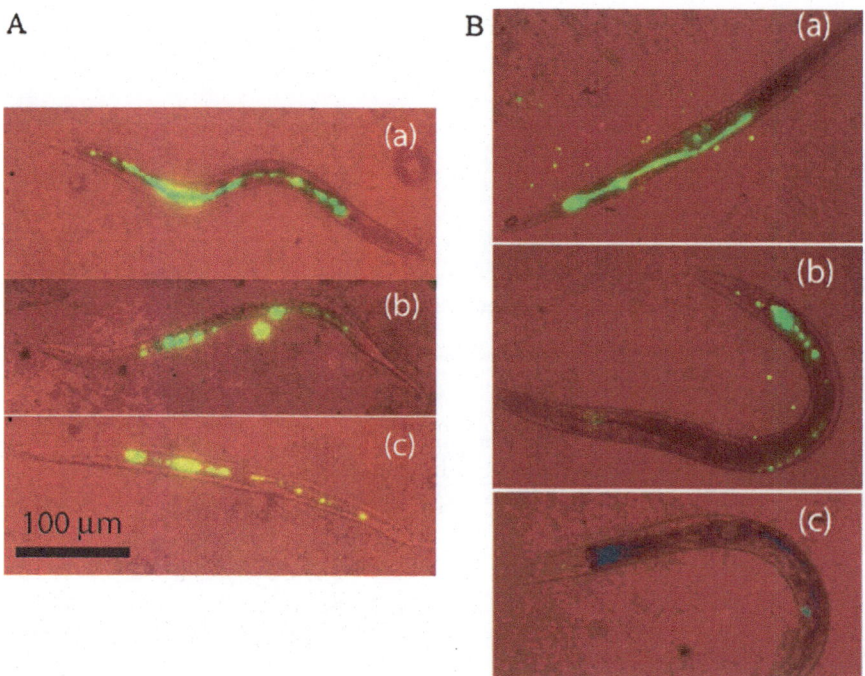

Fig. 8.3 False color two-photon images of phosphor nanoparticles at 980 nm excitation with red representing the bright field and green for the phosphor emission in nematodes [9]
(**a**) The nematodes were deprived of food over a period of 24 h, showing little or no change at (*a*) 0 h, (*b*) 4 h, and (*c*) 24 h. (**b**) The nematodes were given food immediately after being fed with phosphors, showing decreasing amounts of phosphors at (*a*) 0 h, (*b*) 1 h, and (*c*) 2 h

nematodes, implying that the production of Ag-NPs from the $AgNO_3$ may be either nonexistent or below the detection limits in nematodes [8].

8.2.3 Analysis Based on the Fluorescence of ENMs

Certain ENMs, such as the upconversion phosphors NPs, can be imaged by infrared excitation, since this kind of NPs shows the emitted visible fluorescence on the infrared excitation. Based on this physicochemical property, the phosphors NPs were easily visible in the intestines, and most NPs were found beyond the pharynx, extending to the rectum in nematodes (Fig. 8.3) [9]. When the nematodes were deprived of the food (*E. coli* OP50), the phosphors NPs were retained in the body of nematodes, which may be due to the inhibition of excretion as feeding ceases (Fig. 8.3) [9]. In contrast, if the nematodes were given food immediately after being fed with phosphors, this feeding caused the secretion of phosphors NPs in under 2 h (Fig. 8.3) [9]. Thereafter, the nematodes would continue the feeding and appeared unaffected by the prior ingestion of the phosphors NPs (Fig. 8.3) [9].

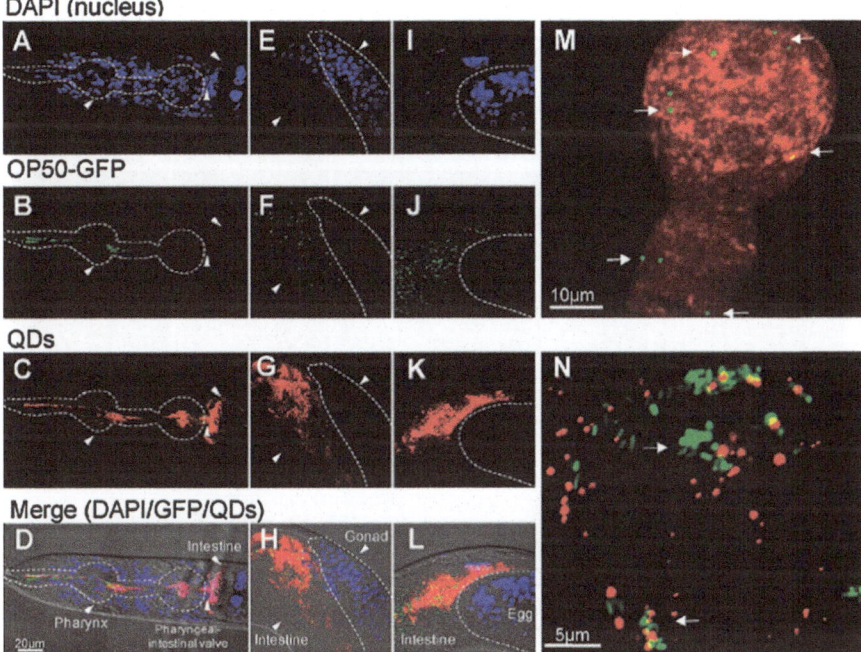

Fig. 8.4 Ingestion and distribution of MPA-CdTe QDs in nematodes [12]

(**a–n**) Laser confocal microscope images of cell nucleus fluorescent dye of DAPI (**a, e, i**), OP50::GFP (**b, f, j**), and MPA-CdTe QDs (**c, g, and k**). Enlarged views of the beginning (**m**) and middle part (**n**) of the intestine are shown by merged images of QDs with OP50::GFP. Arrows indicate the QDs or the OP50::GFP

Based on this physicochemical property, three types of upconversion NPs, including NaYF4:Yb,Er@PEI, NaYF4:Yb,Er@OA, and NaYF4:Yb,Er, were examined, and they show the similar distribution and accumulation in nematodes [10]. All these three types of upconversion NPs were mainly distributed and accumulated in the pharynx and the intestine [10]. In wild-type or *him-5* mutant nematodes, NaYF4:Yb,Tm was also mainly distributed and accumulated in the pharynx and the intestine [11].

Quantum dots (QDs) are another type of ENMs, and we can directly observe them under the commonly used fluorescence microscopy due to their broad optical absorption, bright and tunable fluorescence. After feeding the MPA-CdTe QDs together with the GFP-labeled *E. coli* (OP50::GFP) for 12 h, a large quantity of QDs were ingested together with OP50::GFP through the digestive tract and filled up the pharynx lumen and the intestinal lumen, as well as the pharyngeal-intestinal valve (Fig. 8.4) [12]. In the intestinal lumen, a large amount of QDs were retained in the initial part of the intestine and separated from the OP50::GFP in the middle intestine (Fig. 8.4) [12].

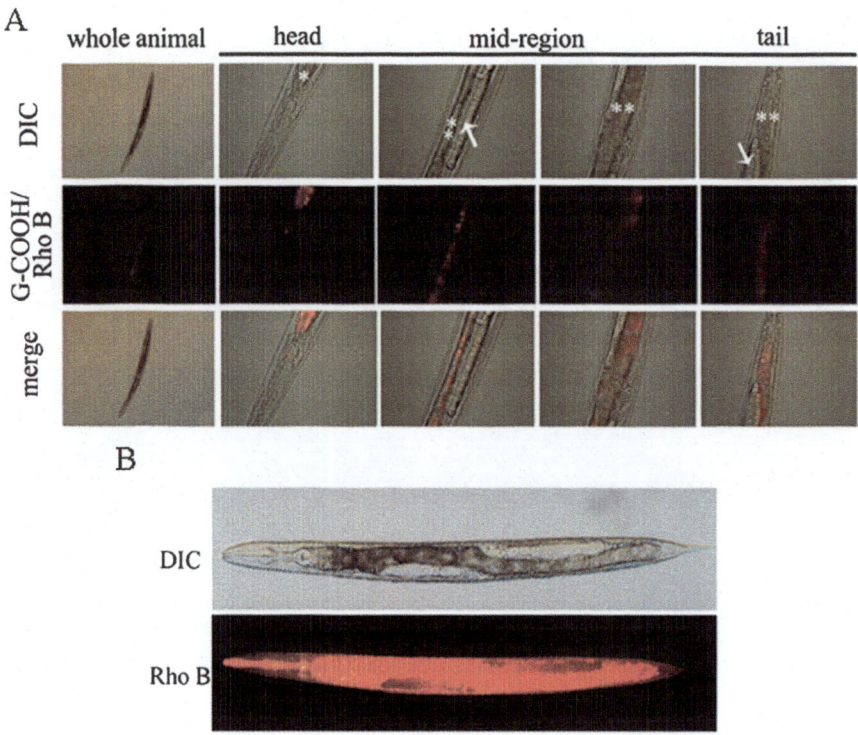

Fig. 8.5 Distribution and translocation of G–COOH in exposed nematodes [7]
(**a**) Distribution and translocation of G–COOH/Rho B in exposed nematodes. Prolonged exposure was performed from L1-larvae to adult day-1. The exposure concentration of G–COOH was 100 mg/L. Arrowheads indicate the gonad at the mid-region of nematodes. The pharynx (*) and the intestine (**) were also indicated. (**b**) Distribution and translocation of Rho B in exposed nematodes

8.2.4   Analysis Based on the Fluorescence Signals of Labeled Probe for the Examined ENMs

If the examined ENMs do not have the fluorescent property themselves, some of the examined ENMs can be labeled with certain fluorescent probe. For example, the graphene materials can be labeled by Rhodamine B (Rho B) through p-p stacking, hydrophobic, and hydrogen-bonding interactions [13]. In nematodes, the Rho B can be loaded on the G–COOH by mixing the Rho B solution (1 mg/mL, 0.3 mL) with the aqueous suspension of G–COOH (0.1 mg/mL, 5 mL) with the aid of shaking overnight. The unbound Rho B will be removed by dialysis against the distilled water over 72 h. After prolonged exposure to G–COOH/Rho B (100 mg/L), the G–COOH/Rho B was mainly distributed in the pharynx and the intestine in nematodes (Fig. 8.5) [7]. No signals for G–COOH/Rho B were detected in the reproductive organs, such as the gonads (Fig. 8.5) [7]. Compared with the distribution of G–COOH/RhoB, exposure to Rho B alone would result in a relatively equal distribution of Rho B fluorescence in the tissues in nematodes (Fig. 8.5) [7].

8.2.5   Transmission Electron Microscopy (TEM) Assay

The TEM technique can help us directly visualize the distribution of ENMs in different cells and especially in different organelles in the examined cells. To perform the TEM assay, the nematodes can be fixed in 0.5% glutaraldehyde and 2% osmium tetroxide. After that, the fixed nematodes are embedded in araldite resin following the infiltration series (30% araldite/acetone for 4 h, 70% araldite/acetone for 5 h, 90% araldite/acetone overnight, and pure araldite for 8 h) [7].

TEM assay at the low resolution is helpful for determining the distribution of ENMs in different tissues or cells. Based on the TEM assay at the low resolution, the biogenic magnetite particles were distributed along the edges at the anterior region (Fig. 8.6) [14]. The high-resolution TEM image in certain cells further confirmed the existence of magnetite particles with the sizes larger than 30–50 nm (Fig. 8.6) [14].

Some of the observed particles with the nanosize in the cells of nematodes during the TEM or scanning electron microscopy (TEM) assay could be further confirmed using energy-dispersive X-ray (EDX) analysis. For example, the upconversion NPs observed in a S.E.M at high spatial resolution could be further identified and confirmed by the existence of EDX spectra of Y K, Yb L, and Er L [9].

TEM technique was especially useful for determining the exact location of ENMs in different organelles in cells. With the TiO_2 nanoparticles (TiO_2-NPs) as an example, some TiO_2-NPs already gradually entered into the intestinal cells after acute exposure to TiO_2-NPs (100 mg/L) (Fig. 8.7) [15]. Nevertheless, acute exposure to TiO_2-NPs (100 mg/L) did not obviously affect the ultrastructure of intestinal cells (Fig. 8.7) [15]. After prolonged exposure to TiO_2-NPs (100 µg/L), a large amount of TiO_2-NPs were distributed and deposited into the intestinal cells, and the TiO_2-NPs were observed to be even located into the mitochondria (Fig. 8.7) [15]. Meanwhile, after prolonged exposure to TiO_2-NPs (100 µg/L), the ultrastructure of intestinal cells was severely disrupted, and some microvilli were lost in TiO_2-NP-exposed nematodes (Fig. 8.7) [15]. Moreover, this disrupted ultrastructure of microvilli in nematodes after prolonged exposure to TiO_2-NPs (100 µg/L) could not be noticeably recovered under the normal condition (Fig. 8.7) [15].

8.2.6   Elemental Analysis

For some certain ENMs, such as some metal NPs or metal oxide NPs, we can determine their distribution and accumulation in the body of nematodes through the elemental analysis. With the TiO_2-NPs as an example, the uptake of TiO_2-NPs in nematodes can also be examined using the inductively coupled plasma-mass spectrometry (ICP-MS) technique. In nematodes, both acute exposure to TiO_2-NPs (100 mg/L) and prolonged exposure to TiO_2-NPs (100 µg/L) could cause the accumulation of a large amount of Ti in the body of nematodes (Fig. 8.7) [15]. Prolonged

(i)   (ii)   (iii)

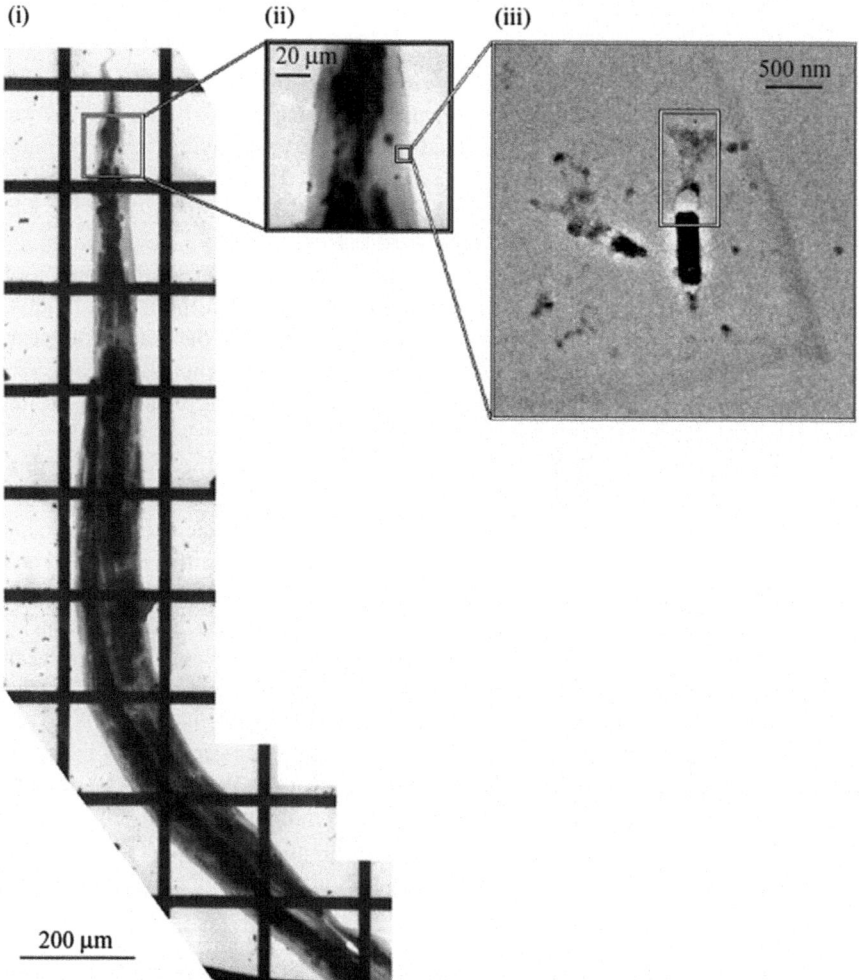

Fig. 8.6 TEM images of a *Caenorhabditis elegans* sample highlighting the location of magnetite discovered within the nematode [14]

exposure to TiO$_2$-NPs (100 µg/L) could result in the more severe accumulation of Ti in the body of nematodes than acute exposure to TiO$_2$-NPs (100 mg/L) (Fig. 8.7) [15]. Moreover, only a small amount of Ti could be detected in TiO$_2$-NPs (100 mg/L) acutely exposed nematodes after transfer to the normal condition for 48 h (Fig. 8.7) [15]. In contrast, still a large amount of Ti existed in the body of nematodes undergoing prolonged exposure to TiO$_2$-NPs (100 µg/L) (Fig. 8.7) [15]. Nevertheless, the ICP-MS technique may meet the difficulty to distinguish the TiO$_2$-NPs and Ti ion in the exposed nematodes.

The ICP-MS technique was also used to determine the effect of *sod-2*, *sod-3*, *mtl-2*, or *hsp-16.48* mutation on the accumulation of TiO$_2$-NPs in the body of nema-

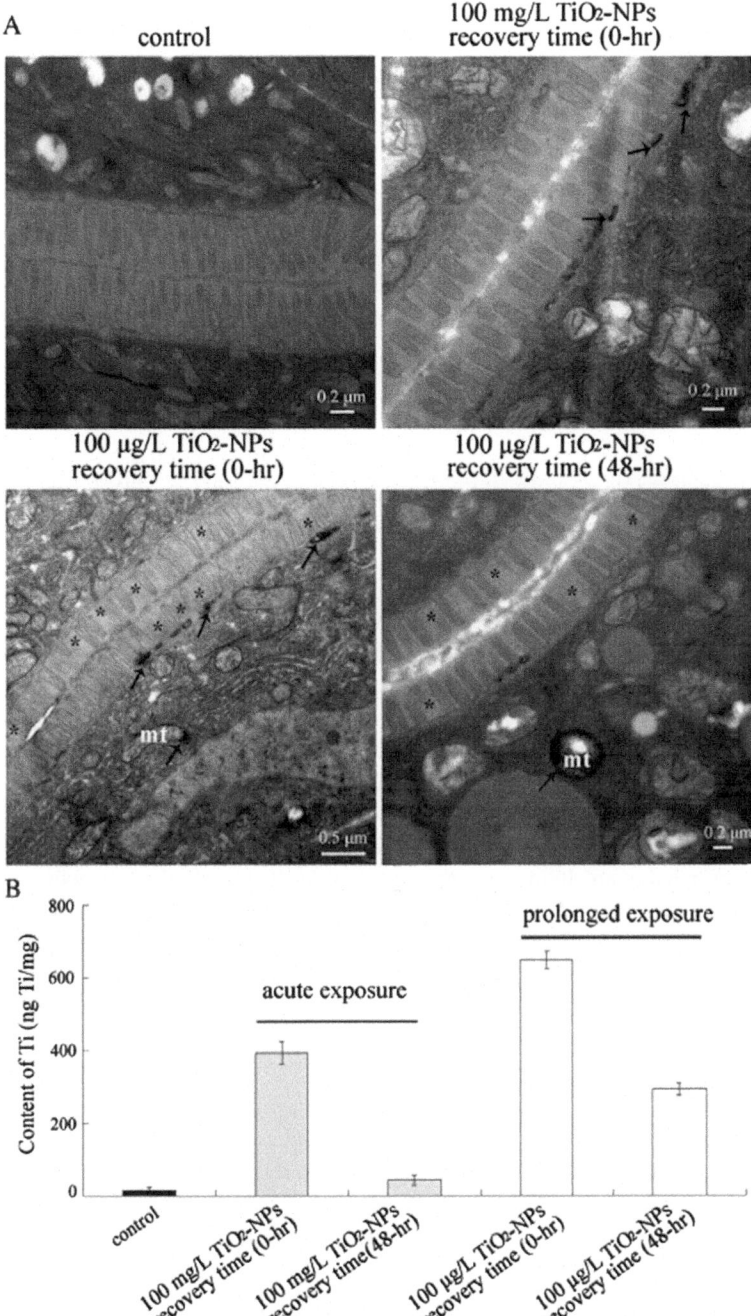

Fig. 8.7 Ultrastructural changes of intestine and uptake of TiO₂-NPs in exposed nematodes after transfer to the normal condition [15]

(a) Ultrastructure of intestine in TiO₂-NP-exposed nematodes after transfer to the normal condition. Asterisks indicate the position without microvilli, and arrowheads indicate the location of TiO₂-NPs. mt, mitochondria. (b) Uptake of Ti in TiO₂-NP-exposed nematodes after transfer to the normal condition. The Ti content was expressed as concentrations of titanium element, ng Ti/mg total protein. Bars represent means ± SEM. Acute exposure concentration for TiO₂-NPs was 100 mg/L, and prolonged exposure concentration for TiO₂-NPs was 100 μg/L

todes. In nematodes, mutation of *sod-2*, *sod-3*, *mtl-2*, or *hsp-16.48* could induce a susceptibility to TiO$_2$-NPs toxicity [16]. Meanwhile, mutation of *sod-2*, *sod-3*, *mtl-2*, or *hsp-16.48* enhanced the uptake of Ti into the body of nematodes [16]. After prolonged exposure to TiO$_2$-NPs (1 μg/L), the contents of Ti in *sod-2*, *sod-3*, *mtl-2*, and *hsp-16.48* mutants were 3.36, 3.26, 2.31, and 2.05 folds of that in wild-type nematodes [16]. Without the exposure to TiO$_2$-NPs (1 μg/L), the Ti content in nematodes was nearly undetectable [16].

8.2.7   Microbeam Synchrotron Radiation X-Ray Fluorescence (M-SRXRF) Technique

SRXRF is a multielemental analysis technique with a high sensitivity. This technique is useful for studying the spatial distribution of major, minor, and even trace elements after electrophoretic separation. Using the m-SRXRF technique, the in situ distribution of copper and other elements in the body of nematodes after exposure to copper nanoparticles (Cu-NPs) was examined in order to determine the absorption and translocation of Cu-NPs. Based on the obtained elemental maps, an obvious increase of Cu level in the head and at a location 1/3 of the way up the body from the tail was detected, although the accumulation of Cu-NPs in a single nematode was not very significant (Fig. 8.8) [17]. A more considerable amount of Cu was detected in other portions of the body, especially in its excretory cells and intestine, when the nematodes were exposed to Cu ion (Fig. 8.8) [17], which implies that the Cu ion may have much higher absorption and accumulation in nematodes. Moreover, Cu exposure led to an alteration in the distribution pattern of other elements as indicated by the increase in K level and the increase in both Fe and Zn content in proximal gonad (Fig. 8.8) [17].

8.2.8   Fourier Transform Infrared Spectroscopy (FT-IR) Mapping Technique

The μFT-IR technique was employed to determine the intake and the distribution of graphite nanoplatelets (GNPs) in the body of nematodes. After exposure to GNPs for 24 h, the GNPs were distributed along the body of nematodes based on the peak centered at 865 cm^{-1}, related to the C–C lattice mode (Fig. 8.9) [18]. Additionally, the GNP-related peak was also observed in the embryos laid by the adult nematodes just after its mounting on the ZnSe window (Fig. 8.9) [18], which suggested the possible translocation of GNPs from the intestine to the gonads. The obtained IR mapping spectroscopy is helpful for the evaluation of distribution and translocation of at least carbon-based ENMs in living organisms, such as the nematodes.

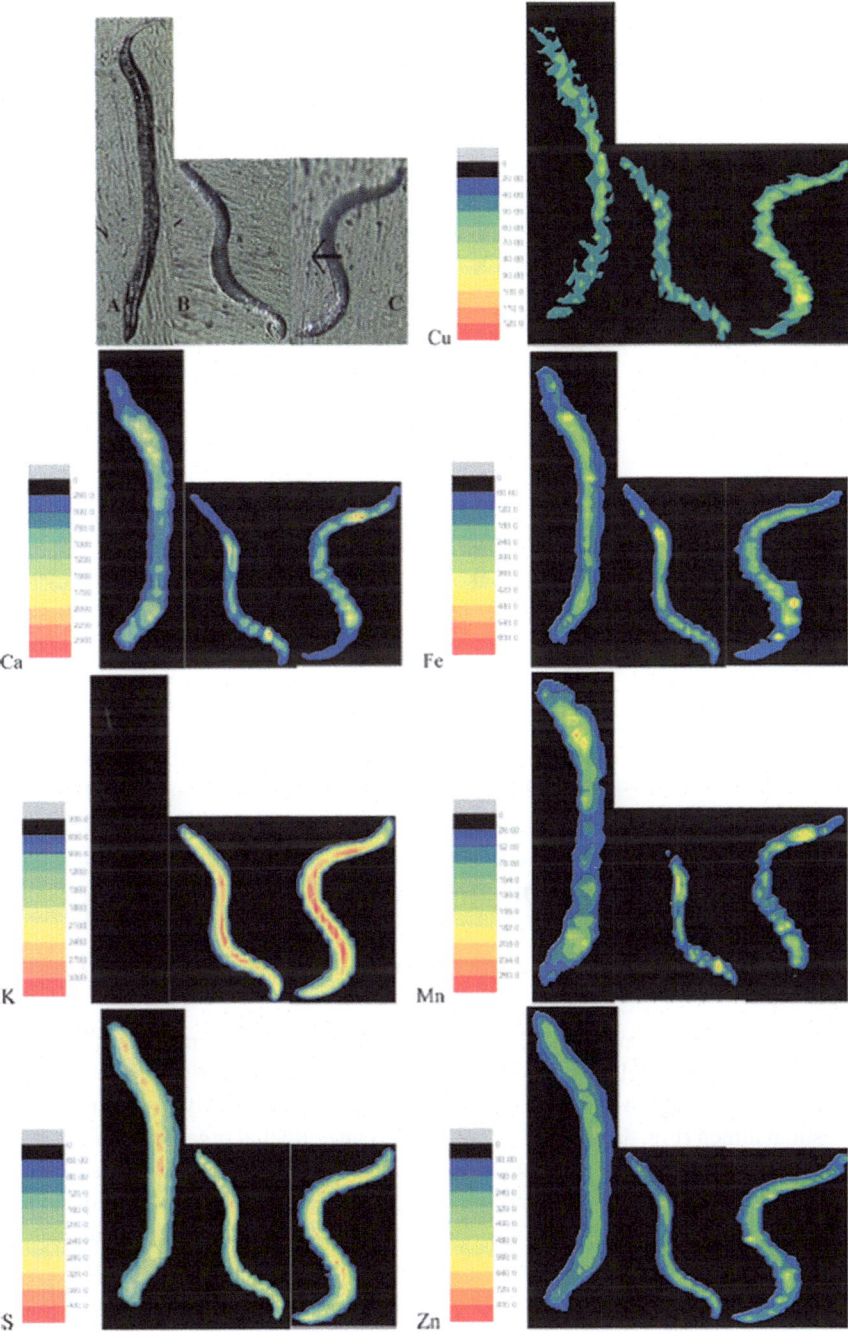

Fig. 8.8 Elemental distribution in the nematodes, which were fed by *E. coli* OP50 + PBS (*A*), *E. coli* OP50 + Cu-NPs (*B*), and *E. coli* OP50 + Cu²⁺ (*C*), respectively [17]

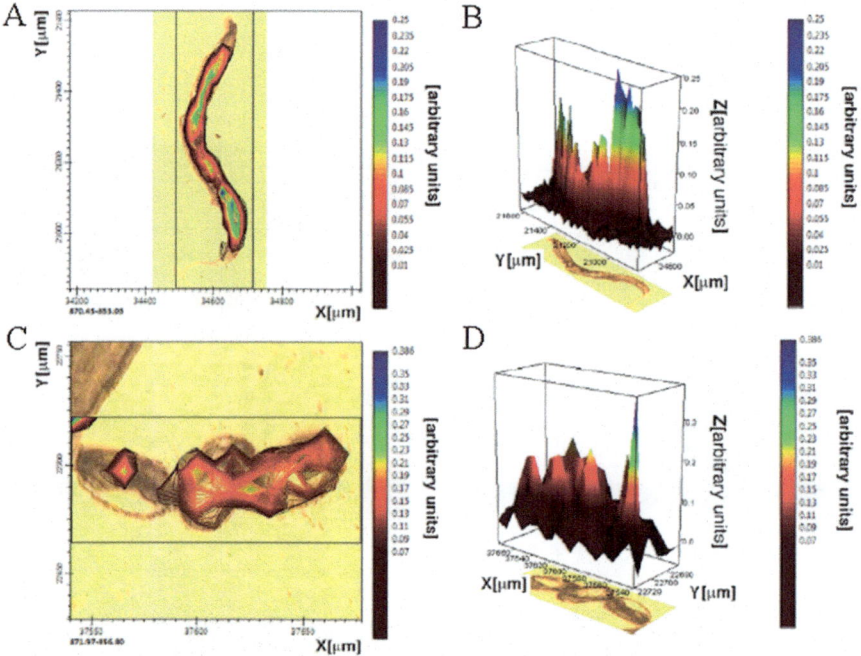

Fig. 8.9 Spectroscopic 2D and 3D images of the graphite nanoplatelets spatially distributed within the adult nematode and the embryos [18]
All spectra were collected in transmission mode with a 3 cm⁻¹ spectral resolution, and the images were obtained by monitoring the 865 cm⁻¹ peak; the absorption intensity was measured and expressed as arbitrary units

8.3 Primary Targeted Organ: Intestine

8.3.1 Location of ENMs in the Intestinal Cells in Nematodes

Using a transgenic strain (CL2120) with the GFP-labeled intestinal cells, the distribution and the accumulation of MPA-CdTe QDs in intestinal cells have been examined. After exposure to QDs (200 nm) for 12 h, QDs were mainly distributed in the intestinal lumen (Fig. 8.10) [12]. The pharynx is an epithelial organ, and the pharyngeal isthmus lies between the first bulb and the second bulb. QDs could attach to the inner surface of the pharyngeal tissues, and QDs could locate at the sidewalls of the pharyngeal isthmus (Fig. 8.10) [12]. The further colocalization suggests that some QDs may also enter into the adjacent cells except for embryos in nematodes (Fig. 8.10) [12]. Moreover, QDs could be excreted from the anus (Fig. 8.10) [12].

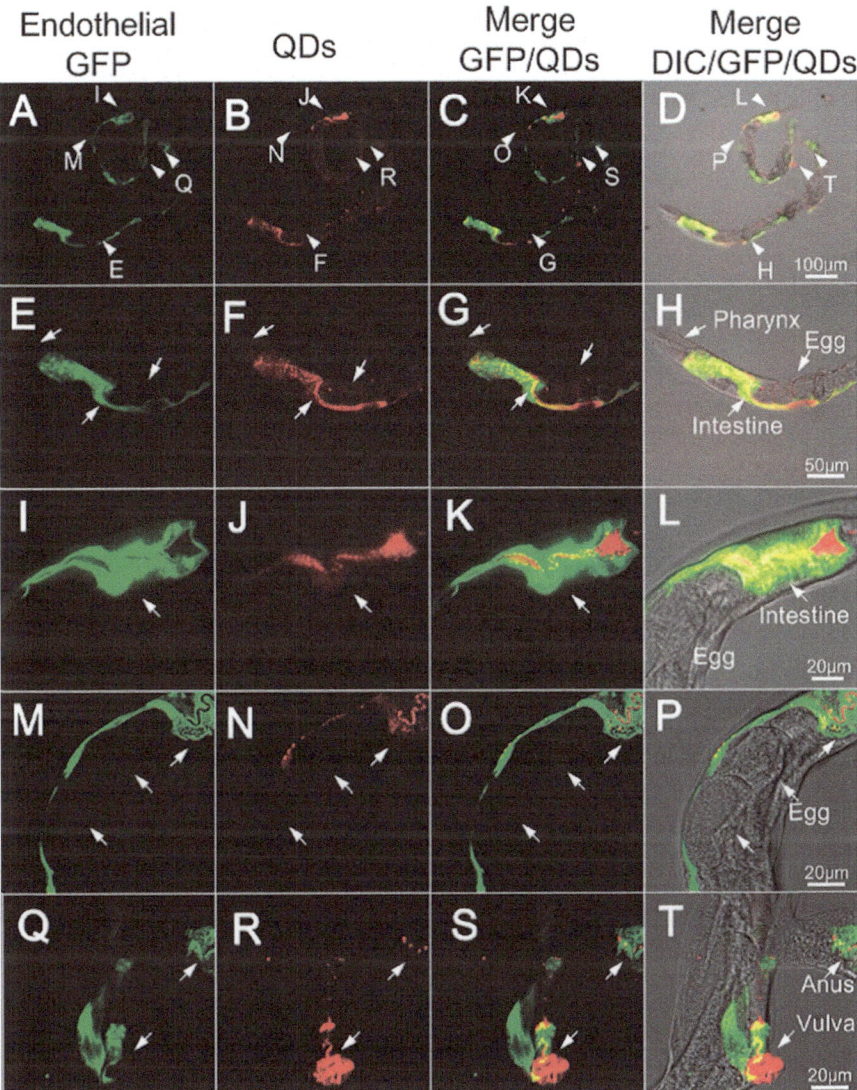

Fig. 8.10 QDs distribution in intestinal GFP-labeled nematodes [12]

(*A–D*) Images of whole-body distribution of MPA-CdTe QDs. (*E–T*) Enlarged views of different body parts. Fluorescent images of GFP-labeled intestinal endothelial cells (A, E, I, M, P) and QDs (B, F, J, N, R) are merged with DIC images subsequently (D, H, L, P, T) to show the exact QDs location in the body of nematodes. Arrows in (A–D) indicate the original positions of enlarged regions, and arrows in (E–T) indicate the colocalized positions

8.3.2   Subcellular Distribution of ENMs in the Intestinal Cells

The subcellular distribution of ENMs in the intestinal cells can be determined by TEM assay. With the G–COOH as an example, exposure to G–COOH was relatively safe for nematodes [7]. After prolonged exposure, although G–COOH could be translocated into the intestinal cells, G–COOH particles were mainly deposited in the small-vesicle structures adjacent to the intestinal microvilli, and only occasionally, the G–COOH particles were found in some large-vesicle structures in the cytosol of intestinal cells (Fig. 8.11) [7]. A moderate amount of G–COOH particles were also distributed in the cytosol in intestinal cells (Fig. 8.11) [7]. In contrast, no G–COOH particles were found within or adjacent to the mitochondria in intestinal cells, and no G–COOH particles were found to be distributed in the germ cells in gonads, the muscles, and the excretory canals [7]. Most of the G–COOH particles were deposited in the intestinal lumen (Fig. 8.11) [7].

Very different from the effects of G–COOH on nematodes, prolonged exposure to graphene oxide (GO) resulted in the severe toxicity on nematodes and GO accumulation in the body in nematodes [19]. After prolonged exposure, a large amount of GO could be translocated into the intestinal cells from the intestinal lumen based on the TEM assay (Fig. 8.12) [19]. In intestinal cells, GO particles were observed to be mainly distributed adjacent to or surrounding the mitochondria in the cytosol (Fig. 8.12) [19]. Meanwhile, the disrupted ultrastructure of microvilli and the loss of many microvilli on intestinal cells were observed in GO-exposed nematodes (Fig. 8.12) [19]. These observations imply the possible direct damage on both the development and the function of intestinal cells by GO exposure. Some of the related evidence is that GO (100 mg/L) induced the significant induction of reactive oxygen species (ROS) production and dysregulation of genes (*gas-1*, *sod-1*, *sod-2*, *sod-3*, *sod-4*, *isp-1*, and *clk-1*) required for the control of oxidative stress [19]. *sod-1* and *sod-4* encode copper/zinc superoxide dismutase, *sod-2* and *sod-3* encode manganese superoxide dismutase, *isp-1* encodes a "Rieske" iron-sulfur protein, *gas-1* encodes a subunit of mitochondrial complex I, and *clk-1* encodes a ubiquinone biosynthesis protein COQ7.

8.3.3   Distribution of ENMs in Certain Organelles in Intestinal Cells

Using some powerful transgenic strains, the biodistribution of G–COOH in specific organelles within the intestinal cells can be further examined in nematodes. In the transgenic strain of *zcIs18*, green fluorescent protein (GFP) is expressed in the cytosol of intestinal cells. With the G–COOH as an example, G–COOH particles were observed to be moderately distributed in the cytosol in intestinal cells in the anterior, mid-, and posterior regions of the *zcIs18* nematodes (Fig. 8.13) [7]. In the transgenic strain of *zcIs17*, GFP is expressed in the mitochondria in intestinal cells. G–COOH

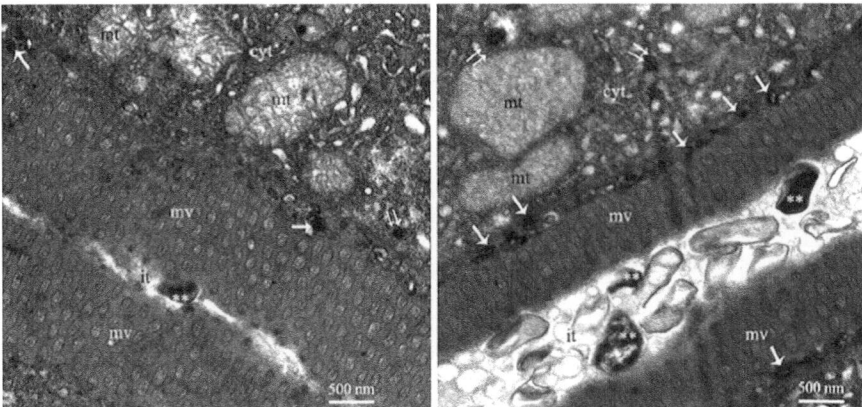

Fig. 8.11 Subcellular distribution of G–COOH in the intestinal cells in nematodes [7] *mt* mitochondria, *cyt* cytosol, *mv* microvilli, *it* intestine. Prolonged exposure was performed from L1-larvae to adult day-1. Exposure concentration of G–COOH was 100 mg/L. Single arrowheads indicate the distribution of G–COOH in the small-vesicle structures adjacent to the intestinal microvilli, double arrowheads indicate the distribution of G–COOH in the cytosol in intestinal cells, and double asterisks indicate the food particles in the intestinal lumen

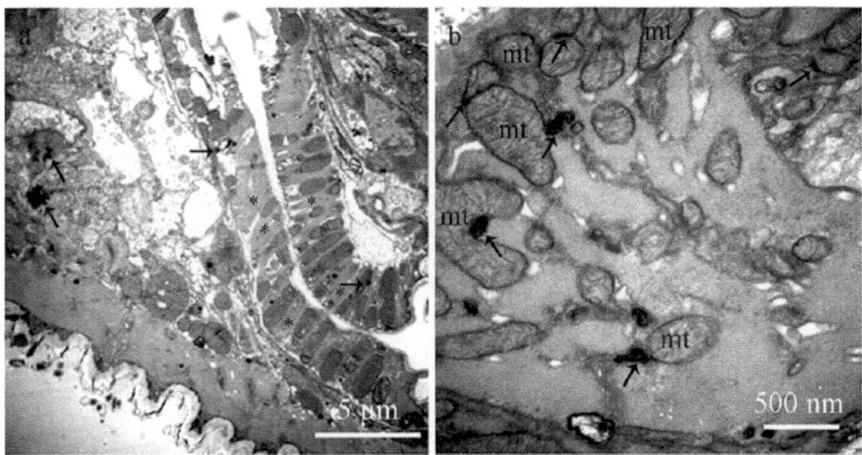

Fig. 8.12 Ultrastructural assay of GO distribution in intestinal cells [19] The left image shows the uptake of GO by intestinal cells in nematodes, and the right image shows the distribution of GO in intestinal cells. Asterisks indicate the position without microvilli, and arrowheads indicate the location of GO. *mt* mitochondria. Prolonged exposure to GO (100 mg/L) was performed from L1-larvae to adult day-1

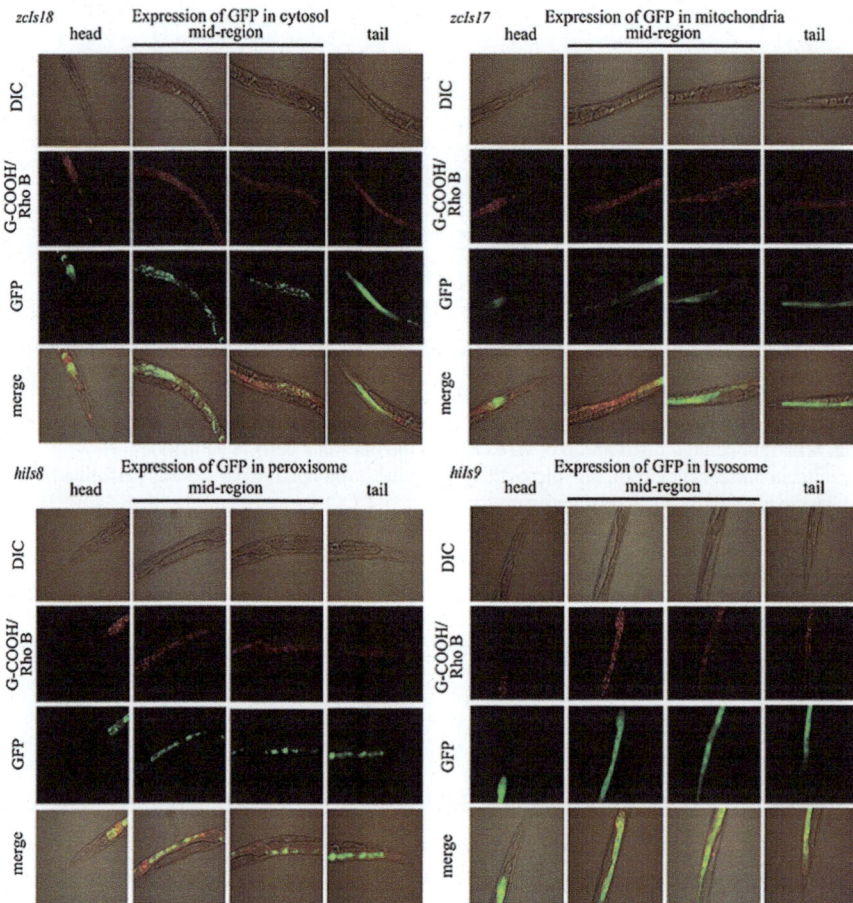

Fig. 8.13 Biodistribution of G–COOH in the cytosol, mitochondria, peroxisome, and lysosome of the intestinal cells in nematodes [7]
Prolonged exposure was performed from L1-larvae to adult day-1. Exposure concentration of G–COOH/Rho B was 100 mg/L

particles were found to be not obviously distributed in the mitochondria in intestinal cells of the *zcIs17* nematodes (Fig. 8.13) [7]. In the transgenic strain of *hiIs8*, GFP is expressed in the peroxisome in intestinal cells. A large amount of G–COOH particles were observed to be distributed in the peroxisome in intestinal cells in the anterior and mid-regions of the *hiIs8* nematodes (Fig. 8.13) [7]. In the transgenic strain of *hiIs9*, GFP is expressed in the lysosome in intestinal cells. A large amount of G–COOH particles were also found to be distributed in the lysosome in intestinal cells in the mid-region and posterior region of the *hiIs9* nematodes (Fig. 8.13) [7].

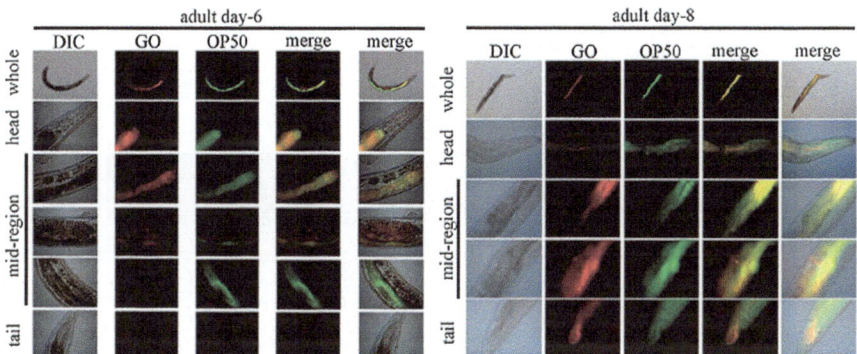

Fig. 8.14 Colocalization of GO with OP50 in GO (1 mg/L)-exposed nematodes at adult day-6 or day-8 [20]

8.3.4 Colocalization of GO with OP50 in Intestinal Lumen of ENM-Exposed Nematodes

In the intestinal lumen, with the aid of OP50::GFP, we can further determine the interaction between certain ENMs and OP50, the food for the nematodes. This can be reflected by the analysis on the colocalization of ENMs, such as GO/Rho B, with OP50::GFP in exposed nematodes [20]. In nematodes at the stage of adult day-8, an obvious colocalization of GO/Rho B with OP50::GFP in the intestinal lumen was observed in GO/Rho B-exposed nematodes, and the colocalization of GO/Rho B with OP50::GFP in the intestinal lumen could be detected at both the anterior region and the posterior region, as well as the region of the defecation structure in the tail, in nematodes (Fig. 8.14) [20]. In contrast, in nematodes at the stage of adult day-6, the colocalization of GO/Rho B with OP50::GFP in the intestinal lumen was mainly detected in the anterior region in nematodes, and seldom of the accumulation of both GO/Rho B and OP50::GFP was observed in the tail region of GO/Rho B-exposed nematodes (Fig. 8.14) [20].

8.4 Secondary Targeted Organ: Reproductive Organs

One of the important evidence for the distribution and translocation of ENMs in the reproductive organs in nematodes is from the observations on multiwalled carbon nanotubes (MWCNTs). After prolonged exposure (from L1-larvae to adult day-1) to MWCNTs/Rho B (10 μg/L or 1 mg/L), a great amount of MWCNTs/Rho B could be translocated and accumulated into the reproductive organs, such as the spermatheca, through the intestinal barrier, although most of the MWCNTs/Rho B were still located and accumulated in the pharynx and the intestine (Fig. 8.15) [21]. In nematodes, the MWCNTs/Rho B could also be translocated into the reproductive

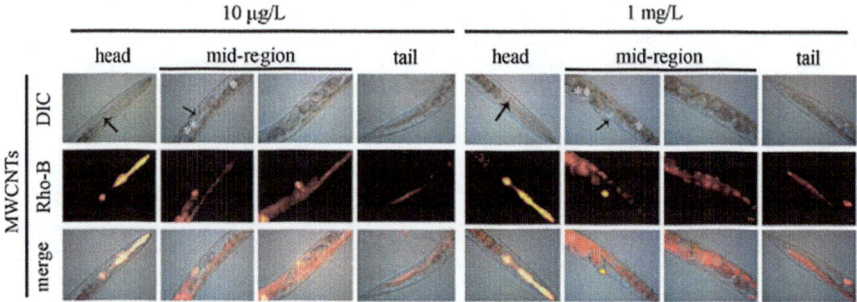

Fig. 8.15 Translocation and distribution of MWCNTs after exposure at different concentrations from L1-larvae to adult day-1 [21]

Arrowheads indicate the pharynx and the spermatheca, respectively, at the head region or mid-region of nematodes. Intestine (*) and embryos (**) in the mid-region are also indicated

organ of the gonad after prolonged exposure [22]. These results imply that the MWCNTs even at the environmentally relevant concentrations can be potentially translocated and accumulated in the secondary targeted organs, such as the reproductive organs, through primary intestinal barrier in nematodes after long-term exposure. More importantly, after prolonged exposure to MWCNTs/Rho B (10 µg/L or 1 mg/L), the MWCNTs/Rho B could be even detected in the embryos in the body of nematodes (Fig. 8.15) [21], which further implies the potential formation of transgenerational toxicity of MWCNTs in nematodes after long-term exposure.

Another line of evidence for the distribution and effect of ENMs in reproductive organs is from the analysis on the distribution and translocation of MPA-CdTe QDs. The evidence first raised is that the QDs could be translocated through the intestinal barrier to the reproductive system, such as the uterus and the vulva, after long-term exposure (Fig. 8.16) [12]. This implies that the embryos may be damaged in such an environment full of the QDs, since the embryos normally pass from the gonad to the uterus and further mature before being released via the vulva after fertilization. Meanwhile, after long-term exposure, difficulty in egg laying, damaged egg left in the vulva, and damaged egg without an intact egg shell were observed in QD-exposed nematodes (Fig. 8.16) [12]. In contrast, no obvious QDs accumulation was observed in vulva, and no deficits in both the self-fertilization and the male mating were detected in QD-exposed nematodes after short-term exposure (Fig. 8.16) [12]. The possible underlying cellular mechanism for the adverse effect of QDs on egg-laying behavior has further been raised. Once the QDs were transferred from the intestine to the gonad and vulva after chronic exposure, the accumulated QDs in the reproductive system will cause the damage on the embryos and lead to the difficulty in egg-laying process [12].

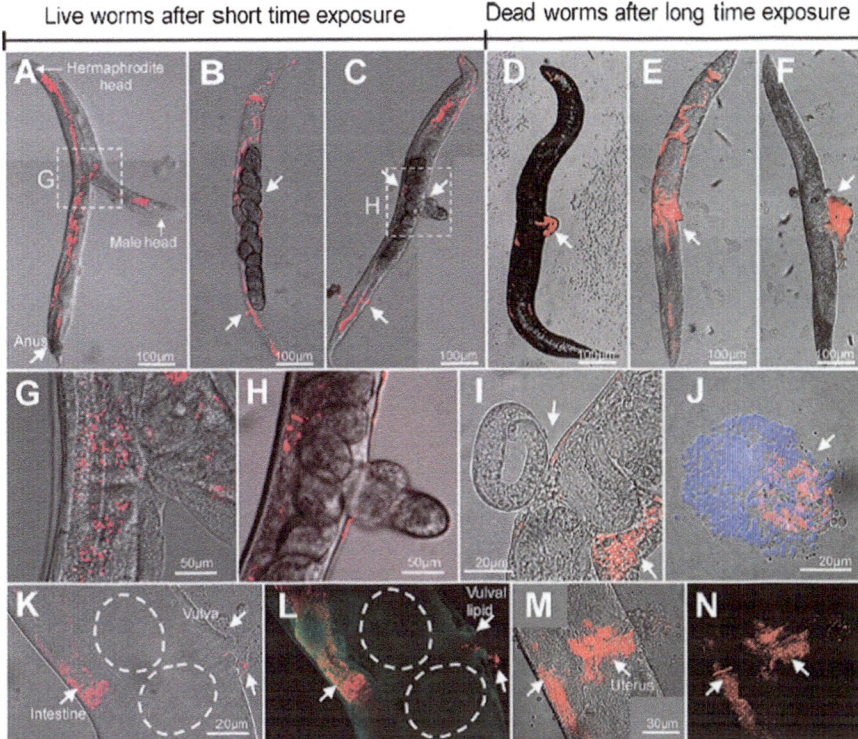

Fig. 8.16 Reproductive behavior and egg-laying difficulty after long-term QDs exposure [12]
(**a**) Mating behavior of hermaphrodite and male nematodes. Boxed region indicates the vulva that
is enlarged as shown in (*G*). (**b–c**) QDs distribution in live nematodes after short-term exposure.
Boxed region in (**c**) indicates the egg-laying process that is enlarged as shown in (*H*). (**d–f**) QDs
distribution in fresh dead worms after long-term exposure (16 days). (**i**) QDs confined in digestive
tract without entering into embryos after short-term exposure. (**j**) Damaged egg after long-time
exposure. (**k** and **m**) Fluorescence/DIC merging images that are focused on the vulva and uterus
features, respectively. (**l** and **n**) Corresponding fluorescence images

8.5 Secondary Targeted Organ: Neurons

So far, only limited evidence has been raised to prove the distribution and translocation of ENMs in the secondary targeted organs of neurons in nematodes. One evidence is from the QDs, and another evidence is from the GO.

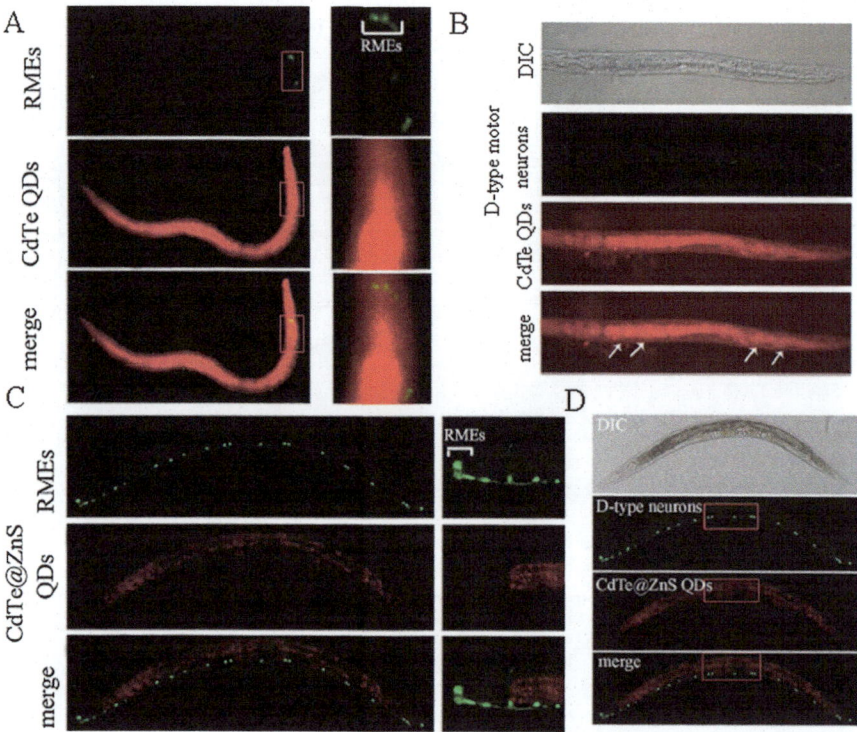

Fig. 8.17 Distribution and translocation of CdTe QDs in nematodes [23, 24]
(**a**) Colocalization of CdTe QDs with RMEs motor neurons. (**b**) Colocalization of CdTe QDs with
D-type motor neurons. (**c**) No colocalization of CdTe@ZnS QDs with RMEs motor neurons was
found in nematodes. (**d**) No colocalization of CdTe@ZnS QDs with D-type motor neurons was
found in nematodes. Prolonged exposure to CdTe QDs or CdTe@ZnS QDs (1 µg/L) was exposed
from L1-larvae to young adult

8.5.1   Evidence from the QDs

Among 26 GABAergic neurons in nematodes, the RMEs neurons (RMED, RMEV,
RMEL, and RMER) in the head region regulate the foraging behavior. After prolonged exposure from L1-larvae to young adult, the CdTe QDs could be distributed
and translocated into the RMEs motor neurons through the intestinal barrier
(Fig. 8.17) [23]. Meanwhile, prolonged exposure to CdTe QDs caused the damage
on the development of RMEs neurons as reflected by the significant decrease in
both the fluorescent intensity and the fluorescent size of cell bodies of RMEs motor
neurons [23]. Prolonged exposure to CdTe QDs also resulted in the formation of
abnormal foraging behavior [23].

Among the GABAergic neurons in nematodes, the ventral cord D-type motor
neurons (6 DD and 13 VD motor neurons) function in inhibiting the contraction of

ventral and dorsal body wall muscles during the locomotion. Once the nematodes lack the functional D-type motor neurons, they would display a phenotype of shrinking behavior. Moreover, after prolonged exposure from L1-larvae to young adult, the CdTe QDs could be distributed and translocated into the D-type motor neurons through the intestinal barrier (Fig. 8.17) [24]. Prolonged exposure to CdTe QDs could also result in the toxicity on the development of D-type motor neurons as reflected by the decrease in the fluorescent intensity of cell bodies, the neuronal loss, and the formation of gaps on both ventral and dorsal cords in D-type motor neurons [24]. Prolonged exposure to CdTe QDs further could cause the formation of shrinking behavior [24]. The bioavailability provides an important cellular basis for the observed neurotoxicity of CdTe QDs on both the development and the functions of RMEs neurons or D-type motor neurons in nematodes.

8.5.2   Evidence from the GO

Another important evidence is from the observation on the distribution and translocation of GO. In nematodes, the defecation behavior is regulated by both the AVL neurons in the head and the DVB neurons in the tail. After chronic exposure (from L1-larvae to adult day-8) to GO (1 mg/L), GO could significantly prolong the mean defecation cycle length and reduce the relative fluorescent puncta sizes of cell body of both AVL and DVB neurons [20].

Meanwhile, the obvious colocalization of GO/Rho B with both the AVL and the DVB neurons could be detected in nematodes after chronic exposure [20], which is helpful for the explanation on the observed damage on both the development and the function of AVL and DVB neurons in GO-exposed nematodes.

8.6   Cellular Metabolism of ENMs in Nematodes

Graphene quantum dots (GQDs), the miniaturized versions of GO sheets, normally have strong two-photon-induced fluorescence; nitrogen-doped GQDs (N-GQDs), one of the member of GQDs, are of potential use in cellular and deep-tissue imaging. Using N-GQDs as an example, the cellular mechanism of ENMs was examined in nematodes. The evidence has been raised that the N-GQDs are relatively safe for nematodes [25]. To observe the dynamic metabolism of N-GQDs, the exposed nematodes were transferred to the normal NGM plants without addition of N-GQDs. At least three stages of the cellular metabolism of N-GQDs were observed in the nematodes. Initially, the N-GQDs were strongly distributed in the pharynx, the intestine,

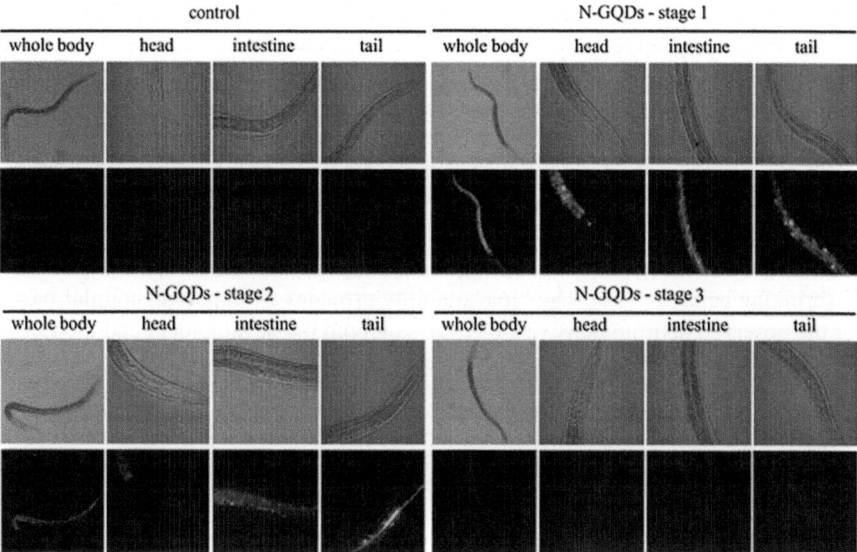

Fig. 8.18 Dynamic metabolism of N-GQDs in nematodes [25]

and the tail regions where the GQDs would be excreted out of the body after the feeding (Fig. 8.18) [25]. At the second stage, the N-GQDs distribution in the pharynx was gradually decreased, and N-GQDs were mainly distributed and accumulated in the intestine and the tail region in nematodes (Fig. 8.18) [25]. At the third stage, no strong N-GQD fluorescent signals would be observed in the nematodes (Fig. 8.18) [25].

8.7   Transgenerational Translocation of Nanomaterials

8.7.1   Formation of Transgenerational Translocation of ENMs in Nematodes

One of the evidence for the formation of transgenerational translocation of ENMs in nematodes is from the study on Ag-NPs. Using a dark-field microscopy combined with a CytoViva VNIR hyperspectral imaging system, the Ag-NPs could not only be accumulated in the body of exposed nematodes but also observed to be translocated into the fertilized embryos (Fig. 8.2) [8].

Another important evidence for the formation of transgenerational translocation of ENMs in nematodes is from the study on QDs. After exposure, the severe distribution and accumulation of CdTe QDs in the pharynx and the intestine could

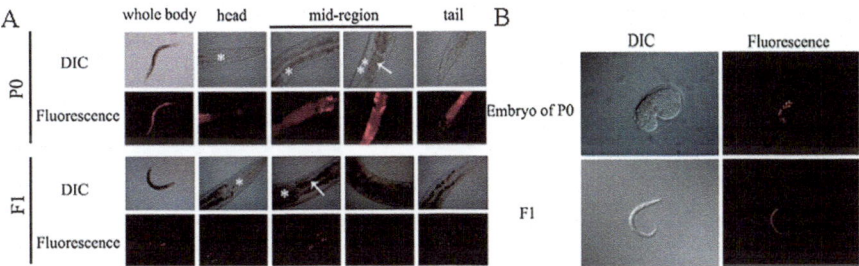

Fig. 8.19 Distribution and translocation of CdTe QDs in exposed nematodes and their progeny [26]

(**a**) CdTe QDs distribution in exposed nematodes and their progeny. Arrowheads indicate the intestine. Pharynx (*) in the head, embryos (*) in the mid-region, and gonad (**) in the mid-region were also indicated. (**b**) CdTe QDs distribution in embryos of exposed nematodes and the animals developed from these embryos. Exposure concentration of CdTe QDs was 20 mg/L. *P0* parents, *F1* the filial generation

be observed in exposed nematodes (Fig. 8.19) [26]. Moreover, in CdTe QDs exposed nematodes, the distribution and accumulation of CdTe QDs in the gonad and the region surrounding the pharynx where most of the neurons are located were also detected (Fig. 8.19) [26], implying the potential translocation of CdTe QDs into the secondary targeted organs through the intestinal barrier. More importantly, the CdTe QDs could be further detected in the embryos of the exposed nematodes and the body of F1 generational nematodes (Fig. 8.19) [26], which suggests the formation of transgenerational translocation of QDs in nematodes.

8.7.2 Molecular Control of the Transgenerational Translocation of ENMs in Nematodes

In nematodes, mutation of *clk-1*, *isp-1*, or *daf-2* could induce a resistance to the toxicity of CdTe QDs in nematodes [26]. *clk-1* encodes a demethoxyubiquinone hydroxylase required for the biosynthesis of ubiquinone, *isp-1* encodes a subunit of mitochondrial complex III, and *daf-2* encodes an insulin receptor in the insulin signaling pathway. Transgenic strain VP303 is a tool for specific intestinal RNA interference (RNAi) of certain genes in nematodes. CdTe QDs distribution in VP303 strain (RNAi control) was similar to that in wild-type nematodes (Fig. 8.20) [26]. Moreover, using this genetic tool, it has been demonstrated that intestine-specific RNAi of *clk-1*, *isp-1*, or *daf-2* only induced the moderate accumulation of CdTe QDs in intestinal lumen of exposed nematodes (Fig. 8.20) [26], implying the effective inhibition of CdTe QDs translocation into secondary targeted organs through the intestinal barrier in CdTe QDs exposed nematodes with intestine-specific RNAi

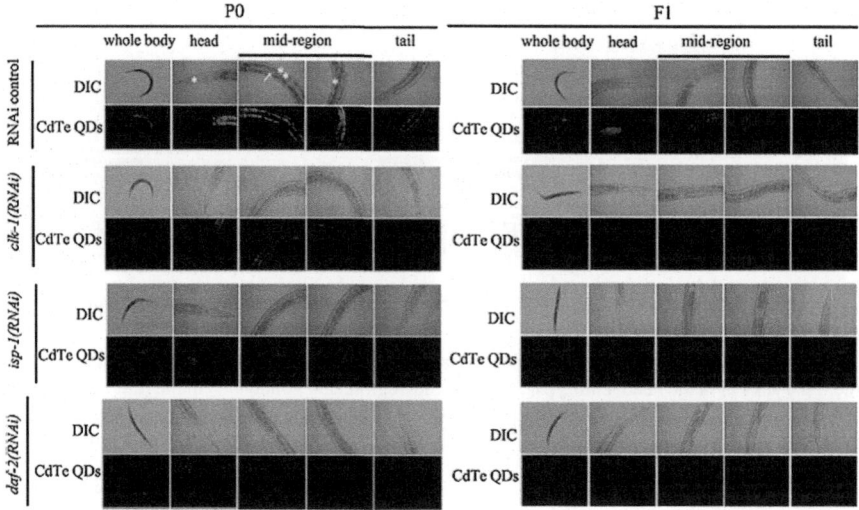

Fig. 8.20 Effects of intestine-specific RNAi of *clk-1*, *isp-1*, or *daf-2* on the distribution and the translocation of CdTe QDs in nematodes [26]
Arrowheads indicate the intestine. Pharynx (*) in the head, embryos (*) in the mid-region, and gonad (**) in the mid-region were also indicated. RNAi control, VP303. Exposure concentration of CdTe QDs was 20 mg/L. *P0* parents, *F1* the filial generation. *QDs* quantum dots

of *clk-1*, *isp-1*, or *daf-2*. Meanwhile, in the progeny of nematodes with intestine-specific RNAi of *clk-1*, *isp-1*, or *daf-2*, no fluorescent signals of CdTe QDs could be detected in the body of nematodes (Fig. 8.20) [26]. Therefore, intestine-specific RNAi of *clk-1*, *isp-1*, or *daf-2* potentially effectively prevents the translocation of CdTe QDs from exposed nematodes to their progeny by maintaining the normal functional state of intestinal barrier.

8.8   Crucial Role of Intestinal Barrier Against the Toxicity of ENMs in Nematodes

8.8.1   Confirmation of Crucial Role of Intestinal Barrier Against the Toxicity of ENMs in Nematodes

In order to confirm the crucial role of intestinal barrier against the toxicity of ENMs, *sod-2* encoding a Mn-SOD protein was expressed specially in the intestinal cells to generate a transgenic strain with the intestinal overexpression of SOD-2. In nematodes, the SOD-2 protein has the key role in protecting the animals from the oxidative stress. Under the normal conditions, intestinal overexpression of SOD-2 did not induce the significant ROS, affect the brood size and the locomotion behavior, and

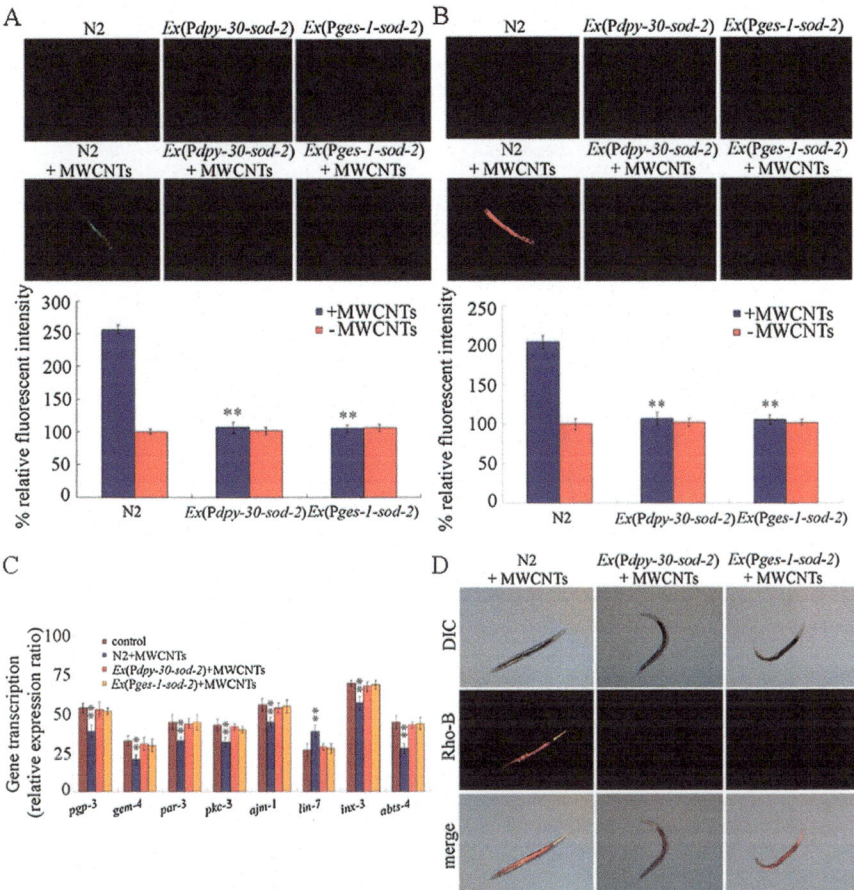

Fig. 8.21 Overexpression of SOD-2 in the intestine suppresses translocation of MWCNTs [22] (**a**) Overexpression of SOD-2 in the intestine prevented the induction of ROS production in the intestines. (**b**) Overexpression of SOD-2 in the intestine inhibited the enhancement of intestinal permeability as indicated by the Nile red intake in MWCNT-exposed nematodes. (**c**) Comparison of expression pattern for genes required for intestinal development between control and MWCNT-exposed nematodes. (**d**) Overexpression of SOD-2 in the intestine suppressed the translocation of MWCNTs into the body of nematodes. Exposure concentration for MWCNTs was 1 mg/L. Bars represent mean ± SEM. **$P < 0.01$

alter the intestinal permeability (Fig. 8.21) [22]. At least three lines evidence has been raised to confirm the crucial role of intestinal barrier against the toxicity of MWCNTs in nematodes. With the aid of this transgenic strain, it has been suggested that intestinal overexpression of SOD-2 could significantly reduce the oxidative stress induced by exposure to MWCNTs (1 mg/L) (Fig. 8.21) [22]. Additionally, intestinal overexpression of SOD-2 could suppress the enhancement of intestinal permeability and recover the dysregulated expressions of genes required for the

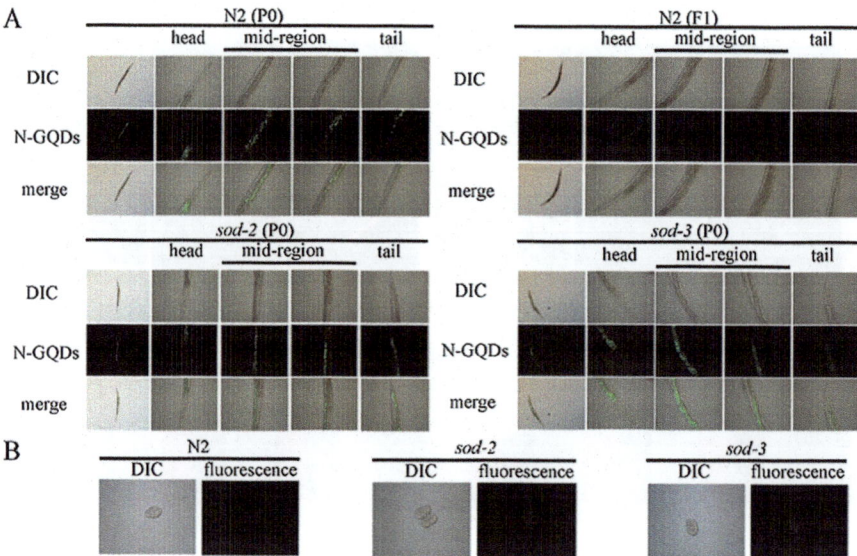

Fig. 8.22 Distribution and translocation of N-GQDs in nematodes [25]
(**a**) Distribution of N-GQDs in exposed nematodes and their progeny. (**b**) Distribution of N-GQDs in embryos of exposed wild-type, *sod-2*, and *sod-3* mutant nematodes. Prolonged exposure to N-GQDs (100 mg/L) was performed from L1-larvae to adult day-1

control of intestinal development induced by exposure to MWCNTs (1 mg/L) (Fig. 8.21) [22]. Moreover, the MWCNTs translocation into the secondary targeted organs through the intestinal barrier was noticeably inhibited in nematodes with the intestinal overexpression of SOD-2 (Fig. 8.21) [22].

8.8.2   Mutation of sod-2 or sod-3 Did Not Affect the Pattern of Distribution of N-GQDs in Nematodes

In wild-type nematodes, N-GQDs were mainly distributed in the pharynx and the intestine after prolonged exposure (Fig. 8.22) [25]. In nematodes, mutation of *sod-2* or *sod-3* encoding Mn-SOD protein can induce a susceptibility to toxicity of environmental toxicants [16, 27]. Similarly, no translocation of N-GQDs into the neurons and the reproductive organs through the biological barrier of the intestine was observed in *sod-2* or *sod-3* mutant nematodes (Fig. 8.22) [25]. In *sod-2* or *sod-3* mutant nematodes, the intestinal permeability is normal, which may be crucial for blocking the N-GQDs translocation into the neurons and the reproductive organs and preventing the induction of N-GQDs toxicity in nematodes.

8.9 Chemical Control of Translocation of ENMs in Nematodes

8.9.1 ZnS Surface Coating Blocks the Translocation of QDs Translocation into Secondary Targeted Organs Through the Intestinal Barrier

ZnS surface coating is an important chemical strategy to modify the physicochemical property and to reduce the toxicity of QDs. As introduced above, the CdTe QDs could be potentially translocated into the RMEs neurons and the D-type motor neurons through the intestinal barrier after prolonged exposure in nematodes [23, 24]. In contrast, it has been demonstrated that ZnS surface coating could effectively block the accumulation and the translocation of CdTe QDs into the RMEs neurons and the D-type motor neurons through the intestinal barrier (Fig. 8.17) [23, 24]. Meanwhile, ZnS surface coating significantly inhibited the toxicity of CdTe QDs on both the development and the functions of RMEs neurons and D-type motor neurons in nematodes [23, 24]. Therefore, the reduction of CdTe QDs neurotoxicity on RMEs neurons and D-type motor neurons by ZnS surface coating may be due to both the blockage of translocation into RMEs neurons and D-type motor neurons and the inhibition in activation of enough damage on intestinal barrier in nematodes.

8.9.2 PEG Surface Modification Blocks the Translocation of MWCNTs or GO Translocation into Secondary Targeted Organs Through the Intestinal Barrier

In nematodes, prolonged exposure (from L1-larvae to adult day-1) to MWCNTs could cause the severe damage on the functions of both primary targeted organs, such as the intestine, and the secondary targeted organs, such as the reproductive organs [22]. Meanwhile, after prolonged exposure, a great amount of MWCNTs could be translocated into the reproductive organs, such as the spermatheca and the gonad, through the intestinal barrier (Fig. 8.23) [22]. In contrast, PEG surface modification effectively prevented the translocation of MWCNTs into the secondary targeted organs, such as the reproductive organs of spermatheca and gonad through the intestinal barrier (Fig. 8.23) [22]. Only a limited amount of MWCNTs-PEG was detected in the pharynx and the intestine in nematodes after prolonged exposure (Fig. 8.23) [22]. Additionally, the amount of MWCNTs-PEG in the intestine of tail region was nearly undetectable (Fig. 8.23) [22].

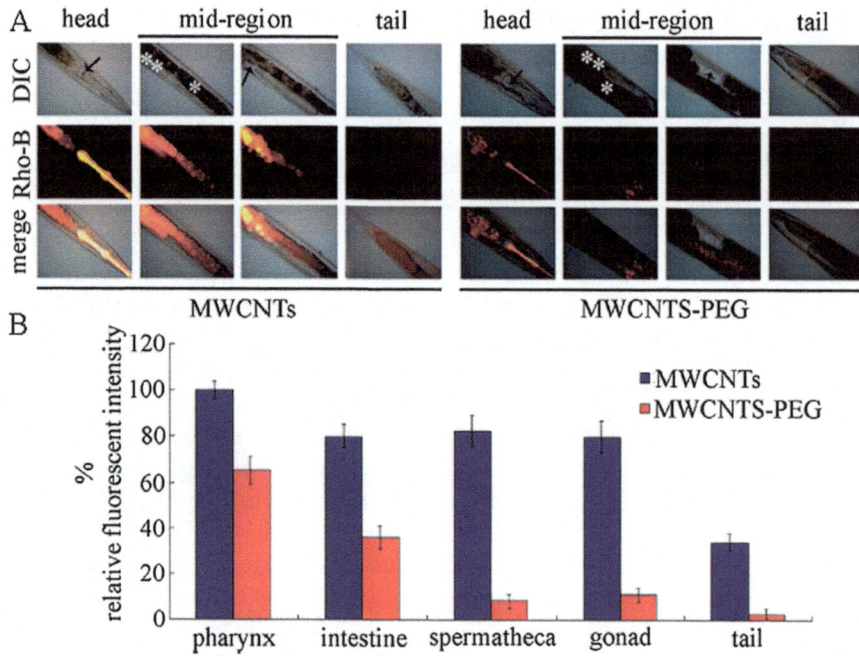

Fig. 8.23 Comparison of translocation and distribution of MWCNTs and MWCNTs-PEG after exposure from L1-larvae to adult [22]

(a) Comparison of translocation and distribution of MWCNTs and MWCNTs-PEG in different regions and organs of nematodes. Arrowheads indicate the pharynx and spermathcca, respectively, in the head region or mid-region of nematodes. The intestine (*) and gonad (**) in the mid-region are also indicated. (b) Comparison of the fluorescent intensities of MWCNTs and MWCNTs-PEG in different regions and organs of nematodes. Exposure concentration for MWCNTs and MWCNTs-PEG was 1 mg/L. Bars represent mean ± SEM

In nematodes, chronic exposure to GO also resulted in the toxicity on the functions of both primary targeted organs, such as the intestine, and the secondary targeted organs, such as the neurons [20]. After chronic exposure, GO could prolong the mean defecation cycle length and adversely affect the development of AVL and DVB neurons controlling the defecation behavior [20]. Chronic exposure to GO could cause the distribution and accumulation of GO into both the AVL and the DVB neurons in nematodes [20]. In contrast, PEG surface modification could significantly reduce the GO toxicity on the functions of both primary targeted organs and the secondary targeted organs [20]. Additionally, PEG surface modification could suppress the increase in mean defecation cycle length induced by GO exposure and inhibit the damage of GO exposure on the development of AVL and DVB neurons [20]. More importantly, it has been observed that the PEG surface modification could prevent the translocation and accumulation of GO in both the AVL and the DVB neurons [20].

8.10 Molecular Signals Regulating the Distribution and Translocation of ENMs

8.10.1 Identification of Molecular Signals Required for Control of ENMs Translocation

In nematodes, a genetic screen with the aim to identify the candidate molecular signals required for the control of toxicity and translocation of GO was performed among 20 *C. elegans* strains with mutations of genes required for stress response or oxidative stress [28]. After prolonged exposure to GO/Rho B (100 mg/L), it has been found that mutation of *hsp-16.48*, *gas-1*, *sod-2*, *sod-3*, *aak-2*, *clk-1*, or *isp-1* noticeably altered the distribution and translocation of GO/Rho B (Fig. 8.24) [28]. Mutation of *hsp-16.48*, *gas-1*, *sod-2*, *sod-3*, or *aak-2* could significantly increase the distribution of GO/Rho B in the intestine and the pharynx and enhance the translocation of GO/Rho B in the spermatheca in nematodes (Fig. 8.24) [28]. In contrast, mutation of *clk-1* or *isp-1* could significantly decrease the distribution of GO/Rho B in the intestine and the pharynx and inhibit the translocation of GO/Rho B in the spermatheca in nematodes (Fig. 8.24) [28]. In *C. elegans*, *hsp-16.48* encodes a heat shock protein, *gas-1* encodes a subunit of mitochondrial complex I, *sod-2* and *sod-3* encode manganese superoxide dismutases (Mn-SODs), *aak-2* encodes a catalytic alpha subunit of AMP-activated protein kinase, *clk-1* encodes a ubiquinone biosynthesis protein COQ7, and *isp-1* encodes a "Rieske" iron-sulfur protein. These identified molecular signals provide an important basis to further deeply elucidate the underlying molecular mechanisms for the translocation of ENMs in organisms.

8.10.2 Molecular Control of GO Translocation by Wnt Signaling

In nematodes, three Wnt ligands, CWN-1, LIN-44, and CWN-2, have been identified to be required for the regulation of GO toxicity in nematodes [29]. Mutation of *cwn-1* or *lin-44* induced a resistance to the GO toxicity on the functions of both primary and secondary targeted organs, whereas mutation of *cwn-2* induced a susceptibility to GO toxicity on the functions of both primary and secondary targeted organs [29]. Meanwhile, mutation of *cwn-2* significantly increased the distribution of GO/Rho B in the intestine and the pharynx and enhanced the translocation of GO/Rho B in the spermatheca in nematodes; however, mutation of *cwn-1* or *lin-44* significantly decreased the distribution of GO/Rho B in the intestine and the pharynx and inhibited the translocation of GO/Rho B in the spermatheca in nematode (Fig. 8.25) [29]. In nematodes, genetic interaction has further indicated that mutation of *cwn-1* or *lin-44* could suppress the susceptibility of *cwn-2* mutant nematodes to GO toxicity [29], implying that CWN-1 and LIN-44 may act downstream of CWN-2 in the regulation of toxicity and translocation of GO in nematodes.

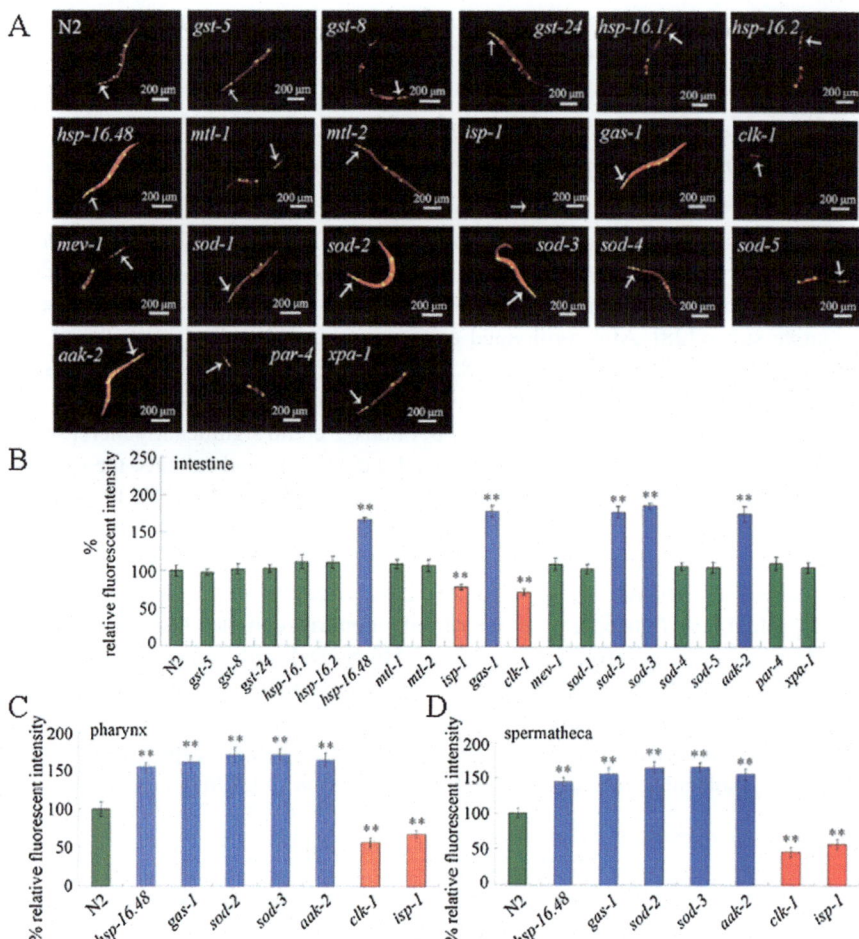

Fig. 8.24 Distributions of GO/Rho B in wild-type and mutant nematodes [28]
(a) Pictures showing the distributions of GO/Rho B in wild-type and mutant nematodes. (b) Comparison of relative fluorescence of GO/Rho B in the intestine between wild-type and mutant nematodes. (c) Comparison of relative fluorescence of GO/Rho B in the pharynx between wild-type and mutant nematodes. (d) Comparison of relative fluorescence of GO/Rho B in the spermatheca between wild-type and mutant nematodes. Arrowheads indicate the pharynx. The intestine (**) and the spermatheca (*) are also indicated. GO exposure was performed from L1-larvae to young adult. Exposure concentration of GO was 100 mg/L. Bars represent means ± SEM. $**P < 0.01$ vs. N2

8.10.3 Molecular Control of Ag-NPs Translocation by Intracellular Trafficking Signaling

To determine the possible role of endocytosis and lysosomal function in uptake and subsequent toxicity of Ag-NPs, the sensitivity of three endocytosis-deficient mutants

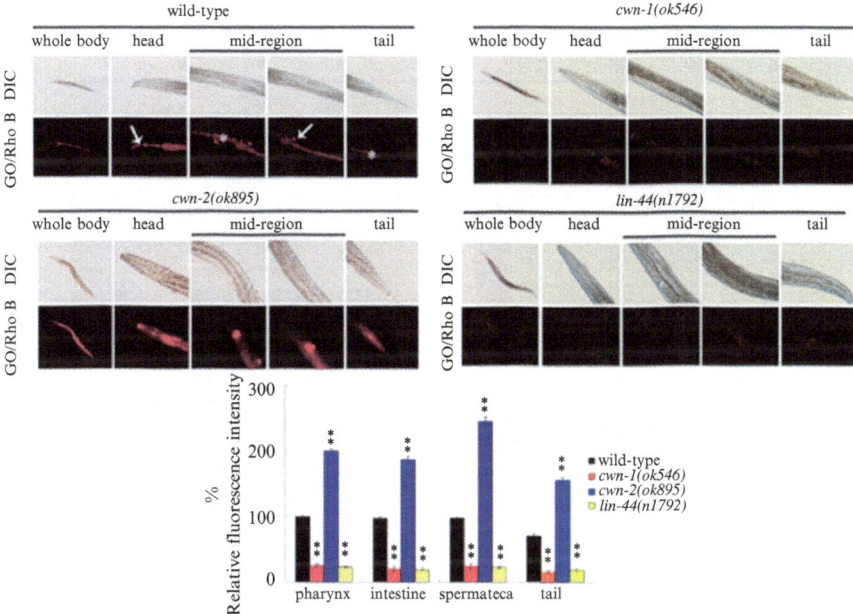

Fig. 8.25 Distribution and translocation of GO in the body of wild-type, *cwn-1*, *cwn-2*, and *lin-44* mutant nematodes [29]
Arrowheads indicate pharynx in the head and spermatheca in the mid-region, respectively. Asterisks indicate intestine in the mid-region and tail, respectively. GO/Rho B exposure concentration was 100 mg/L. Bars represent means ± SD. **P < 0.01 vs wild type

(*rme-1*, *rme-6*, and *rme-8*) and two lysosomal function deficient mutants (*cup-5* and *glo-1*) to toxicity and translocation of Ag-NPs was analyzed in nematodes. Among these examined mutants, the lysosome and lysosome-related organelle mutants showed the sensitivity to the growth-inhibiting effects of CIT-Ag-NPs [30]. In contrast, one of the endocytosis-deficient mutants (*rme-6*) took up less silver and exhibited the resistance to toxicity of CIT-Ag-NPs [30]. Besides these, treatment of clathrin-mediated endocytosis inhibitor (chlorpromazine) could also reduce the toxicity of CIT-Ag-NPs [30].

8.11   Chemical Metabolism and Degradation of ENMs in the Body of Animals

In the recent years, some important techniques provide important basis for the determination of the in situ metabolism and degradation of ENMs in the alimentary system animals. In 2011, an experimental example on this aspect has been raised on in vivo QDs behavior in nematodes [12]. One of such techniques is the microbeam synchrotron radiation X-ray fluorescence (μ-SRXRF) technique, which can provide

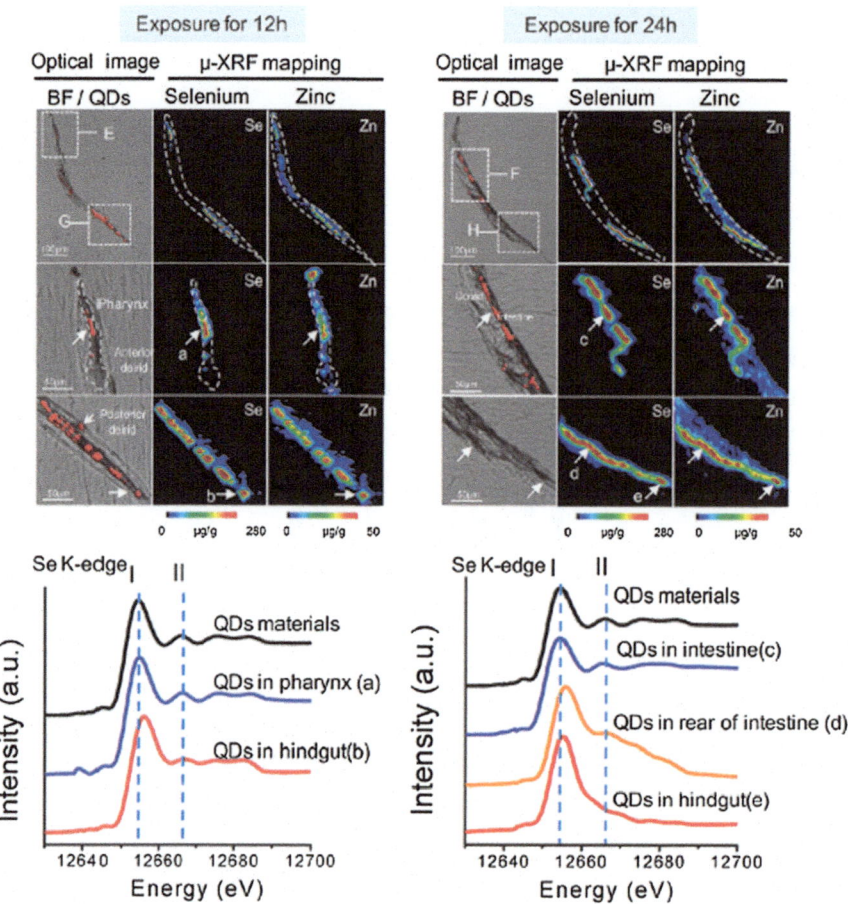

Fig. 8.26 In situ elemental analysis on metabolism and degradation of QDs in nematodes [12] Microbeam X-ray fluorescence (μ-XRF) mappings of Se and Zn elements in QDs. Quantity of each individual element is plotted on the respective scale below. The dashed lines delimit the boundary of the worm body. The large white dashed lines delimit boundary of the pharynx. Arrows indicate the beam positions of μ-XANES spectra

precise information on the subcellular spatial resolution and the simultaneous distributions of various elements with the high sensitivity [12].

Moreover, the technique of μ-XANES can allow the analysis of the physicochemical changes of chemical species in vivo. The technique of μ-SRXRF can be further combined with the technique of μ-XANES with the aim to investigate the metabolic process of nanomaterials in the physiological microenvironment. This combinational use of μ-SRXRF and μ-XANES was used to determine the metabolism of QDs within the digestive tract in nematodes [12]. It has been found that, after exposure for 12 h, the fluorescence images and the elemental distribution of QDs matched perfectly (Fig. 8.26) [12]. In contrast, after exposure for 24 h, no fluorescence signals (fluorescent quenching) were detected at the retral part of the intestine, where Se and Zn elements are largely present (Fig. 8.26) [12]. The Se K-edge XANES of CdSe QDs are mainly used to monitor the unoccupied p states of the Se element. In nematodes, all the XANES spectra are similar to the original QDs with only slight differences at the positions where XRF mapping matches well with fluorescence imaging (Fig. 8.26) [12]. However, the XANES of QDs after digestion where QDs fluorescence quenched exhibited obvious differences, such as (1) intensity of peak I increases strongly, (2) position of peak I shifts to higher energy by about 2 eV, and (3) intensity of shoulder II gradually decreases and disappears (Fig. 8.26) [12]. The peak I positions were quite close to that of the characteristic peak of Na_2SeO_3, suggesting that the Se^{2-} in the CdSe core may be oxidized to the $SeO3^{2-}$ [12]. Meanwhile, the changes of Se XANES spectra at the retral part of the intestine may be attributed to the collapse of core/sell structure after digestion [12]. That is, the protection structure of the CdSe core by ZnS shell may be destroyed, and such an oxidation of CdSe core gives rise to dramatic decrease of QDs fluorescence [12]. These results highlight the possibility of partial degradation and fluorescence quenching of QDs. Furthermore, QDs exposure led to alterations in the distribution patterns of other physiologically important elements, including Ca and K [12]. The obtained data in nematodes have demonstrated that the synchrotron radiation related imaging techniques may be a more accurate way to determine and understand the QDs fate in the organisms. An underlying chemical mechanism for the observed metabolism of QDs has been raised in nematodes [12]. Under the normal conditions, the surface coating may effectively reduce the oxidation of CdSe cores in the QDs. However, due to the oxidation of inner selenium and the possible release of free Cd and Se ions, the CdSe@ZnS core/shell QDs can be degraded. And then, their fluorescence would be quenched, and the toxicity of QDs would be induced in organisms.

8.12 Perspectives

We here first introduce several techniques with the potential to visualize and track the distribution and especially the translocation of certain ENMs in the body of nematodes after exposure. These employed techniques largely help us observe and understand the dynamic behavior of ENMs once intake into the body of nematodes. Moreover, these techniques provide the basis for understanding how the ENMs interact with the cells or tissues in the body of organisms. With the research progress and the participation of more chemists, more powerful techniques will be used in the study on nanotoxicology by opening more useful windows. With the aid of these techniques, we should pay attention to not only the cellular and physiological mechanisms of nanotoxicity formation but also the chemical and molecular mechanisms of nanotoxicity formation.

In this chapter, we further systematically introduce the progress on the distribution and translocation of different ENMs in different targeted organs and the underlying cellular, molecular, and chemical metabolisms in nematodes. Nevertheless, the knowledge on the translocation of ENMs into secondary targeted organs through the biological barrier of primary targeted organs is still limited. That is, the underlying molecular mechanisms for translocation of ENMs into secondary targeted organs are still largely unclear in nematodes. Additionally, the underlying cellular and molecular mechanisms for intracellular transport of ENMs are also largely unknown in nematodes. Moreover, the underlying cellular and molecular mechanisms for transgenerational toxicity and translocation of ENMs need to be further deeply elucidated in organisms. Meanwhile, the exact molecular mechanism for the clearance of ENMs out of the body also needs to be further determined in nematodes.

References

1. Leung MC, Williams PL, Benedetto A, Au C, Helmcke KJ, Aschner M, Meyer JN (2008) *Caenorhabditis elegans*: an emerging model in biomedical and environmental toxicology. Toxicol Sci 106:5–28
2. Zhao Y-L, Wu Q-L, Li Y-P, Wang D-Y (2013) Translocation, transfer, and in vivo safety evaluation of engineered nanomaterials in the non-mammalian alternative toxicity assay model of nematode *Caenorhabditis elegans*. RSC Adv 3:5741–5757
3. Wang D-Y (2016) Biological effects, translocation, and metabolism of quantum dots in nematode *Caenorhabditis elegans*. Toxicol Res 5:1003–1011
4. Pluskota A, Horzowski E, Bossinger O, von Mikecz A (2009) In *Caenorhabditis elegans* nanoparticle-bio-interactions become transparent: silica-nanoparticles induce reproductive senescence. PLoS One 4(8):e6622
5. McGhee JD (2007) The *C. elegans* intestine. WormBook. https://doi.org/10.1895/wormbook.1.133.1
6. Chisholm AD, Hardin J (2005) Epidermal morphogenesis. WormBook. https://doi.org/10.1895/wormbook.1.35.1
7. Yang J-N, Zhao Y-L, Wang Y-W, Wang H-F, Wang D-Y (2015) Toxicity evaluation and translocation of carboxyl functionalized graphene in *Caenorhabditis elegans*. Toxicol Res 4:1498–1510
8. Meyer JN, Lord CA, Yang XY, Turner EA, Badireddy AR, Marinakos SM, Chilkoti A, Wiesner MR, Auffan M (2010) Intracellular uptake and associated toxicity of silver nanoparticles in *Caenorhabditis elegans*. Aquat Toxicol 100:140–150
9. Lim SF, Riehn R, Ryu WS, Khanarian N, Tung C, Tank D, Austin RH (2006) *In vivo* and scanning electron microscopy imaging of upconverting nanophosphors in *Caenorhabditis elegans*. Nano Lett 6(2):169–174
10. Chen J, Guo C, Wang M, Huang L, Wang L, Mi C, Li J, Fang X, Mao C, Xu S (2011) Controllable synthesis of NaYF4: Yb, Er upconversion nanophosphors and their application to *in vivo* imaging of *Caenorhabditis elegans*. J Mater Chem 21(8):2632
11. Zhou J, Yang Z, Dong W, Tang R, Sun L, Chun-Hua Yan C (2011) Bioimaging and toxicity assessments of near-infrared upconversion luminescent NaYF4:Yb,Tm nanocrystals. Biomaterials 32:9059–9067
12. Qu Y, Li W, Zhou Y, Liu X, Zhang L, Wang L, Li Y, Iida A, Tang Z, Zhao Y, Chai Z, Chen C (2011) Full assessment of fate and physiological behavior of quantum dots utilizing *Caenorhabditis elegans* as a model organism. Nano Lett 11:3174–3183
13. Zhang L, Xia J, Zhao Q, Liu L, Zhang Z (2010) Functional graphene oxide as a nanocarrier for controlled loading and targeted delivery of mixed anticancer drugs. Small 6:537–544

14. Cranfield CG, Dawe A, Karloukovski V, Dunin-Borkowski RE, de Pomerai D, Dobson J (2004) Biogenic magnetite in the nematode *Caenorhabditis elegans*. Proc R Soc Lond B 271(Suppl):S436–S439

15. Zhao Y-L, Wu Q-L, Tang M, Wang D-Y (2014) The *in vivo* underlying mechanism for recovery response formation in nano-titanium dioxide exposed *Caenorhabditis elegans* after transfer to the normal condition. Nanomedicine 10:89–98

16. Wu Q-L, Zhao Y-L, Li Y-P, Wang D-Y (2014) Susceptible genes regulate the adverse effects of TiO_2-NPs at predicted environmental relevant concentrations on nematode *Caenorhabditis elegans*. Nanomedicine 10:1263–1271

17. Gao Y, Liu N, Chen C, Luo Y, Li Y, Zhang Z, Zhao Y, Zhao B, Iida A, Chai Z (2008) Mapping technique for biodistribution of elements in a model organism, *Caenorhabditis elegans*, after exposure to copper nanoparticles with microbeam synchrotron radiation X-ray fluorescence. J Anal At Spectrom 23:1121–1124

18. Zanni E, De Bellis G, Bracciale MP, Broggi A, Santarelli ML, Sarto MS, Palleschi C, Uccelletti D (2012) Graphite nanoplatelets and *Caenorhabditis elegans*: insights from an *in vivo* model. Nano Lett 12:2740–2744

19. Wu Q-L, Yin L, Li X, Tang M, Zhang T, Wang D-Y (2013) Contributions of altered permeability of intestinal barrier and defecation behavior to toxicity formation from graphene oxide in nematode *Caenorhabditis elegans*. Nanoscale 5(20):9934–9943

20. Wu Q-L, Zhao Y-L, Fang J-P, Wang D-Y (2014) Immune response is required for the control of *in vivo* translocation and chronic toxicity of graphene oxide. Nanoscale 6:5894–5906

21. Nouara A, Wu Q-L, Li Y-X, Tang M, Wang H-F, Zhao Y-L, Wang D-Y (2013) Carboxylic acid functionalization prevents the translocation of multi-walled carbon nanotubes at predicted environmental relevant concentrations into targeted organs of nematode *Caenorhabditis elegans*. Nanoscale 5:6088–6096

22. Wu Q-L, Li Y-X, Li Y-P, Zhao Y-L, Ge L, Wang H-F, Wang D-Y (2013) Crucial role of biological barrier at the primary targeted organs in controlling translocation and toxicity of multi-walled carbon nanotubes in nematode *Caenorhabditis elegans*. Nanoscale 5:11166–11178

23. Zhao Y-L, Wang X, Wu Q-L, Li Y-P, Wang D-Y (2015) Translocation and neurotoxicity of CdTe quantum dots in RMEs motor neurons in nematode *Caenorhabditis elegans*. J Hazard Mater 283:480–489

24. Zhao Y-L, Wang X, Wu Q-L, Li Y-P, Tang M, Wang D-Y (2015) Quantum dots exposure alters both development and function of D-type GABAergic motor neurons in nematode *Caenorhabditis elegans*. Toxicol Res 4:399–408

25. Zhao Y-L, Liu Q, Shakoor S, Gong JR, Wang D-Y (2015) Transgenerational safe property of nitrogen-doped graphene quantum dots and the underlying cellular mechanism in *Caenorhabditis elegans*. Toxicol Res 4:270–280

26. Liu Z-F, Zhou X-F, Wu Q-L, Zhao Y-L, Wang D-Y (2015) Crucial role of intestinal barrier in the formation of transgenerational toxicity in quantum dots exposed nematodes *Caenorhabditis elegans*. RSC Adv 5:94257–94266

27. Li Y-X, Wang W, Wu Q-L, Li Y-P, Tang M, Ye B-P, Wang D-Y (2012) Molecular control of TiO_2-NPs toxicity formation at predicted environmental relevant concentrations by Mn-SODs proteins. PLoS One 7(9):e44688

28. Wu Q-L, Zhao Y-L, Li Y-P, Wang D-Y (2014) Molecular signals regulating translocation and toxicity of graphene oxide in nematode *Caenorhabditis elegans*. Nanoscale 6:11204–11212

29. Zhi L-T, Ren M-X, Qu M, Zhang H-Y, Wang D-Y (2016) Wnt ligands differentially regulate toxicity and translocation of graphene oxide through different mechanisms in *Caenorhabditis elegans*. Sci Rep 6:39261

30. Maurer LL, Yang X, Schindler AJ, Taggart RK, Jiang C, Hsu-Kim H, Sherwood DR, Meyer JN (2015) Intracellular trafficking pathways in silver nanoparticle uptake and toxicity in *Caenorhabditis elegans*. Nanotoxicology 11:1–5

Chapter 9
Confirmation of Nanomaterials with Low-Toxicity or Non-toxicity Property

Abstract In the nanoscience field, besides the preparation of engineered nanomaterials (ENMs) with excellent physicochemical properties, to obtain the ENMs with the safe property for environmental organisms and human beings is also another important aim. In this chapter, we summarize the progress on the assessment and confirmation of low-toxicity or relative non-toxicity property of some important ENMs (such as graphite, graphene quantum dots (GQDs), carboxyl-functionalized graphene (G–COOH), Gd@C82(OH)22, fluorescent nanodiamond (FND), halloysite clay nanotubes (HNTs), and titanium dioxide nanoparticles (TiO$_2$-NPs)) using *Caenorhabditis elegans*. With the TiO$_2$-NPs as an example, the relative safe property of TiO$_2$-NPs at realistic concentrations was further discussed. Moreover, the recovery response of toxicity of TiO$_2$-NPs after transfer to the normal conditions and the underlying mechanisms in nematodes were also introduced.

Keywords Low-toxicity · Non-toxicity · Nanomaterials · *Caenorhabditis elegans*

9.1 Introduction

In the nanoscience field, one of the important aims is to prepare the engineered nanomaterials (ENMs) with both the excellent physicochemical properties and the safe property for environmental organisms and human beings. Some of the successfully prepared ENMs may exhibit the safe property once released into the environment to allow the exposure of them to environmental organisms and human beings. At least in the past 10 years, many reports have been published to demonstrate the sensitivity of *Caenorhabditis elegans* to environmental toxicants, including heavy metals, organic pollutants, ENMs, nanoplastics, and even fine particulate matter (PM$_{2.5}$) [1–10]. Thus, *C. elegans* may be a wonderful in vivo assay model for the evaluation of ecotoxicological effects of certain ENMs for environmental organisms. Meanwhile, *C. elegans* will be a useful assay model for the confirmation of low-toxicity or relative non-toxicity property of certain ENMs.

In this chapter, we summarize the progress on the evaluation and confirmation of low-toxicity or relative non-toxicity property of some important ENMs using in vivo assay system in *C. elegans*. We mainly introduced (1) graphite, (2) graphene

© Springer Nature Singapore Pte Ltd. 2018
D. Wang, *Nanotoxicology in* Caenorhabditis elegans,
https://doi.org/10.1007/978-981-13-0233-6_9

quantum dots (GQDs), (3) carboxyl-functionalized graphene (G–COOH), (4) Gd@ C82(OH)22 (a water-soluble endohedral metallofullerene derivative), (5) fluorescent nanodiamond (FND), (6) halloysite clay nanotubes (HNTs), and (7) titanium dioxide nanoparticles (TiO$_2$-NPs).

9.2   Confirmation of ENMs with Low-Toxicity or Non-toxicity Property

9.2.1   Graphite

Graphite nanoplatelets (GNPs), one of the important graphene-related ENMs, can be potentially used in drug delivery, photothermal anticancer, bioimaging, DNA sequencing, tissue engineering, antibacterial, and biosensors [11–20]. Using in vivo acute exposure assay model of *C. elegans*, the relative safety property of GNPs for organisms after short-term exposure has been proven.

After acute exposure to GNPs (3 h) on nematode growth medium (NGM) agar plates in the presence of *E. coli* OP50 strain, the GNPs at the concentration of 100 or 250 mg/L were not able to increase the mortality of nematodes (Fig. 9.1) [21]. Acute exposure to GNPs at concentrations of 50–250 mg/L also did not influence the reproduction as indicated by the endpoint of brood size (Fig. 9.1) [21]. Moreover, acute exposure to GNPs at the concentration of 100 or 250 mg/L did not significantly affect the lifespan of nematodes (Fig. 9.1) [21]. Throughout the entire assessment, the fullerol nanoparticles (100 mg/L) were used as a positive control. Acute exposure to fullerol nanoparticles (100 mg/L) could significantly decrease the percentage of living animals, reduce the brood size, and suppress the lifespan of nematodes (Fig. 9.1) [21].

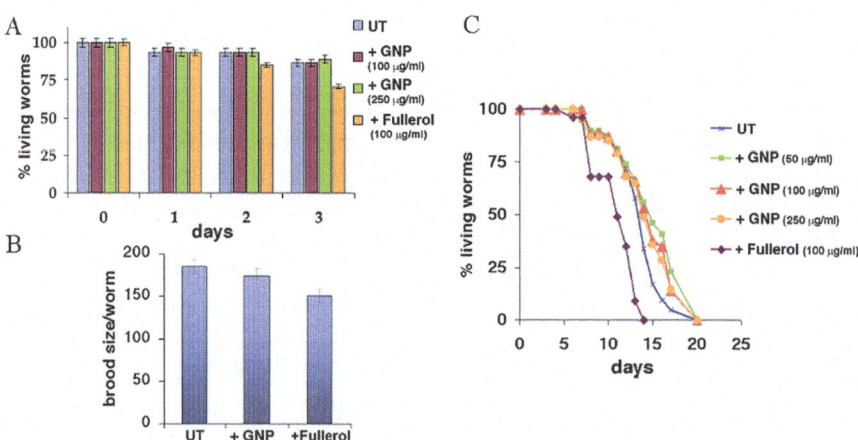

Fig. 9.1   Effect of acute exposure to GNPs on nematodes [21]
(**a**) Effect of GNPs on mortality. (**b**) Effect of GNPs on brood size. (**c**) Effect of GNPs on lifespan. Fullerol nanoparticles suspension (100 mg/L) was used as a positive control

9.2.2 GQDs

In the recent several years, GQDs acting as a new member of the graphene-related ENMs have received the great attention. Actually, GQDs are the miniaturized versions of GO. GQDs are potentially applied in cellular imaging, photovoltaic devices, biosensors, catalysis, and DNA cleavage [22–29]. Nitrogen-doped graphene quantum dots (N-GQDs) can be prepared using a one-pot solvothermal approach based on the GO sheets [30]. N-GQDs have the excellent physicochemical properties for cellular and deep-tissue imaging [30].

9.2.2.1 Effects of N-GQD Exposure on Nematodes

Prolonged exposure from L1-larvae to adult day-1 was performed to assess the possible long-term effects of N-GQDs on wild-type nematodes. Firstly, prolonged exposure to N-GQDs at concentrations of 0.1–100 mg/L would not cause the formation of lethal phenotype [31]. Secondly, prolonged exposure to N-GQDs at concentrations of 0.1–100 mg/L could not obviously affect the lifespan of wild-type nematodes (Fig. 9.2) [31]. Thirdly, prolonged exposure to N-GQDs at concentrations of 0.1–100 mg/L could not cause the damage on the functions of reproductive organs and neurons as indicated by the endpoints of brood size and locomotion behavior in wild-type nematodes (Fig. 9.2) [31]. Fourthly, prolonged exposure to N-GQDs at concentrations of 0.1–100 mg/L also could not induce the significant induction of intestinal reactive oxygen species (ROS) production in wild-type nematodes (Fig. 9.2) [31], implying the normal physiological function of intestinal cells. Therefore, these results imply the relative safe property of examined N-GQDs on wild-type nematodes after prolonged exposure.

9.2.2.2 Effects of N-GQD Exposure on Progeny of Exposed Nematodes

Prolonged exposure to N-GQDs may be not only safe for the exposed wild types but also safe for the F1 progeny of the exposed wild-type nematodes. In order to examine the effect of N-GQDs exposure on the progeny of exposed nematodes, the eggs of N-GQD-exposed nematodes were transferred to new NGM plates without the addition of N-GQDs. In the progeny of N-GQD-exposed wild-type nematodes, no induction of lethality, no obvious alteration in lifespan, brood size, and locomotion behavior, and no significant induction of intestinal ROS production were detected [31]. These results imply that the normal physiological functions of primary and secondary targeted organs can be maintained in both the N-GQD-exposed nematodes and their F1 progeny. More importantly, these results exclude the possibility that the possible toxicity of N-GQDs may be amplified and formed in the progeny of exposed wild-type nematodes.

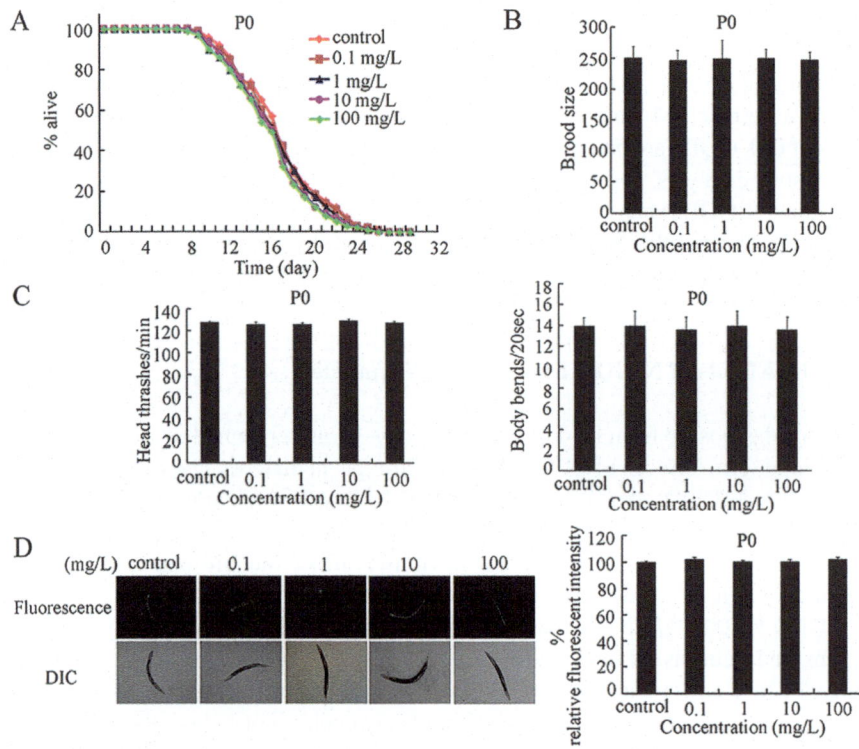

Fig. 9.2 Effects of N-GQD exposure on animals [31]
(**a**) Effect of N-GQDs on lifespan. (**b**) Effect of N-GQDs on brood size. (**c**) Effect of N-GQDs on locomotion behavior. (**d**) Effect of N-GQDs on induction of intestinal ROS production. Prolonged exposure to GQDs was performed from L1-larvae to adult day-1. Bars represent means ± standard error of the mean (SEM)

9.2.2.3 Molecular Mechanisms for the Observed Safe Property of N-GQDs in Nematodes

The underlying molecular basis with the concern on the control of oxidative stress has been raised for the observed safe property of N-GQDs. In *C. elegans*, induction of the oxidative stress is crucial for the toxicity formation of ENMs [2, 3]. In wild-type nematodes, exposure to N-GQDs (100 mg/L) did not alter the transcriptional expressions of genes (*mev-1*, *gas-1*, *isp-1*, and *clk-1*) encoding the subunits of mitochondrial complexes and components on electron transport chain [31], implying that exposure to N-GQDs did not activate the molecular machinery inducing the oxidative stress. *mev-1* encodes a subunit of mitochondrial complex II, *gas-1* encodes a subunit of mitochondrial complex I, *isp-1* encodes a subunit of mitochondrial complex III, and *clk-1* encodes a demethoxyubiquinone hydroxylase. Additionally, exposure to N-GQDs (100 mg/L) did not change the transcriptional expressions of genes (*sod-1*, *sod-2*, *sod-3*, *sod-4*, and *sod-5*) encoding superoxide dismutases (SODs) or genes (*ctl-1*, *ctl-2*, and *ctl-3*) encoding catalases in nematodes [31], implying that exposure to N-GQDs did not induce a certain stress with the potential in inducing the response of SODs or catalases expression. *sod-1* encodes a copper/zinc SOD, *sod-2* and *sod-3* encode Mn-SODs, and *sod-4* and *sod-5* encode extracellular copper/zinc SODs. The transcriptional expressions of these examined genes were also normal in the F1 progeny of nematodes exposed to N-GQDs (100 mg/L) [31].

9.2.2.4   Cellular Mechanisms for the Observed Safe Property of N-GQDs in Nematodes

Bioavailability plays a pivotal role in the toxicity induction of ENMs [2, 3]. One of the important cellular mechanisms for the observed safe property of N-GQDs is that N-GQDs were mainly distributed in the intestinal lumen in nematodes after prolonged exposure [31]. That is, the N-GQDs could not be translocated into the secondary targeted organs, such as neurons and reproductive organs, through the intestinal barrier in nematodes. Meanwhile, no green fluorescence signals of N-GQDs were detect in the F1 progeny of N-GQD-exposed nematodes, since N-GQDs could not be translocated into the reproductive organs [31]. Intestinal permeability is a crucial factor affecting the patterns of distribution and translocation of ENMs in nematodes [2, 3]. The second cellular mechanism for the observed safe property of N-GQDs is that prolonged exposure to N-GQDs (100 mg/L) could not obviously alter the intestinal permeability as indicated by the distribution of Nile red signals in nematodes [31], implying the maintenance of normal physiological function of intestinal barrier in N-GQD-exposed wild-type nematodes. Besides these, it has been further shown that the N-GQDs exposure did not alter the defecation behavior as indicated by the endpoint of mean defecation cycle length and the structure of neurons controlling the defecation behavior in nematodes [31], implying the normal defecation behavior in N-GQD-exposed nematodes. In nematodes, both AVL neurons in the head and the DVB neurons in the tail are required for the control of defecation behavior [32].

9.2.2.5   Effects of N-GQD Exposure on *sod-2* or *sod-3* Mutant Nematodes

In *C. elegans*, mutation of *sod-2* or *sod-3* usually causes a susceptibility to environmental toxicants [33]. An important question is that whether prolonged exposure to N-GQDs is still safe in *sod-2* or *sod-3* mutant nematodes with the susceptible property to environmental toxicants. Prolonged exposure to N-GQDs (100 mg/L) did not result in the reduction in lifespan, the reduction in brood size, the decrease in locomotion behavior, and the induction of intestinal ROS production in *sod-2* or *sod-3* mutant nematodes and their progeny (Fig. 9.3) [31], suggesting the normal physiological functions of both primary and secondary targeted organs were maintained in N-GQD-exposed *sod-2* or *sod-3* mutant nematodes and their progeny. Meanwhile, prolonged

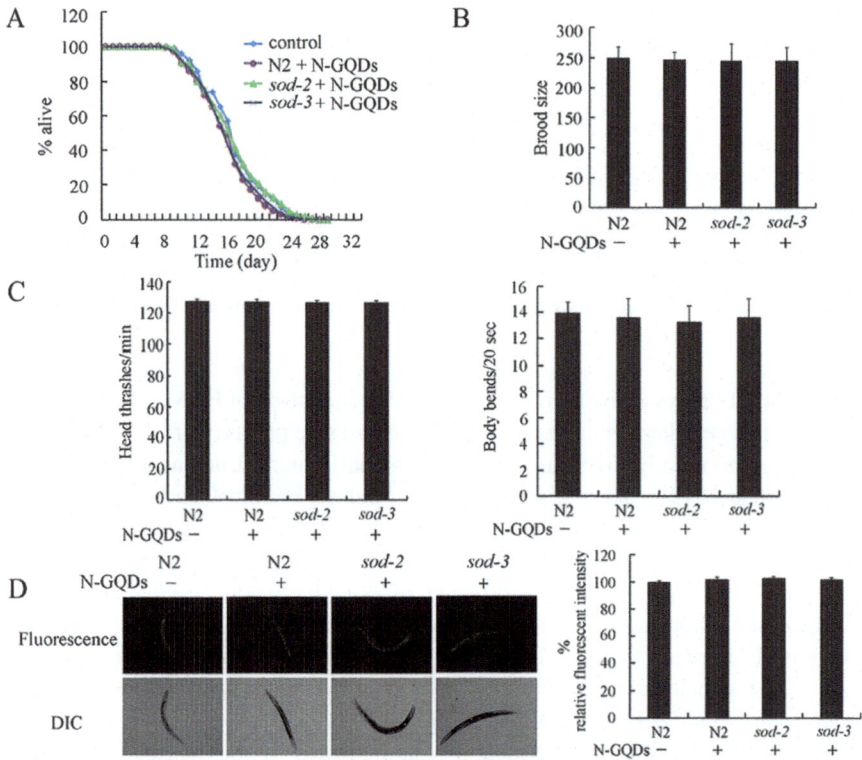

Fig. 9.3 Effects of N-GQD exposure on *sod-2* and *sod-3* mutant nematodes [31]
(**a**) Effects of N-GQDs on lifespans of *sod-2* and *sod-3* mutant nematodes. (**b**) Effects of N-GQDs on brood sizes of *sod-2* and *sod-3* mutant nematodes. (**c**) Effects of N-GQDs on locomotion behavior in *sod-2* and *sod-3* mutant nematodes. (**d**) Effects of N-GQDs on induction of intestinal ROS production in *sod-2* and *sod-3* mutant nematodes. Prolonged exposure to N-GQDs (100 mg/L) was performed from L1-larvae to adult day-1. Bars represent means ± standard error of the mean (SEM)

exposure to N-GQDs (100 mg/L) could also not be able to obviously affect both the intestinal permeability and the defecation behavior in *sod-2* or *sod-3* mutant nematodes and their progeny [31]. Moreover, N-GQDs were still mainly distributed in the intestinal lumen in *sod-2* or *sod-3* mutant nematodes, and no accumulation of N-GQDs was detected in the progeny of *sod-2* or *sod-3* mutant nematodes [31].

9.2.3   G–COOH

G–COOH is another important member of graphene-related family. G–COOH has a high carboxyl ratio and additional ethanoic acid groups on the sp^3-hybridized carbon. G–COOH is considered to be potentially applied in biosensors, anti-infection, gas adsorption, supercapacitor, and radiofrequency ablation of cancer cells [34–40].

9.2.3.1   Effects of Prolonged Exposure to G–COOH on Nematodes

Five endpoints were selected to evaluate the possible effect of prolonged exposure (from L1-larvae to adult day-1) to G–COOH on nematodes, and they were lifespan, body length, induction of intestinal ROS production, locomotion behavior, and brood size. After the prolonged exposure, G–COOH at all the examined concentrations (0.01–100 mg/L) could not cause toxicity on lifespan, body length, locomotion behavior, and brood size and induce the significant induction of intestinal ROS production (Fig. 9.4) [41]. As a positive control, prolonged exposure to GO (100 mg/L) significantly reduced the lifespan, decreased the body length, induced the significant induction of intestinal ROS production, decreased the locomotion behavior, and reduced the brood size in nematodes (Fig. 9.4) [41]. That is, the nematodes exposed to G–COOH at the examined concentrations can maintain the normal physiological functions of different organs.

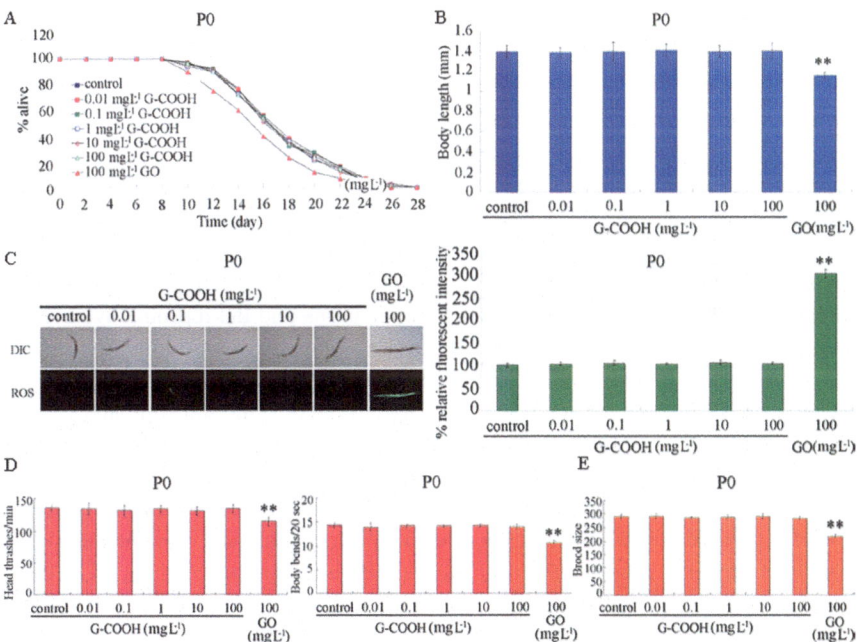

Fig. 9.4   Effects of prolonged exposure to G–COOH on nematodes [41]. (**a**) Effects of G–COOH on the lifespan. (**b**) Effects of G–COOH on the body length. (**c**) Effects of G–COOH on the induction of intestinal ROS production. The left shows the pictures, and the right shows the comparison of the relative fluorescence intensity in the intestine. (**d**) Effects of G–COOH on locomotion behavior. (**e**) Effects of G–COOH on the brood size. Prolonged exposure was performed from L1-larvae to adult day-1. Bars represent means ± SEM **$P < 0.01$

9.2.3.2   Effect of Chronic Exposure to G–COOH on Nematodes

The effects of chronic exposure to G–COOH (from L1-larvae to adult day-10, approximately 13 days) were also examined. With the endpoint of locomotion behavior as the toxicity assessment endpoint, chronic exposure to G–COOH (100 mg/L) also could not cause obvious toxicity on the locomotion behavior [41]. As a positive control, chronic exposure to GO (1 mg/L) could result in the significant decrease in the locomotion behavior in nematodes [41].

9.2.3.3   Effects of Prolonged Exposure to G–COOH on the Progeny
of the Exposed Nematodes

Similarly to the observations in the progeny of N-GQD-exposed nematodes as introduced above, no toxicity on lifespan, body length, locomotion behavior, and brood size and no significant induction of intestinal ROS production were observed in the progeny of G–COOH (1–100 mg/L)-exposed nematodes [41]. Different from this, prolonged exposure to GO (100 mg/L) caused the formation of toxicity on lifespan, body length, locomotion behavior, and brood size and significant induction of intestinal ROS production in the progeny of exposed nematodes [41].

9.2.3.4   Cellular Basis for the Safe Property of G–COOH in Nematodes

The pattern of distribution and distribution of G–COOH is the primary cellular contributor for the observed safe property of G–COOH in nematodes. It has been found that G–COOH/Rho B was mainly distributed in the pharynx and the intestine in nematodes after prolonged exposure to G–COOH/Rho B (100 mg/L), and no signals of GCOOH/Rho B were observed in the neurons and the reproductive organs [41]. Additionally, no signals of G–COOH/Rho B were detected in the progeny of nematodes exposed to G–COOH/Rho B (100 mg/L) [41]. These observations suggest that the G–COOH particles were mainly confined to the intestine and cannot have the potential to be translocated into the secondary targeted organs through the intestinal barrier in nematodes.

The second important cellular mechanism for the observed safe property of G–COOH is that the normal intestinal permeability was maintained in nematodes after prolonged exposure to G–COOH (0.01–100 mg/L) and their progeny [41]. The third important cellular mechanism for the observed safe property of G–COOH is that the normal defecation behavior and the normal development of AVL and DVB neurons controlling the defecation behavior were sustained in nematodes after prolonged exposure to G–COOH (0.01–100 mg/L) and their progeny [41]. These two cellular contributors also well explain the observation on the pattern of distribution and distribution of G–COOH introduced above. In contrast, after prolonged exposure to GO (100 mg/L), the intestinal permeability was enhanced, and the mean defecation cycle length was increased in nematodes [41]. Additionally, prolonged exposure to

GO (100 mg/L) induced the developmental deficits in AVL and DVB neurons as indicated by a reduction in the relative size of fluorescent puncta in the cell bodies [41].

9.2.3.5 Molecular Basis for the Safe Property of G–COOH in Nematodes

One of the important molecular mechanisms for the observed safe property of G–COOH is that prolonged exposure to G–COOH (100 mg/L) could not significantly affect the transcriptional expressions of genes (*mev-1*, *clk-1*, *gas-1*, and *isp-1*) encoding subunits on mitochondrial complexes and components on electron transport chain [41], implying that prolonged exposure to G–COOH may be not able to activate the action of molecular machinery controlling the oxidative stress in nematodes. Another important molecular mechanism for the observed safe property of G–COOH is that prolonged exposure to G–COOH (100 mg/L) could not significantly influence the transcriptional expressions of genes (*sod-2* and *sod-3*) encoding Mn-SODs [41], which suggests that prolonged exposure to G–COOH may not have the potential to form at least an enough stress to induce the increase in Mn-SODs expression in nematodes. The third raised molecular mechanism for the observed safe property of G–COOH is that prolonged exposure to G–COOH (100 mg/L) did not significantly alter the transcriptional expressions of genes (*gem-4*, *nhx-2*, *pkc-3*, *par-3*, *par-6*, *dlg-1*, *ajm-1*, *nfm-1*, *abts-4*, *erm-1*, *eps-8*, *act-5*, and *ifb-2*) required for the control of intestinal development [41]. *gem-4*, *nhx-2*, *pkc-3*, *par-3*, *par-6*, *erm-1*, *eps-8*, *act-5*, and *ifb-2* are required for the development of microvilli on intestinal cells, *nfm-1* and *abts-4* are required for the development of basolateral domain in intestine, and *dlg-1* and *ajm-1* are required for the development of apical junction in intestine [42].

9.2.4 Gd@C82(OH)22

Gd@C82(OH)22 is a new metallofullerenol with an average size of 22 nm. Some reports have suggested its values in the application of cancer therapy, drug-like nanomedicine, and antineoplastic activity [43–49].

9.2.4.1 Effects of Gd@C82(OH)22 Exposure on Nematodes

The endpoints of lifespan, growth, feeding rate, and reproductive capacity were selected to assess the possible safe property in nematodes. After exposure to Gd@C82(OH)22 at concentrations of 0.01–10 mg/L for one generation, no alteration in lifespan was observed in exposed nematodes [50]. Additionally, after exposure for one generation, Gd@C82(OH)22 (0.01–10 mg/L) also did not significantly affect the growth of nematodes as indicated by the endpoint of body length, and no

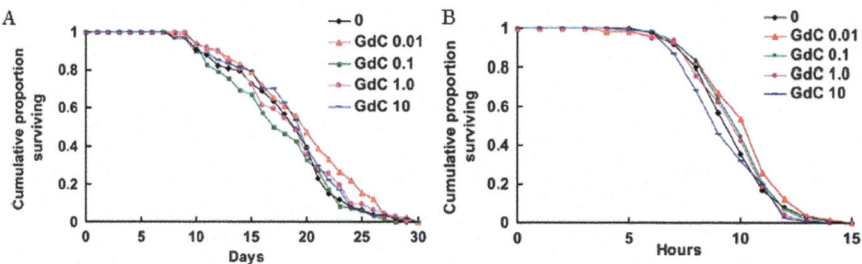

Fig. 9.5 Effects of long-term exposure to Gd@C82(OH)22 on lifespan (**a**) and thermotolerance (**b**) in nematodes with treatment for six generations [50]
The nematodes were exposed to Gd@C82(OH)22 nanoparticles for six generations. For the thermotolerance assay, the nematodes were subjected to heat shock after exposure to Gd@C82(OH)22 nanoparticles for six generations

abnormalities in the development could be detected in all experimental groups during the lifespan [50]. Moreover, the pharyngeal pumping rate was also not changed by exposure to Gd@C82(OH)22 at concentrations of 0.01–10 mg/L for one generation [50], suggesting the normal feeding rate in Gd@C82(OH)22-exposed nematodes. More importantly, the reproductive capacity was also normal in nematodes after exposure to Gd@C82(OH)22 (0.01–10 mg/L) for one generation [50].

Besides these, after exposure to Gd@C82(OH)22 at concentrations of 0.01–10 mg/L for successive six generations, still not obvious alteration in lifespan was detected in exposed nematodes (Fig. 9.5) [50].

9.2.4.2   Effects of Gd@C82(OH)22 Exposure on Thermotolerance in Nematodes

Considering the fact that there is close correlation between the stress resistance and the longevity in nematodes [51–54], the effect of Gd@C82(OH)22 exposure on thermotolerance at 35 °C was further examined. The nematodes were first exposed to Gd@C82(OH)22 and then performed the heat shock at 35 °C for 72 h. After exposure to Gd@C82(OH)22 (0.01–10 mg/L) for one generation, it has been found that pretreatment with Gd@C82(OH)22 could not significantly enhance the adverse effect of heat shock on lifespan in nematodes [50]. Similarly, after exposure to Gd@C82(OH)22 (0.01–10 mg/L) for successive six generations, Gd@C82(OH)22 still did not significantly enhance the adverse effect of heat shock on lifespan in nematodes [50].

9.2.5   FND

In the recent years, the nanodiamonds have received the great attention for their potential applications in bioimaging, drug delivery, regenerative medicine, thermal conductivity, catalysts, gene delivery, biosensors, environmental remediation,

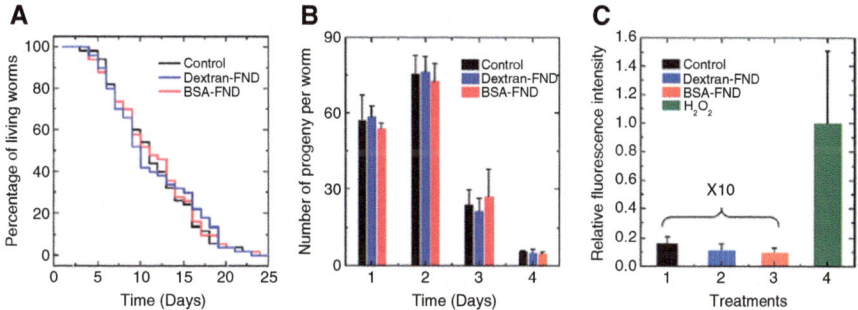

Fig. 9.6 Effects of FNDs exposure on lifespan (**a**), brood size (**b**), and ROS level (**c**) in nematodes [66]
Exposure was performed from L4-larvae for 3 h. During the ROS level assay, a hydrogen peroxide treatment was used as the positive control

antitumor immunotherapy, and antioxidant activity [55–65]. The safe property of two types of FNDs (dextran-coated FND (dextran-FND) and BSA-coated FND (BSA-FND)) was examined after short-term exposure in nematodes.

9.2.5.1   Effects of FNDs Exposure on Lifespan, Brood Size, and ROS Level in Nematodes

After exposure from L4-larvae for 3 h, both dextran-FND (1 mg/mL) and BSA-FND (1 mg/mL) did not significantly affect the lifespan and the number of progeny per worm from adult day-1 to day-4 (Fig. 9.6) [66]. Additionally, after exposure from L4-larvae for 3 h, both dextran-FND (1 mg/mL) and BSA-FND (1 mg/mL) could not cause the significant increase in ROS level (Fig. 9.6) [66]. In contrast, exposure to hydrogen peroxide (a positive control) induced the significant increase in ROS level in nematodes (Fig. 9.6) [66].

9.2.5.2   Effects of FNDs Exposure on Expressions of DAF-16::GFP and GCS-1::GFP in Nematodes

In *C. elegans*, both insulin signaling and SKN-1/Nrf signaling are required for the control of oxidative stress and stress response [67–71]. *daf-16* encodes a FOXO transcriptional factor in the insulin signaling pathway. The environmental stress, such as the heat shock, can induce the significant nuclear translocation of DAF-16::GFP [66]. However, exposure to both dextran-FND (1 mg/mL) and BSA-FND (1 mg/mL) for 3 h did not affect the fluorescence intensity of DAF-16::GFP and induces the obvious nuclear translocation of DAF-16::GFP [66]. In nematodes, GCS-1 acts as a direct target for SKN-1/Nrf during the control of oxidative stress or stress response [72, 73]. The environmental toxicants, such as the heavy metal of As, could induce the significant increase in GCS-1::GFP expression [66]. However, exposure to both dextran-FND (1 mg/mL) and BSA-FND (1 mg/mL) for 3 h could not induce the significant increase in GCS-1::GFP expression in nematodes [66]. Nevertheless, in this study, the exposure was performed only for 3 h, and the potential toxicity of long-term exposure to FNDs on nematodes is still uncertain.

9.2.6   HNTs

HNTs are one of the promising natural nanomaterials with the potential used in antitumor, gene delivery, targeted drug delivery, photocatalysis, food packaging, bioimaging, magnetic resonance imaging, tissue engineering, environmental remediation, enzyme immobilization, and biosensors [74–85].

In nematodes, short-term exposure to HNTs at concentrations of 0.05–1 mg/mL did not significantly affect both the brood size and the lifespan in nematodes [86]. Nevertheless, HNTs exposure at concentrations of 0.05–1 mg/mL could obviously reduce the body length [86]. Therefore, HNTs may be safe for environmental organisms at certain aspects.

9.3   Confirmation of the Relative Safe Property of Titanium Dioxide Nanoparticles (TiO_2-NPs) at Realistic Concentrations

TiO_2-NPs are among the most widely used ENMs and used in paints, plastics, inks, paper, creams, cosmetics, drugs, and foods. Meanwhile, many reports have indicated the in vitro and in vivo toxicity of TiO_2-NPs on environmental organisms and human beings [87–89].

In *C. elegans*, some publications have further demonstrated the toxicity of TiO_2-NPs (10 nm) at environmentally relevant concentrations after long-term exposure [90–92]. In contrast, acute exposure to TiO_2-NPs (10 nm) for 24 h only at concentrations in the range of mg/L induced the increase in mortality, the reduction in growth, and the decrease in reproductive capacity [93–95].

Normally, the duration from young man to old man is 42 years for the human beings according to the suggestion from WHO. Actually, 1 day in nematodes is approximately comparable to 4.2 years of human [95], since the population of dead animals increases sharply after adult day-10 during the developmental [96]. In other words, 5.71 h in nematodes is comparable to 1 year of human, and 0.48 h in nematodes is comparable to 1 month of human [95]. It has been reported that acute exposure to 10–25 mg/L TiO_2-NPs from sugar-coated chewing gum could not result in the adverse effects on GES-1 cells [97]. In nematodes, acute exposure to TiO_2-NPs (25 mg/L) for 24 h resulted in the toxicity on both reproduction and locomotion behavior and induced significant intestinal ROS production in wild-type nematodes, and these toxic effects could be enhanced by mutation of *sod-2*, *sod-3*, *mtl-2*, or *hsp-16.48* [95]. However, even in *sod-2*, *sod-3*, *mtl-2*, or *hsp-16.48* mutant nematodes, acute exposure to TiO$_2$-NPs (25 mg/L) for 0.48 h or 5.71 h could not alter both the brood size and the locomotion behavior (Fig. 9.7) [95]. Additionally, acute exposure to TiO$_2$-NPs (25 mg/L) for 0.48 h or 5.71 h also could not induce the significant intestinal ROS production in *sod-2*, *sod-3*, *mtl-2*, or *hsp-16.48* mutant nematodes (Fig. 9.7) [95]. That is, acute exposure to TiO$_2$-NPs (25 mg/L) for 0.48 h or 5.71 h cannot obviously affect the functions of both primary and secondary targeted organs for TiO$_2$-NPs in nematodes. These results imply that short-term intake of TiO$_2$-NPs at concentrations ≤25 mg/L may be relatively safe for environmental organisms and human beings.

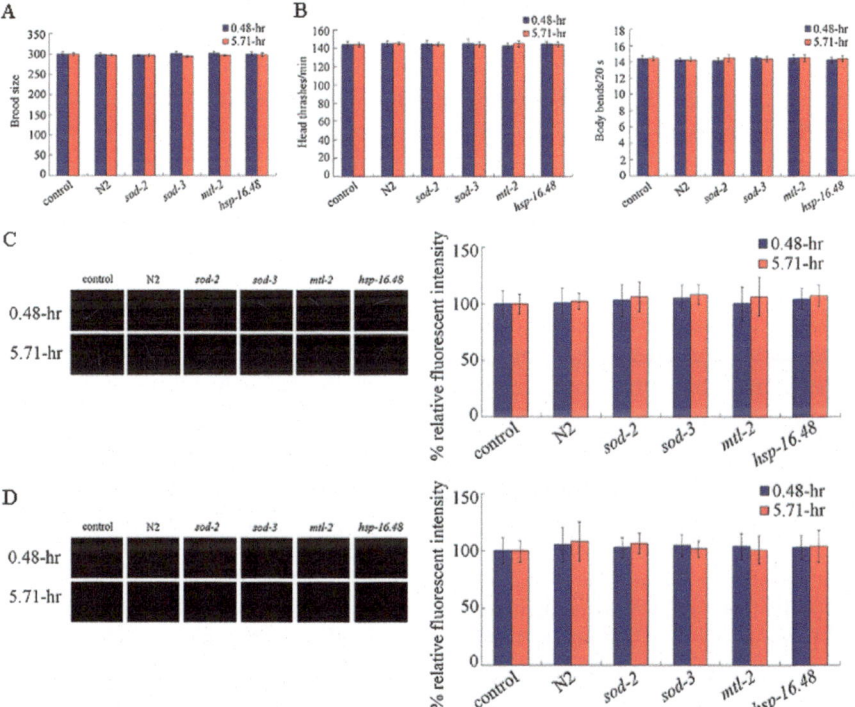

Fig. 9.7 Effects of acute exposure to TiO$_2$-NPs for 0.48 h and 5.71 h on wild-type and mutant nematodes [95]
(**a**) Comparison of brood size between wild type and mutants in nematodes acutely exposed to TiO$_2$-NPs. (**b**) Comparison of locomotion behavior between wild type and mutants in nematodes acutely exposed to TiO$_2$-NPs. (**c**) Comparison of intestinal autofluorescence between wild type and mutants in nematodes acutely exposed to TiO$_2$-NPs. (**d**) Comparison of ROS production between wild type and mutants in nematodes acutely exposed to TiO$_2$-NPs. Bars represent mean ± SEM.

9.4 Recovery Response of Toxicity of TiO₂-NPs After Transfer to the Normal Conditions in Nematodes

9.4.1 Recovery Response of Toxicity at Different Aspects of TiO₂-NPs After Transfer to the Normal Condition in Nematodes

It has been indicated that acute exposure to TiO$_2$-NPs (10 nm) in the range of mg/L significantly suppressed the development, decreased the locomotion behavior, suppressed the pumping rates, and induced the significant intestinal ROS production in nematodes [91]. The concentrations of 10–100 mg/L were selected for acute exposure (from young adults for 24 h) to TiO$_2$-NPs, and the concentrations of 1–100 μg/L were selected for prolonged exposure (from L1-larvae to adult day-1) to TiO$_2$-NPs [32].

The nematodes were first acutely exposed to TiO$_2$-NPs, and then transferred to the normal conditions. After transfer to the normal condition for 24 h, only the body length of TiO$_2$-NP (50 mg/L)-exposed nematodes could return to the control level, and the body length of TiO$_2$-NP (100 mg/L)-exposed nematodes was only moderately recovered [32]. After transfer to the normal condition for 24 h, the head thrashes and the body bends of TiO$_2$-NP (10–50 mg/L)-exposed nematodes could return to control levels (Fig. 9.8) [32]. After transfer to the normal condition for 24 h, the pumping rate of TiO$_2$-NP (50 mg/L)-exposed nematodes could be recovered to the control level [32]. Additionally, after transfer to the normal condition for 24 h, the intestinal ROS production of TiO$_2$-NP (100 mg/L)-exposed nematodes could not be recovered to the control levels [32].

After transfer to the normal condition for 48 h, the body length of TiO$_2$-NP (100 mg/L)-exposed nematodes could return to the control level [32]. After transfer to the normal condition for 48 h, the head thrashes and the body bends of TiO$_2$-NP (100 mg/L)-exposed nematodes could return to the control levels (Fig. 9.8) [32]. After transfer to the normal condition for 48 h, the pumping rate of TiO$_2$-NP (100 mg/L)-exposed nematodes could also return to the control level [32]. Additionally, after transfer to the normal condition for 48 h, the intestinal ROS production of TiO$_2$-NP (100 mg/L)-exposed nematodes could return to the control levels [32].

The prolonged exposure to TiO$_2$-NPs was first performed, and then the nematodes were transferred to the normal conditions. After transfer to the normal condition for 24 h, the body lengths of TiO$_2$-NP (10 or 100 μg/L)-exposed nematodes could not return to the control levels [32]. After transfer to the normal condition for 24 h, the head thrashes and the body bends of TiO$_2$-NP (1–100 μg/L)-exposed nematodes were all not recovered to the control levels (Fig. 9.8) [32]. After transfer to the normal condition for 24 h, the pumping rates of TiO$_2$-NP (10–100 μg/L)-exposed nematodes could not return to the control levels [32]. Additionally, after transfer to the normal condition for 24 h, the intestinal ROS production of TiO$_2$-NP (100 μg/L)-exposed nematodes was only moderately recovered to the control levels [32].

After transfer to the normal condition for 48 h, only the body length of TiO$_2$-NP (10 μg/L)-exposed nematodes could be recovered to the control level [32]. After

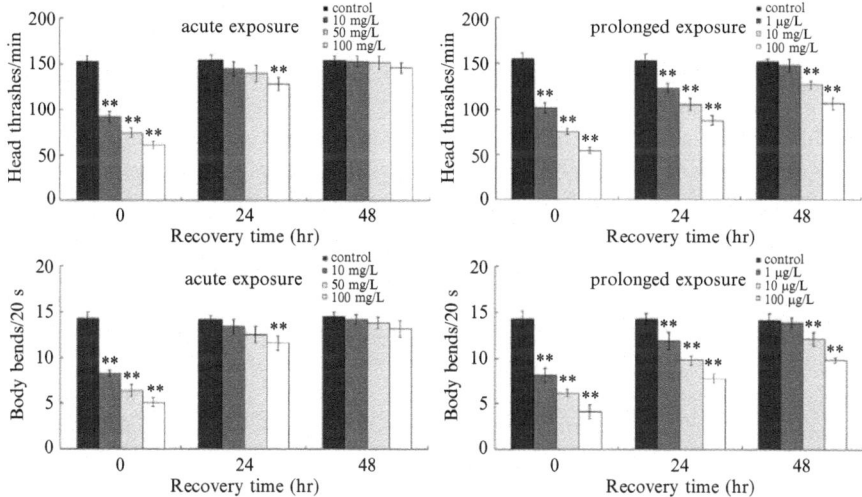

Fig. 9.8 Locomotion behavior during the recovery response of nematodes exposed to TiO₂-NPs [32]

Bars represent means ± SEM **$P < 0.01$

transfer to the normal condition for 48 h, only the head thrashes and the body bends of TiO₂-NP (1 μg/L)-exposed nematodes could return to the control levels (Fig. 9.8) [32]. After transfer to the normal condition for 48 h, only the pumping rate of TiO₂-NP (10 μg/L)-exposed nematodes could return to the control level [32]. Additionally, after transfer to the normal condition for 48 h, the intestinal ROS production of TiO₂-NP (100 μg/L)-exposed nematodes was still not completely recovered to the control levels [32].

9.4.2   Cellular Mechanisms for the Formation of Recovery Response of TiO₂-NPs Toxicity After Transfer to the Normal Condition in Nematodes

The first important cellular mechanism for the formation of recovery response of TiO₂-NPs toxicity after transfer to the normal condition is that the deficit in intestinal permeability caused by acute exposure to TiO₂-NPs (100 mg/L) did not return to the control level after transfer to the normal condition for 24 h, whereas the deficit in intestinal permeability caused by acute exposure to TiO₂-NPs (100 mg/L) could be recovered to the control level after transfer to the normal condition for 48 h [32]. Different from this, the deficit in intestinal permeability caused by prolonged exposure to TiO₂-NPs (100 μg/L) could not be recovered to the control level after transfer to the normal condition for 24 or 48 h [32]. The analysis on the structural changes of intestinal barrier using a transmission electron microscope (TEM) further confirms these observations. Acute exposure to TiO_2-NPs (100 mg/L) did not obviously alter the ultrastructure of intestinal cells; however, the ultrastructure of intestinal cells was severely disrupted, and some microvilli were lost in nematodes after prolonged exposure to TiO_2-NPs (100 μg/L) [32]. Moreover, the damage on ultrastructure of microvilli in nematodes after prolonged exposure to TiO_2-NPs (100 μg/L) was not noticeably recovered under the normal condition [32], which suggests that the different developmental or permeable state of intestinal barrier during the recovery response may be very important for the formation of recovery response of TiO_2-NPs toxicity after transfer to the normal condition.

The second important cellular mechanism for the formation of recovery response of TiO_2-NPs toxicity after transfer to the normal condition is that the deficit in the defecation behavior as reflected by the endpoint of mean defecation cycle length caused by acute exposure to TiO_2-NPs (100 mg/L) could be recovered to the control level after transfer to the normal condition for 24 or 48 h [32]. Different from this, the deficit in the defecation behavior and the development of AVL and DVB neurons controlling the defecation behavior caused by prolonged exposure to TiO_2-NPs (100 μg/L) could not be recovered to the control level after transfer to the normal condition for 24 or 48 h [32].

The third important cellular mechanism for the formation of recovery response of TiO_2-NPs toxicity after transfer to the normal condition is that the severe uptake of TiO_2-NPs after acute exposure to TiO_2-NPs (100 mg/L) could be largely recovered to the control level after transfer to the normal condition for 48 h [32]. However, the severe uptake of TiO_2-NPs after prolonged exposure to TiO_2-NPs (100 μg/L) could be only moderately recovered to the control level after transfer to the normal condition for 48 h [32]. That is, after prolonged exposure to TiO_2-NPs (100 μg/L), the normal condition may be not able to successfully help the nematodes to excrete the ingested TiO_2-NPs from the body during the recovery response duration.

9.5 Perspectives

In this chapter, we introduced some experimental examples performed in nematodes with the concerns on different ENMs to discuss their possible safe property on environmental organisms and/or human beings. The relatively high sensitivity of *C. elegans* makes the data obtained in this model animal convincing for the conclusions on the safe property for some ENMs at certain aspects. The data obtained from the nematodes will be largely helpful for our understanding of the safe property of certain ENMs on organisms at the level of whole animal. In contrast, the data obtained from the nematodes may be difficult to reflect the possible effects or safe property of certain ENMs on many organs which existed in human beings and mammals, since these organs cannot be found in nematodes.

Nevertheless, most of the data on the safe property of certain ENMs are only based on the experimental analysis in wild-type nematodes. First of all, genetic mutations of many genes can result in the susceptibility to environmental toxicants or environmental stresses. Besides the performance in wild-type nematodes, what's about the performance of these ENMs in genetic mutants with the susceptibility to environmental toxicants or environmental stresses? With the MWCNTs–COOH as an example, prolonged exposure to MWCNTs–COOH at environmentally relevant concentrations was safe to wild-type nematodes [98]. In genetic mutants with the susceptibility to environmental toxicants or environmental stresses, can this safe property of MWCNTs–COOH be still maintained for animals? Moreover, most of the data obtained in nematodes on the safe property of certain ENMs are based on a premise that the biological barrier is normal or well sustained. A series of evidence has demonstrated that the crucial cellular mechanism for the safe property of certain ENMs is that the examined ENMs cannot be translocated into the secondary targeted organs through the primary intestinal barrier in nematodes. We need to further pay attention to the possible effects of ENMs on environmental organisms or human beings once a large amount of them would be allowed to enter into the secondary targeted organs through the primary biological barrier in environmental organisms or human beings with the deficit in primary biological barrier.

With the concerns on the safe property of ENMs to environmental organisms, we suggest that it is very important to pay attention to the safe property of certain ENMs at environmentally relevant concentrations on environmental organisms after long-term or multigenerational exposure. With the concerns on the safe property of ENMs to human health, we then suggest that it is necessary to further pay attention to the safe property of certain ENMs at realistic concentrations on human beings after long-term intake and the possible transgenerational effects of certain ENMs.

References

1. Leung MC, Williams PL, Benedetto A, Au C, Helmcke KJ, Aschner M, Meyer JN (2008) *Caenorhabditis elegans*: an emerging model in biomedical and environmental toxicology. Toxicol Sci 106:5–28
2. Zhao Y-L, Wu Q-L, Li Y-P, Wang D-Y (2013) Translocation, transfer, and in vivo safety evaluation of engineered nanomaterials in the non-mammalian alternative toxicity assay model of nematode *Caenorhabditis elegans*. RSC Adv 3:5741–5757
3. Wang D-Y (2016) Biological effects, translocation, and metabolism of quantum dots in nematode *Caenorhabditis elegans*. Toxicol Res 5:1003–1011
4. Luo X, Xu S, Yang Y, Li L, Chen S, Xu A, Wu L (2016) Insights into the ecotoxicity of silver nanoparticles transferred from *Escherichia coli* to *Caenorhabditis elegans*. Sci Rep 6:36465
5. Li J, Li D, Yang Y, Xu T, Li P, He D (2016) Acrylamide induces locomotor defects and degeneration of dopamine neurons in *Caenorhabditis elegans*. J Appl Toxicol 36:60–67
6. Zuo Y, Hu Y, Lu W, Cao J, Wang F, Han X, Lu W, Liu A (2016) Toxicity of 2,6-dichloro-1,4-benzoquinone and five regulated drinking water disinfection by-products for the *Caenorhabditis elegans* nematode. J Hazard Mater 321:456–463
7. Du H, Wang M, Wang L, Dai H, Wang M, Hong W, Nie X, Wu L, Xu A (2015) Reproductive toxicity of endosulfan: implication from germ cell apoptosis modulated by mitochondrial dysfunction and genotoxic response genes in *Caenorhabditis elegans*. Toxicol Sci 145(1):118–127

8. Wu Q-L, Han X-X, Wang D, Zhao F, Wang D-Y (2017) Coal combustion related fine particulate matter (PM$_{2.5}$) induces toxicity in *Caenorhabditis elegans* by dysregulating microRNA expression. Toxicol Res 6:432–441

9. Yang R-L, Rui Q, Kong L, Zhang N, Li Y, Wang X, Tao J, Tian P, Ma Y, Wei J, Li G, Wanng D (2016) Metallothioneins act downstream of insulin signaling to regulate toxicity of outdoor fine particulate matter (PM$_{2.5}$) during spring festival in Beijing in nematode *Caenorhabditis elegans*. Toxicol Res 5:1097–1105

10. Zhao L, Qu M, Wong G, Wang D-Y (2017) Transgenerational toxicity of nanopolystyrene particles in the range of μg/L in nematode *Caenorhabditis elegans*. Environ Sci Nano 4:2356–2366

11. Xu J, Dou Y, Wei Z, Ma J, Deng Y, Li Y, Liu H, Dou S (2017) Recent progress in graphite intercalation compounds for rechargeable metal (Li, Na, K, Al)-ion batteries. Adv Sci 4(10):1700146

12. Zhang B, Wei P, Zhou Z, Wei T (2016) Interactions of graphene with mammalian cells: molecular mechanisms and biomedical insights. Adv Drug Deliv Rev 105(Pt B):145–162

13. Shi L, Chen J, Teng L, Wang L, Zhu G, Liu S, Luo Z, Shi X, Wang Y, Ren L (2016) The antibacterial applications of graphene and its derivatives. Small 12(31):4165–4184

14. Chen YW, Su YL, Hu SH, Chen SY (2016) Functionalized graphene nanocomposites for enhancing photothermal therapy in tumor treatment. Adv Drug Deliv Rev 105(Pt B):190–204

15. Lin J, Chen X, Huang P (2016) Graphene-based nanomaterials for bioimaging. Adv Drug Deliv Rev 105(Pt B):242–254

16. Shim G, Kim MG, Park JY, Oh YK (2016) Graphene-based nanosheets for delivery of chemotherapeutics and biological drugs. Adv Drug Deliv Rev 105(Pt B):205–227

17. Ambrosi A, Chua CK, Latiff NM, Loo AH, Wong CH, Eng AY, Bonanni A, Pumera M (2016) Graphene and its electrochemistry – an update. Chem Soc Rev 45(9):2458–2493

18. Shin SR, Li YC, Jang HL, Khoshakhlagh P, Akbari M, Nasajpour A, Zhang YS, Tamayol A, Khademhosseini A (2016) Graphene-based materials for tissue engineering. Adv Drug Deliv Rev 105(Pt B):255–274

19. Heerema SJ, Dekker C (2016) Graphene nanodevices for DNA sequencing. Nat Nanotechnol 11(2):127–136

20. Yoo JM, Kang JH, Hong BH (2015) Graphene-based nanomaterials for versatile imaging studies. Chem Soc Rev 44(14):4835–4852

21. Zanni E, De Bellis G, Bracciale MP, Broggi A, Santarelli ML, Sarto MS, Palleschi C, Uccelletti D (2012) Graphite nanoplatelets and *Caenorhabditis elegans*: insights from an *in vivo* model. Nano Lett 12:2740–2744

22. Wu JB, Lin ML, Cong X, Liu HN, Tan PH (2018) Raman spectroscopy of graphene-based materials and its applications in related devices. Chem Soc Rev. https://doi.org/10.1039/c6cs00915h

23. Li K, Zhao X, Wei G, Su Z (2017) Recent advance in the cancer bioimaging with graphene quantum dots. Curr Med Chem. https://doi.org/10.2174/0929867324666170223154145

24. Schroeder KL, Goreham RV, Nann T (2016) Graphene quantum dots for theranostics and bioimaging. Pharm Res 33(10):2337–2357

25. Gan Z, Xu H, Hao Y (2016) Mechanism for excitation-dependent photoluminescence from graphene quantum dots and other graphene oxide derivates: consensus, debates and challenges. Nanoscale 8(15):7794–7807

26. Du Y, Guo S (2016) Chemically doped fluorescent carbon and graphene quantum dots for bioimaging, sensor, catalytic and photoelectronic applications. Nanoscale 8(5):2532–2543

27. Zheng XT, Ananthanarayanan A, Luo KQ, Chen P (2015) Glowing graphene quantum dots and carbon dots: properties, syntheses, and biological applications. Small 11(14):1620–1636

28. Li L, Wu G, Yang G, Peng J, Zhao J, Zhu JJ (2013) Focusing on luminescent graphene quantum dots: current status and future perspectives. Nanoscale 5(10):4015–4039

29. Shen J, Zhu Y, Yang X, Li C (2012) Graphene quantum dots: emergent nanolights for bioimaging, sensors, catalysis and photovoltaic devices. Chem Commun 48(31):3686–3699

30. Liu Q, Guo B, Rao Z, Zhang B, Gong JR (2013) Strong two-photon-induced fluorescence from photostable, biocompatible nitrogen-doped graphene quantum dots for cellular and deep-tissue imaging. Nano Lett 13(6):2436–2341

31. Zhao Y-L, Liu Q, Shakoor S, Gong JR, Wang D-Y (2015) Transgenerational safe property of nitrogen-doped graphene quantum dots and the underlying cellular mechanism in *Caenorhabditis elegans*. Toxicol Res 4:270–280

32. Zhao Y-L, Wu Q-L, Tang M, Wang D-Y (2014) The *in vivo* underlying mechanism for recovery response formation in nano-titanium dioxide exposed *Caenorhabditis elegans* after transfer to the normal condition. Nanomedicine 10:89–98

33. Li Y-X, Wang W, Wu Q-L, Li Y-P, Tang M, Ye B-P, Wang D-Y (2012) Molecular control of TiO_2-NPs toxicity formation at predicted environmental relevant concentrations by Mn-SODs proteins. PLoS One 7(9):e44688

34. Qian W, Yan C, He D, Yu Y, Yuan L, Liu M, Luo G, Deng J (2018) pH-triggered charge-reversible of glycol chitosan conjugated carboxyl graphene for enhancing photothermal ablation of focal infection. Acta Biomater. https://doi.org/10.1016/j.actbio.2018.01.022

35. Lalitha M, Lakshmipathi S (2017) Gas adsorption efficacy of graphene sheets functionalised with carboxyl, hydroxyl and epoxy groups in conjunction with Stone-Thrower-Wales (STW) and inverse stone-thrower-Wales (ISTW) defects. Phys Chem Chem Phys 19(45):30895–30913

36. Chiu NF, Fan SY, Yang CD, Huang TY (2017) Carboxyl-functionalized graphene oxide composites as SPR biosensors with enhanced sensitivity for immunoaffinity detection. Biosens Bioelectron 89(Pt 1):370–376

37. Wu Q, Sun Y, Zhang D, Li S, Wang X, Song D (2016) Magnetic field-assisted SPR biosensor based on carboxyl-functionalized graphene oxide sensing film and Fe3O4-hollow gold nano-hybrids probe. Biosens Bioelectron 86:95–101

38. Xie B, Chen Y, Yu M, Shen X, Lei H, Xie T, Zhang Y, Wu Y (2015) Carboxyl-assisted synthesis of nitrogen-doped graphene sheets for supercapacitor applications. Nanoscale Res Lett 10(1):1031

39. Sasidharan A, Sivaram AJ, Retnakumari AP, Chandran P, Malarvizhi GL, Nair S, Koyakutty M (2015) Radiofrequency ablation of drug-resistant cancer cells using molecularly targeted carboxyl-functionalized biodegradable graphene. Adv Healthc Mater 4(5):679–684

40. Liang B, Fang L, Yang G, Hu Y, Guo X, Ye X (2013) Direct electron transfer glucose biosensor based on glucose oxidase self-assembled on electrochemically reduced carboxyl graphene. Biosens Bioelectron 43:131–136

41. Yang J-N, Zhao Y-L, Wang Y-W, Wang H-F, Wang D-Y (2015) Toxicity evaluation and translocation of carboxyl functionalized graphene in *Caenorhabditis elegans*. Toxicol Res 4:1498–1510

42. McGhee JD (2007) Intestine. WormBook. https://doi.org/10.1895/wormbook.1.133.1

43. Liu J, Kang SG, Wang P, Wang Y, Lv X, Liu Y, Wang F, Gu Z, Yang Z, Weber JK, Tao N, Qin Z, Miao Q, Chen C, Zhou R, Zhao Y (2018) Molecular mechanism of Gd@C82(OH)22 increasing collagen expression: implication for encaging tumor. Biomaterials 152:24–36

44. Kang SG, Araya-Secchi R, Wang D, Wang B, Huynh T, Zhou R (2014) Dual inhibitory pathways of metallofullerenol Gd@C82(OH)22 on matrix metalloproteinase-2: molecular insight into drug-like nanomedicine. Sci Rep 4:4775

45. Song Y, Jin J, Li J, He R, Zhang M, Chang Y, Chen K, Wang Y, Sun B, Xing G (2014) Gd@C82(OH)22 nanoparticles constrain macrophages migration into tumor tissue to prevent metastasis. J Nanosci Nanotechnol 14(6):4022–4028

46. Meng J, Liang X, Chen X, Zhao Y (2013) Biological characterizations of [Gd@C82(OH)22]n nanoparticles as fullerene derivatives for cancer therapy. Integr Biol 5(1):43–47

47. Kang SG, Zhou G, Yang P, Liu Y, Sun B, Huynh T, Meng H, Zhao L, Xing G, Chen C, Zhao Y, Zhou R (2012) Molecular mechanism of pancreatic tumor metastasis inhibition by Gd@C82(OH)22 and its implication for de novo design of nanomedicine. Proc Natl Acad Sci USA 109(38):15431–15436

48. Wang B, Yang D, Sun B, Wei X, Guo H, Liu X, Ying G, Niu R, Zhang N, Ma Y (2011) An anti-tumor nanoparticle, [Gd@C82(OH)22]n, induces macrophage activation. J Nanosci Nanotechnol 11(3):2321–2329
49. Chen C, Xing G, Wang J, Zhao Y, Li B, Tang J, Jia G, Wang T, Sun J, Xing L, Yuan H, Gao Y, Meng H, Chen Z, Zhao F, Chai Z, Fang X (2005) Multihydroxylated [Gd@C82(OH)22] n nanoparticles: antineoplastic activity of high efficiency and low toxicity. Nano Lett 5(10):2050–2057
50. Zhang W, Sun B, Zhang L, Zhao B, Nie G, Zhao Y (2011) Biosafety assessment of Gd@ C82(OH)22 nanoparticles on *Caenorhabditis elegans*. Nanoscale 3:2636–2641
51. Zhang L, Jie G, Zhang J, Zhao B (2009) Significant longevity-extending effects of EGCG on *Caenorhabditis elegans* under stress. Free Radic Biol Med 46(3):414–421
52. Beckman KB, Ames BN (1997) Oxidative decay of DNA. J Biol Chem 272(32):19633–19636
53. Lithgow GJ, White TM, Melov S, Johnson TE (1995) Thermotolerance and extended life-span conferred by single-gene mutations and induced by thermal stress. Proc Natl Acad Sci USA 92(16):7540–7544
54. Muñoz MJ, Riddle DL (2003) Positive selection of *Caenorhabditis elegans* mutants with increased stress resistance and longevity. Genetics 163(1):171–180
55. Santacruz-Gomez K, Sarabia-Sainz A, Acosta-Elias M, Sarabia-Sainz M, Janetanakit W, Khosla N, Melendrez R, Pedroza-Montero M, Lal R (2018) Antioxidant activity of hydrated carboxylated nanodiamonds and its influence on water γ-radiolysis. Nanotechnology. https://doi.org/10.1088/1361-6528/aaa80e
56. Sotoma S, Hsieh FJ, Chen YW, Tsai PC, Chang HC (2018) Highly stable lipid-encapsulation of fluorescent nanodiamonds for bioimaging applications. Chem Commun 54(8):1000–1003
57. Huang H, Liu M, Jiang R, Chen J, Mao L, Wen Y, Tian J, Zhou N, Zhang X, Wei Y (2017) Facile modification of nanodiamonds with hyperbranched polymers based on supramolecular chemistry and their potential for drug delivery. J Colloid Interface Sci 513:198–204
58. Basu S, Pacelli S, Wang J, Paul A (2017) Adoption of nanodiamonds as biomedical materials for bone repair. Nanomedicine 12(24):2709–2713
59. Song N, Cui S, Hou X, Ding P, Shi L (2017) Significant enhancement of thermal conductivity in nanofibrillated cellulose films with low mass fraction of nanodiamond. ACS Appl Mater Interfaces 9(46):40766–40773
60. Chen TM, Tian XM, Huang L, Xiao J, Yang GW (2017) Nanodiamonds as pH-switchable oxidation and reduction catalysts with enzyme-like activities for immunoassay and antioxidant applications. Nanoscale 9(40):15673–15684
61. Lim DG, Rajasekaran N, Lee D, Kim NA, Jung HS, Hong S, Shin YK, Kang E, Jeong SH (2017) Polyamidoamine-decorated nanodiamonds as a hybrid gene delivery vector and siRNA structural characterization at the charged interfaces. ACS Appl Mater Interfaces 9(37):31543–31556
62. Whitlow J, Pacelli S, Paul A (2017) Multifunctional nanodiamonds in regenerative medicine: recent advances and future directions. J Control Release 261:62–86
63. Lee H, Lee C, Kim JH (2017) Response to comment on "Activation of persulfate by graphitized nanodiamonds for removal of organic compounds". Environ Sci Technol 51(9):5353–5354
64. Ong SY, Chipaux M, Nagl A, Schirhagl R (2017) Shape and crystallographic orientation of nanodiamonds for quantum sensing. Phys Chem Chem Phys 19(17):10748–10752
65. Suarez-Kelly LP, Campbell AR, Rampersaud IV, Bumb A, Wang MS, Butchar JP, Tridandapani S, Yu L, Rampersaud AA, Carson WE 3rd. (2017) Fluorescent nanodiamonds engage innate immune effector cells: a potential vehicle for targeted anti-tumor immunotherapy. Nanomedicine 13(3):909–920
66. Mohan N, Chen C, Hsieh H, Wu Y, Chang H (2010) *In vivo* imaging and toxicity assessments of fluorescent nanodiamonds in *Caenorhabditis elegans*. Nano Lett 10:3692–3699
67. Zhao Y-L, Yang R-L, Rui Q, Wang D-Y (2016) Intestinal insulin signaling encodes two different molecular mechanisms for the shortened longevity induced by graphene oxide in *Caenorhabditis elegans*. Sci Rep 6:24024

68. Ren M-X, Zhao L, Lv X, Wang D-Y (2017) Antimicrobial proteins in the response to graphene oxide in *Caenorhabditis elegans*. Nanotoxicology 11(4):578–590

69. Zhuang Z-H, Li M, Li H, Luo L-B, Gu W-D, Wu Q-L, Wang D-Y (2016) Function of RSKS-1-AAK-2-DAF-16 signaling cascade in enhancing toxicity of multi-walled carbon nanotubes can be suppressed by *mir-259* activation in *Caenorhabditis elegans*. Sci Rep 6:32409

70. Zhao Y-L, Zhi L-T, Wu Q-L, Yu Y-L, Sun Q-Q, Wang D-Y (2016) p38 MAPK-SKN-1/ Nrf signaling cascade is required for intestinal barrier against graphene oxide toxicity in *Caenorhabditis elegans*. Nanotoxicology 10(10):1469–1479

71. Li W-J, Wang D-Y, Wang D-Y (2018) Regulation of the response of *Caenorhabditis elegans* to simulated microgravity by p38 mitogen-activated protein kinase signaling. Sci Rep 8:857

72. Inoue H, Hisamoto N, An JH, Oliveira RP, Nishida E, Blackwell TK, Matsumoto K (2005) The *C. elegans* p38 MAPK pathway regulates nuclear localization of the transcription factor SKN-1 in oxidative stress response. Genes Dev 19(19):2278–2283

73. Wang J, Robida-Stubbs S, Tullet JM, Rual JF, Vidal M, Blackwell TK (2010) RNAi screening implicates a SKN-1-dependent transcriptional response in stress resistance and longevity deriving from translation inhibition. PLoS Genet 6(8):e1001048

74. Li K, Zhang Y, Chen M, Hu Y, Jiang W, Zhou L, Li S, Xu M, Zhao Q, Wan R (2017) Enhanced antitumor efficacy of doxorubicin-encapsulated halloysite nanotubes. Int J Nanomedicine 13:19–30

75. Long Z, Zhang J, Shen Y, Zhou C, Liu M (2017) Polyethyleneimine grafted short halloysite nanotubes for gene delivery. Mater Sci Eng C Mater Biol Appl 81:224–235

76. Hu Y, Chen J, Li X, Sun Y, Huang S, Li Y, Liu H, Xu J, Zhong S (2017) Multifunctional halloysite nanotubes for targeted delivery and controlled release of doxorubicin in-vitro and in-vivo studies. Nanotechnology 28(37):375101

77. Peng H, Liu X, Tang W, Ma R (2017) Facile synthesis and characterization of ZnO nanoparticles grown on halloysite nanotubes for enhanced photocatalytic properties. Sci Rep 7(1):2250

78. Makaremi M, Pasbakhsh P, Cavallaro G, Lazzara G, Aw YK, Lee SM, Milioto S (2017) Effect of morphology and size of halloysite nanotubes on functional pectin bionanocomposites for food packaging applications. ACS Appl Mater Interfaces 9(20):17476–17488

79. Zhou T, Jia L, Luo YF, Xu J, Chen RH, Ge ZJ, Ma TL, Chen H, Zhu TF (2016) Multifunctional nanocomposite based on halloysite nanotubes for efficient luminescent bioimaging and magnetic resonance imaging. Int J Nanomedicine 11:4765–4776

80. Fakhrullin RF, Lvov YM (2016) Halloysite clay nanotubes for tissue engineering. Nanomedicine 11(17):2243–2246

81. Zeng G, He Y, Zhan Y, Zhang L, Pan Y, Zhang C, Yu Z (2016) Novel polyvinylidene fluoride nanofiltration membrane blended with functionalized halloysite nanotubes for dye and heavy metal ions removal. J Hazard Mater 317:60–72

82. Tully J, Yendluri R, Lvov Y (2016) Halloysite clay nanotubes for enzyme immobilization. Biomacromolecules 17(2):615–621

83. Zhang H, Ren T, Ji Y, Han L, Wu Y, Song H, Bai L, Ba X (2015) Selective modification of halloysite nanotubes with 1-pyrenylboronic acid: a novel fluorescence probe with highly selective and sensitive response to hyperoxide. ACS Appl Mater Interf 7(42):23805–23811

84. Lvov Y, Wang W, Zhang L, Fakhrullin R (2016) Halloysite clay nanotubes for loading and sustained release of functional compounds. Adv Mater 28(6):1227–1250

85. Jana S, Das S, Ghosh C, Maity A, Pradhan M (2015) Halloysite nanotubes capturing isotope selective atmospheric CO_2. Sci Rep 5:8711

86. Fakhrullina GI, Akhatova FS, Lvovb YM, Fakhrullin RF (2015) Toxicity of halloysite clay nanotubes in vivo: a *Caenorhabditis elegans* study. Environ Sci Nano 2:54–59

87. Wu T, Tang M (2018) The inflammatory response to silver and titanium dioxide nanoparticles in the central nervous system. Nanomedicine 13(2):233–249

88. Liu K, Lin X, Zhao J (2013) Toxic effects of the interaction of titanium dioxide nanoparticles with chemicals or physical factors. Int J Nanomedicine 8:2509–2520

89. Shi H, Magaye R, Castranova V, Zhao J (2013) Titanium dioxide nanoparticles: a review of current toxicological data. Part Fibre Toxicol 10:15
90. Wu Q-L, Wang W, Li Y-X, Li Y-P, Ye B-P, Tang M, Wang D-Y (2012) Small sizes of TiO₂-NPs exhibit adverse effects at predicted environmental relevant concentrations on nematodes in a modified chronic toxicity assay system. J Hazard Mater 243:161–168
91. Wu Q-L, Nouara A, Li Y-P, Zhang M, Wang W, Tang M, Ye B-P, Ding J-D, Wang D-Y (2013) Comparison of toxicities from three metal oxide nanoparticles at environmental relevant concentrations in nematode *Caenorhabditis elegans*. Chemosphere 90:1123–1131
92. Wu Q-L, Zhao Y-L, Li Y-P, Wang D-Y (2014) Susceptible genes regulate the adverse effects of TiO₂-NPs at predicted environmental relevant concentrations on nematode *Caenorhabditis elegans*. Nanomedicine 10:1263–1271
93. Wang H, Wick RL, Xing B (2009) Toxicity of nanoparticulate and bulk ZnO, Al₂O₃ and TiO₂ to the nematode *Caenorhabditis elegans*. Environ Pollut 157:1171–1177
94. Khare P, Sonane M, Pandey R, Ali S, Gupta KC, Satish A (2011) Adverse effects of TiO₂ and ZnO nanoparticles in soil nematode, *Caenorhabditis elegans*. J Biomed Nanotechnol 7:116–117
95. Rui Q, Zhao Y-L, Wu Q-L, Tang M, Wang D-Y (2013) Biosafety assessment of titanium dioxide nanoparticles in acutely exposed nematode *Caenorhabditis elegans* with mutations of genes required for oxidative stress or stress response. Chemosphere 93(10):2289–2296
96. Shen L-L, Xiao J, Ye H-Y, Wang D-Y (2009) Toxicity evaluation in nematode *Caenorhabditis elegans* after chronic metal exposure. Environ Toxicol Pharmacol 28:125–132
97. Chen X, Cheng B, Yang Y, Cao A, Liu J, Du L, Liu Y, Zhao Y, Wang H (2013) Characterization and preliminary toxicity assays of nano-titanium dioxide additive in sugar-coated chewing gum. Small 9(9–10):1765–1774
98. Nouara A, Wu Q-L, Li Y-X, Tang M, Wang H-F, Zhao Y-L, Wang D-Y (2013) Carboxylic acid functionalization prevents the translocation of multi-walled carbon nanotubes at predicted environmental relevant concentrations into targeted organs of nematode *Caenorhabditis elegans*. Nanoscale 5:6088–6096

Chapter 10
Surface Chemical Modification to Reduce the Toxicity of Nanomaterials

Abstract In the field of nanoscience, a series of surface chemical modifications or coatings have been designed to acquire the new physicochemical properties or to enhance the certain aspects of physicochemical properties. Meanwhile, some of the surface chemical modifications or coatings also have the beneficial effects in reducing the potential toxicity of nanomaterials. In this chapter, we focus on three important surface chemical modifications, polyethylene glycol (PEG) modification, ZnS or fetal bovine serum (FBS) surface coating, and carboxyl (–COOH) modification, to introduce the beneficial effects of these surface chemical modifications in reducing the toxicity of nanomaterials in nematodes.

Keywords Surface modification · Surface coating · Toxicity reduction · Nanomaterials · *Caenorhabditis elegans*

10.1 Introduction

In the field of nanoscience, the main effort to reduce the toxicity of engineered nanomaterials (ENMs) is to design the suitable modification on the surface of examined ENMs [1–4]. Actually, reduction of the potential toxicity of certain ENMs is normally not the only research aim for surface chemical modifications. The aims to design certain surface chemical modifications usually contain both the toxicity reduction and the acquirement of new physicochemical properties or the enhancement of certain aspects of physicochemical properties [5–14]. One of the chemical modifications is to design chemical binding of certain group(s) [15, 16]. Another important chemical modification is to design certain form of coating to cover the examined ENMs [17, 18]. So far, besides the acquirement of new physicochemical properties or the enhancement of certain aspects of physicochemical properties, the increasing evidence on nanotoxicology has prompted the attention for chemists or material scientists the necessity to consider thoroughly the safety of ENMs potentially used in different industrial and medical applications.

In this chapter, we mainly focus on three important chemical modifications, polyethylene glycol (PEG) modification, ZnS or fetal bovine serum (FBS) surface

© Springer Nature Singapore Pte Ltd. 2018
D. Wang, *Nanotoxicology in* Caenorhabditis elegans,
https://doi.org/10.1007/978-981-13-0233-6_10

coating, and carboxyl (–COOH) modification, to introduce the different aspects of certain surface chemical modifications to reduce the potential toxicity of ENMs in nematodes.

10.2   PEG Modification

PEG surface modification is a frequently employed strategy to acquire the new physicochemical properties and to reduce the toxicity of ENMs [19–23]. More importantly, it has been indicated that the surface PEGylation could provide a "stealth" characteristic for ENMs; otherwise it would be identified as foreign materials by the body of organisms [23, 24].

10.2.1   PEG Modification to Reduce the Toxicity of Graphene Oxide (GO)

Along with the worldwide GO production, it is possible that a large amount of GO could be released into the environment, and GO would pose a significant environmental and health risk. PEG surface modification is a normally used chemical strategy to reduce the GO toxicity [25, 26].

10.2.1.1   Beneficial Function of PEG Modification in Reducing GO Toxicity in Nematodes

Chronic exposure to GO-PEG (1 mg/L) from L1-larvae to adult day-8 did not cause the toxicity on locomotion behavior as reflected by the endpoints of head thrash and body bend in nematodes [27] (Fig. 10.1). Similarly, chronic exposure to GO-PEG (1 mg/L) also did not result in the noticeable inductions of either intestinal autofluorescence or intestinal reactive oxygen species (ROS) production in nematodes [27] (Fig. 10.1). In contrast, chronic exposure to GO (1 mg/L) led to the significant decrease in locomotion behavior and the significant induction of both intestinal autofluorescence and intestinal ROS production in nematodes [27] (Fig. 10.1). These results demonstrate that PEG surface modification can be helpful for maintaining the normal functional state of both primary and secondary targeted organs in GO-exposed nematodes.

10.2.1.2   PEG Surface Modification Suppresses both the GO Deposition and the OP50 Accumulation in the Intestine

In nematodes, chronic exposure to GO from L1-larvae to adult day-8 caused the severe GO deposition in the body, as well as the severe OP50 accumulation in the intestine, which is considered to be closely associated with the possible decrease in

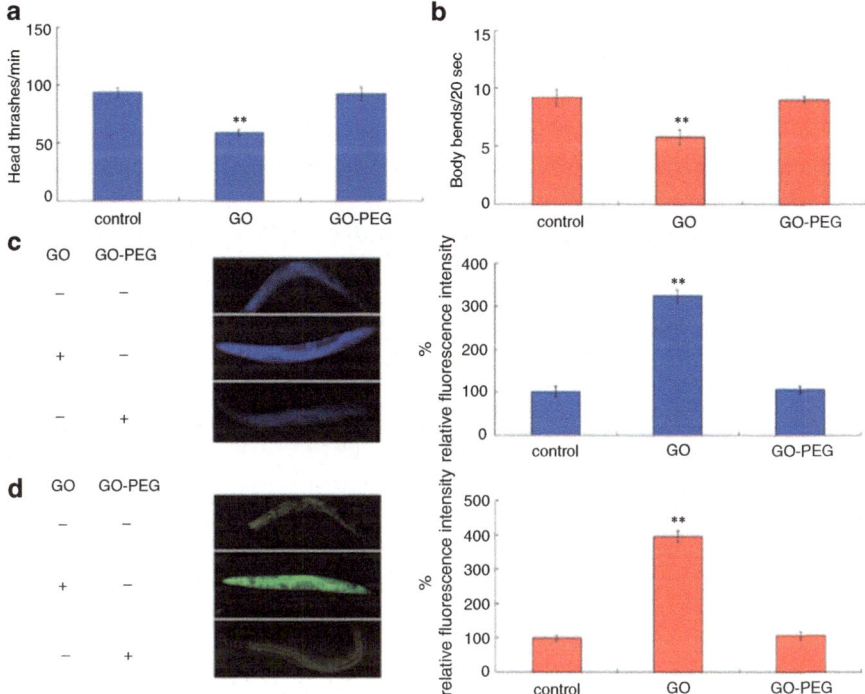

Fig. 10.1 PEG surface modification reduced chronic GO toxicity [27]
(**a**) Effect of PEG surface modification on head thrash of GO (1 mg/L)-exposed nematodes. (**b**) Effect of PEG surface modification on body bend of GO (1 mg/L)-exposed nematodes. (**c**) Effect of PEG surface modification on induction of intestinal autofluorescence in GO (1 mg/L)-exposed nematodes. (**d**) Effects of PEG surface modification on induction of intestinal ROS production in GO (1 mg/L)-exposed nematodes. Exposure was performed from L1-larvae to adult day-8. Bars represent means ± SEM **$P < 0.01$

innate immune response of animals [27]. In contrast, PEG surface modification noticeably suppressed the GO deposition in the body, including the intestine, in nematodes [27]. More importantly, PEG surface modification obviously inhibited the OP50 accumulation in the intestine in nematodes [27]. Meanwhile, PEG surface modification significantly recovered the expression patterns of dysregulated genes encoding antimicrobial peptides or p38 MAPK signaling pathway induced by GO chronic exposure [27]. That is, PEG surface modification may have the potential to prevent the OP50 accumulation and to maintain the normal intestinal innate immune response in GO-exposed nematodes.

10.2.1.3 PEG Surface Modification Is Helpful for Maintaining Normal Developmental and Functional States of AVL and DVB Neurons

In nematodes, AVL neurons in the head and DVB neurons in the tail are required for the control of defecation behavior [28]. GO exposure could cause the significant increase in mean defecation cycle length in nematodes [27, 29]. Additionally, GO exposure resulted in the developmental deficit in both the AVL neurons and the DVB neurons as indicated by the decrease in fluorescent size of AVL or DVB neurons in GO-exposed nematodes [27]. In contrast, PEG surface modification could be helpful for nematodes to maintain the normal developmental state of both the AVL neurons and the DVB neurons [27]. Moreover, the normal mean defecation cycle length was observed in PEG-GO-exposed nematodes after chronic exposure from L1-larvae to adult day-8 [27].

A cellular mechanism has been raised to explain these observations. In GO-exposed nematodes, the obvious colocalization of Rho B-labeled GO with AVL or DVB neurons has been detected [27]. Different from this, PEG surface modification inhibited the colocalization of Rho B-labeled GO with AVL or DVB neurons [27].

10.2.1.4   Molecular Basis for PEG Modification to Reduce GO Toxicity in Nematodes

An important molecular mechanism raised to explain the beneficial effect of PEG surface modification to reduce the GO toxicity is that PEG modification may reduce the GO toxicity by influencing the functions of specific lncRNAs in nematodes [30]. After the comparison of 34 candidate long noncoding RNAs (lncRNAs) in control, GO- or GO-PEG-exposed nematodes, expression patterns of some dysregulated lncRNAs induced by GO exposure could be reversed by PEG surface modification [30]. PEG surface modification could increase the expressions of *linc-37*, *linc-5*, *linc- 24*, *linc-14*, *XLOC_013642*, *XLOC_010849*, and *XLOC_004416* and decrease the expression of *XLOC_007959* in GO-exposed nematodes [30].

Another corresponding molecular mechanism has also been raised to explain the observations introduced above. In nematodes, GO exposure dysregulated the expression levels of some genes required for the control of defecation behavior [27, 29]. In contrast, PEG surface modification could be helpful for nematodes to maintain the normal expression patterns of genes required for the control of defecation behavior including those dysregulated by GO exposure [27].

10.2.2   PEG Surface Modification to Reduce the Toxicity of Multiwalled Carbon Nanotubes (MWCNTs)

10.2.2.1   Beneficial Function of PEG Surface Modification Against the MWCNT Toxicity

MWCNT is another important carbon-based ENM, which has been widely used in different fields. In nematodes, prolonged exposure (from L1-larvae to adult day-1) to MWCNTs at environmental relevant concentrations, such as 0.1–1 µg/L, could cause the toxicity in inducing intestinal autofluorescence or ROS production and in decreasing locomotion behavior (Fig. 10.2) [31]. These results imply that long-term exposure to MWCNTs could potentially cause the damage on the functions of both

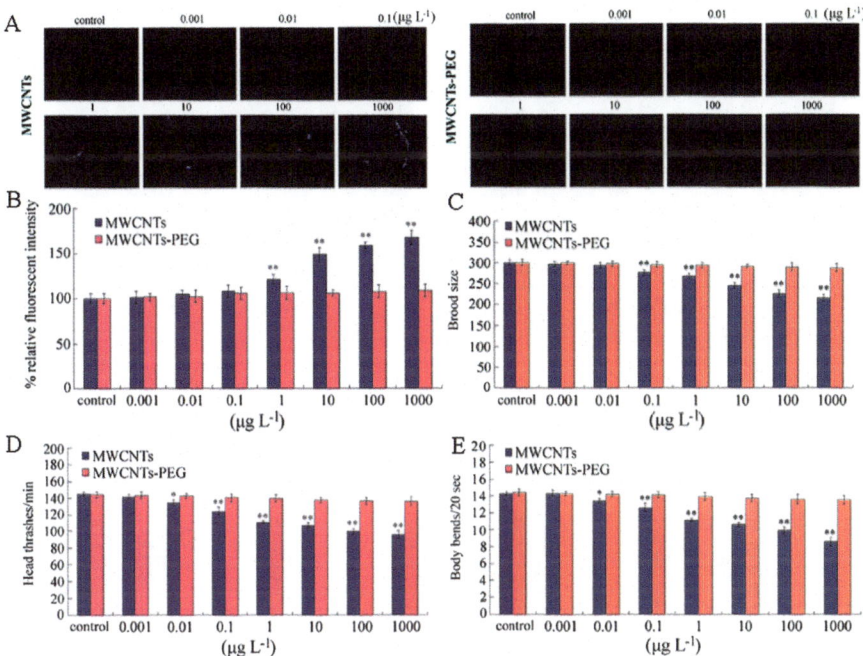

Fig. 10.2 Beneficial effect PEG surface modification in reducing MWCNT toxicity [31]
(**a**) Pictures showing intestinal autofluorescences of MWCNTs and MWCNT-PEG-exposed nematodes. (**b**) Comparison of intestinal autofluorescences in MWCNT-PEG-exposed nematodes from those in MWCNT-exposed nematodes. (**c**) Comparison of brood sizes in MWCNT-PEG-exposed nematodes from those in MWCNT-exposed nematodes. (**d**) Comparison of head thrashes in MWCNT-PEG-exposed nematodes from those in MWCNT-exposed nematodes. (**e**) Comparison of body bends in MWCNT-PEG-exposed nematodes from those in MWCNT-exposed nematodes. Exposure was performed from L1-larvae to adult day-1. Bars represent mean ± SEM *$P < 0.05$, **$P < 0.01$.

primary targeted organs, such as the intestine, and secondary targeted organs, such as the neurons, in nematodes. In contrast, PEG surface modification could effectively reduce the potential toxicity induced by prolonged exposure to MWCNTs at concentrations of 0.1–1000 μg/L as indicated by the endpoints of intestinal autofluorescence, intestinal ROS production, locomotion behavior, and brood size (Fig. 10.2) [31]. That is, the normal physiological function of both primary targeted organs and secondary targeted organs could be detected in nematodes exposed to MWCNTs-PEG.

10.2.2.2   Beneficial Effect of PEG Surface Modification in Inhibiting Accumulation and Translocation of MWCNTs in Nematodes

After prolonged exposure, a larger amount of MWCNTs were accumulated in the pharynx, the intestine, and the tail region (Fig. 10.3) [31]. Meanwhile, a great amount of MWCNTs were translocated into the spermatheca and the gonad through

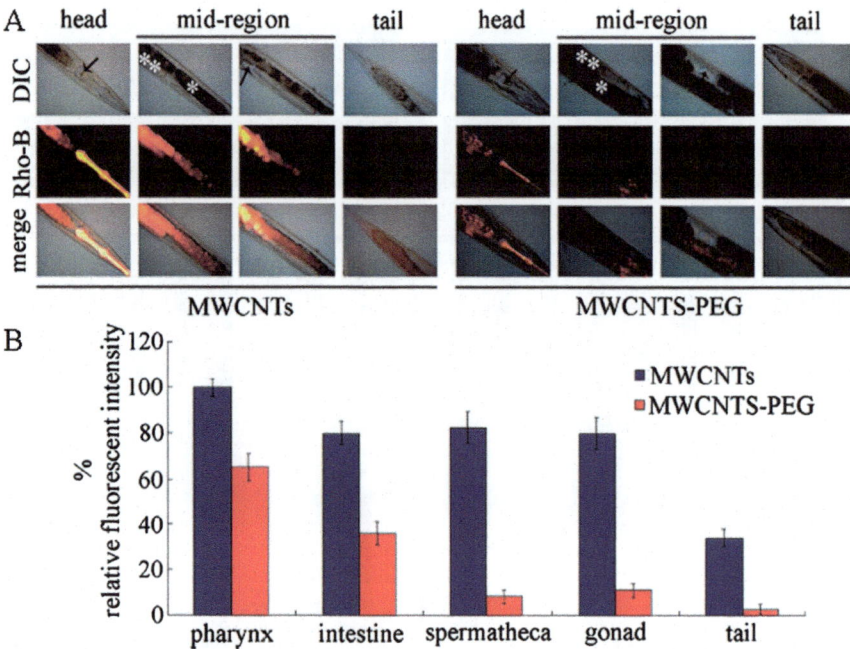

Fig. 10.3 Comparison of translocation and distribution of MWCNTs and MWCNTs-PEG [31]
(**a**) Comparison of translocation and distribution of MWCNTs and MWCNTs-PEG in different regions and organs of nematodes. Arrowheads indicate the pharynx and spermatheca, respectively, in the head region or mid-region of nematodes. The intestine (*) and gonad (**) in the mid-region are also indicated. (**b**) Comparison of the fluorescent intensities of MWCNTs and MWCNTs-PEG in different regions and organs of nematodes. Exposure concentration for MWCNTs and MWCNTs-PEG was 1 mg/L. Exposure was performed from L1-larvae to adult day-1. Bars represent mean ± SEM.

the intestinal barrier (Fig. 10.3) [31]. A certain amount of MWCNTs were also translocated into the region surrounding the pharynx where most of neurons are located (Fig. 10.3) [31]. Different from these, PEG surface modification obviously reduced the accumulation of MWCNTs in the pharynx, the intestine, and the tail region (Fig. 10.3) [31]. More importantly, PEG surface modification could effectively block the translocation of MWCNTs into the spermatheca, the gonad, and the region surrounding the pharynx (Fig. 10.3) [31]. These results imply an important cellular mechanism for PEG surface modification in reducing MWCNT toxicity. That is, PEG surface modification may reduce the MWCNT toxicity by inhibiting the accumulation of MWCNTs in the pharynx, the intestine, and the tail region and blocking the translocation of MWCNTs into secondary targeted organs in nematodes.

10.2.2.3 Maintenance of Normal Intestinal Permeability in Reducing MWCNT Toxicity Induced by PEG Surface Modification

Based on the assessment data using lipophilic fluorescent dye of Nile red, prolonged exposure to MWCNTs (1 mg/L) enhanced the intestinal permeability of nematodes [31]. Additionally, prolonged exposure to MWCNTs (1 mg/L) altered the expression patterns of some genes required for the control of intestinal development and function. These genes contain *pgp-3*, *gem-4*, *par-3*, *pkc-3*, *ajm-1*, *inx-3*, *abts-4*, and *lin-7* [31]. In nematodes, *pkc-3*, *par-3*, and *pgp-3* are required for the development of microvilli on intestinal cells, *inx-3* and *abts-4* are required for the development of a basolateral domain of the intestine, and *ajm-1* and *lin-7* are associated with development of an apical junction of the intestine [32]. In nematodes, *pgp-3* encodes a transmembrane protein, *gem-4* encodes a Ca^{2+}-dependent phosphatidylserine binding protein, *par-3* encodes a PDZ domain-containing protein, *pkc-3* encodes an atypical protein kinase, *ajm-1* encodes a member of apical junction molecule class, *lin-7* encodes a protein that contains a PDZ domain and an L27 domain, *inx-3* encodes a gap protein, and *abts-4* encodes an anion transporter.

After PEG surface modification, the normal intestinal permeability was observed in MWCNT-PEG (1 mg/L)-exposed nematodes [31]. Moreover, PEG surface modification could rescue the dysregulation of some genes (*pgp-3*, *gem-4*, *par-3*, *pkc-3*, *ajm-1, inx-3, abts-4*, and *lin-7*) induced by prolonged exposure to MWCMTs (1 mg/L). These observations provide another important cellular mechanism for PEG surface modification in reducing MWCNT toxicity. These observations are also useful for explaining the beneficial function of PEG surface modification in inhibiting the accumulation and the translocation of MWCNTs as introduced above.

10.3 ZnS Surface Coating

In the field of nanoscience, quantum dots (QDs) have received the tremendous attentions due to their fascinating optical and electronic properties [33, 34]. Considering the potential application of QDs in medical imaging, drug delivery, targeted therapy, nanodiagnostics, gene technology, and biosensor detection [35–40], QDs would be released into the environment and available for human beings and environmental organisms. Both in vitro and in vivo evidence has suggested the potential QDs toxicity for organisms [41, 42]. To reduce the potential toxicity of QDs, ZnS surface coating is a normally used chemical strategy to generate a core-shell structure together with the inner QDs [42, 43].

10.3.1 Beneficial Effect of ZnS Surface Coating in Reducing CdTe QD Toxicity on Both Development and Function of RME Motor Neurons

In nematodes, γ-aminobutyric acid (GABA) is an amino acid neurotransmitter and acts primarily at the neuromuscular synapses [44]. Among the GABAergic neurons, RME motor neurons (RMED, RMEV, RMEL, and RMER) are located in the head region and control the foraging behavior of nematodes [44]. After prolonged exposure (from L1-larvae to young adults), CdTe QDs at concentrations ≥0.1 μg/L induced the abnormal foraging behavior and the deficits in the development of RME motor neurons [45]. In contrast, ZnS surface coating could suppress the toxicity of CdTe QDs (0.1–1 μg/L) on foraging behavior (Fig. 10.4) [45]. Meanwhile, ZnS surface coating obviously inhibited the toxicity of CdTe QDs (1 μg/L) on the development of RME motor neurons as indicated by the morphological alteration, relative fluorescent intensity, and the relative fluorescent size of cell body of RMEV neuron (Fig. 10.4) [45]. Therefore, ZnS surface coating can reduce the neurotoxicity of CdTe QDs on both the development and the function of RME motor neurons.

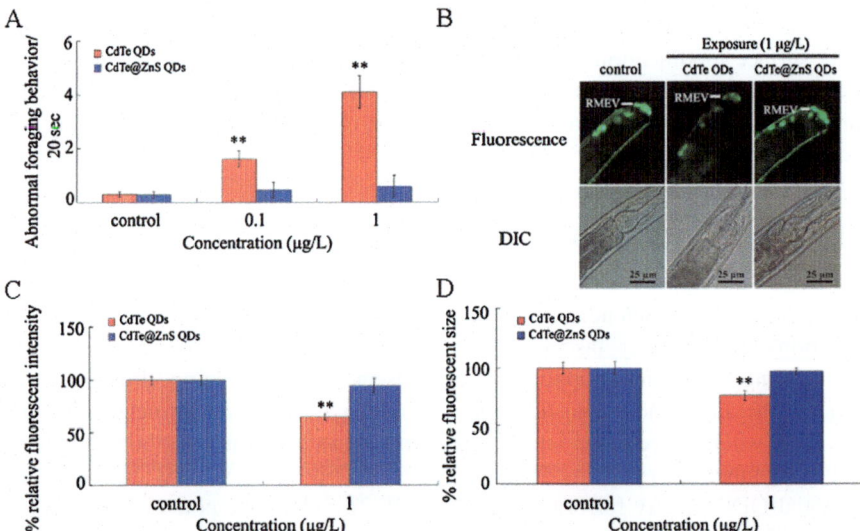

Fig. 10.4 ZnS surface coating reduces the toxicity of CdTe QDs on development and function of RME motor neurons [45]

(**a**) Comparison of effects on foraging behavior between CdTeQDs and CdTe@ZnS QDs. (**b**) Pictures showing the effects of CdTe@ZnS QDs on development of RME motor neurons. (**c**) Comparison of effects on fluorescent intensity of cell bodies in RMEV motor neurons between CdTe QDs and CdTe@ZnS QDs. (**d**) Comparison of effects on fluorescent size of cell bodies in RMEV motor neurons between CdTeQDs and CdTe@ZnS QDs. CdTe QDs was exposed from L1-larvae to young adult. Bars represent mean ± SEM **$P < 0.01$

10.3.2 ZnS Surface Coating Inhibits Translocation and Accumulation of CdTe QDs in RME Motor Neurons

In nematodes, prolonged exposure to CdTe QDs (1 µg/L) could cause the translocation of CdTe QDs into the RME motor neurons as indicated by the colocalization signals between CdTe QDs and RME motor neurons and the accumulation of CdTe QDs in the RME motor neurons (Fig. 10.5) [45]. In contrast, ZnS surface coating effectively blocked the translocation of CdTe QDs into the RME motor neurons (Fig. 10.5) [45]. In CdTe@ZnS QD-exposed nematodes, the CdTe@ZnS QDs

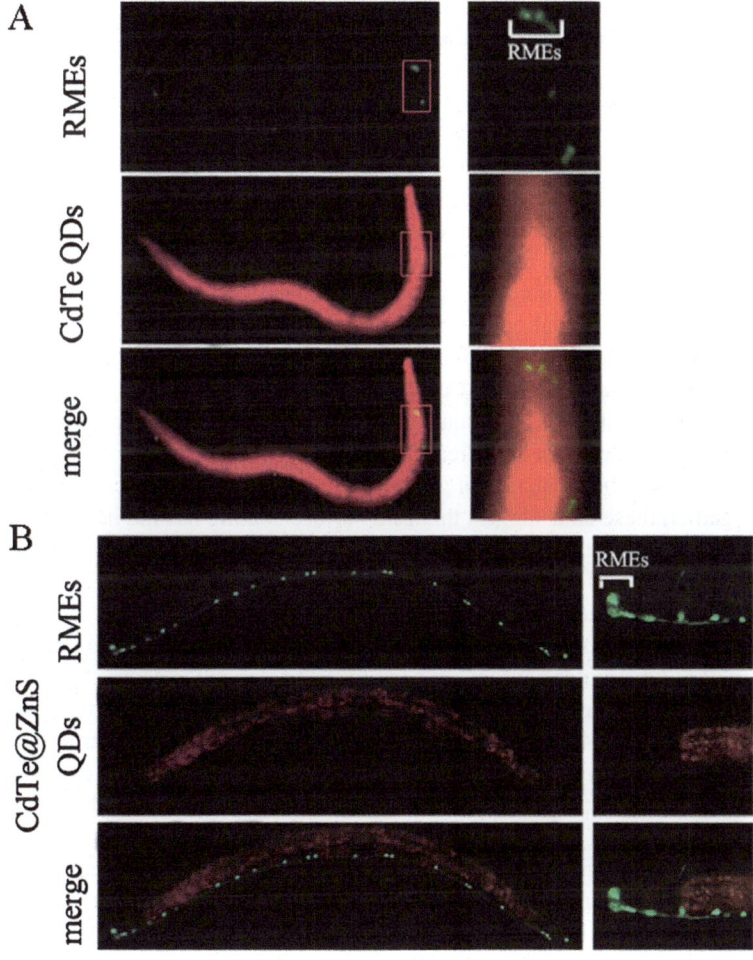

Fig. 10.5 Distribution and translocation of CdTe QDs (**a**) and CdTe@ZnS QDs (**b**) in RME motor neurons [45]

signals were mainly distributed in the intestine, and the fluorescence intensity of CdTe@ZnS QDs signals in the intestine was much lower than that of CdTe QDs signals in the intestine (Fig. 10.5) [45]. These observations provide an important cellular basis for the beneficial effect of ZnS surface coating in reducing the neurotoxicity of CdTe QDs on the development and the function of RME motor neurons.

10.3.3   Beneficial Effect of ZnS Surface Coating in Reducing Toxicity of CdTe QDs on the Development and Function of D-Type Motor Neurons

In nematodes, among the 26 GABAergic neurons, ventral cord D-type neurons, including 6 DD and 13 VD motor neurons, function in inhibiting the contraction of ventral and dorsal body wall muscles during the locomotion [44]. When the nematodes lack the functional D-type motor neurons, it would display a shrinking behavior after touching [44]. In nematodes, prolonged exposure (from L1-larvae to young adults) to CdTe QDs at concentrations ≥0.1 µg/L induced the significant shrinker behavior [46]. Moreover, prolonged exposure to CdTe QDs at concentrations ≥0.1 µg/L caused the deficits in D-type motor neurons as indicated by the decrease in fluorescence intensity of cell bodies, the neuronal loss, and the formation of gaps on both ventral and dorsal cords in D-type motor neurons [46]. In contrast, ZnS surface coating prevented the formation of shrinker behavior in nematodes exposed to CdTe QDs at concentrations ≤1 µg/L (Fig. 10.6) [46]. Additionally, ZnS surface coating effectively inhibited the damage of CdTe QDs on the development of D-type motor neurons, including the decrease in fluorescence intensity of cell bodies, the neuronal loss, and the formation of gaps on both ventral and dorsal cords (Fig. 10.6) [46]. Together, these data suggest that ZnS surface coating has the beneficial effect in reducing the neurotoxicity of CdTe QDs on both the development and function of D-type motor neurons in nematodes.

An underlying molecular basis was raised to explain the beneficial effect of ZnS surface coating in reducing the neurotoxicity CdTe QDs on the development and the function of D-type motor neurons. In nematodes, prolonged exposure to CdTe QDs (1 µg/L) resulted in the significant decrease in transcriptional expressions of *unc-25*, *unc-30*, and *unc-47* [46]. In contrast, it has been shown that ZnS surface coating could prevent the decreasing tendency of *unc-25*, *unc-30*, and *unc-47* expression in CdTe QD (1 µg/L)-exposed nematodes (Fig. 10.6) [46]. *unc-25* encodes a glutamic acid decarboxylase, a biosynthesis enzyme for the GABA [47]. *unc-30* encodes a homeodomain transcription factor that is required for the GABA neuron identity [48]. *unc-47* encodes a GABA transporter with the function in pumping the GABA into synaptic vesicles [49].

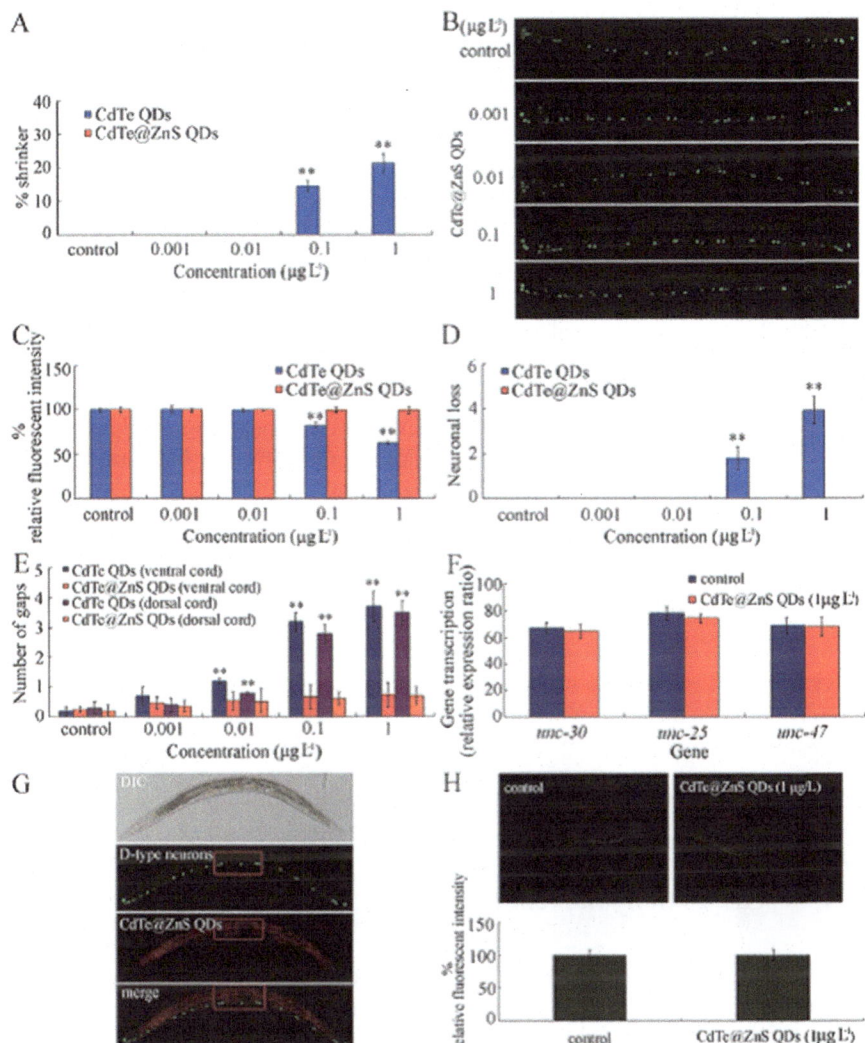

Fig. 10.6 Beneficial effect of ZnS surface coating in reduced neurotoxicity of CdTe QDs on development and function of D-type motor neurons [46]

(a) Comparison of effects on the shrinking behavior formation between CdTe QDs and CdTe@ZnS QDs. (b) Images showing effects of CdTe@ZnS QDs on the development of D-type motor neurons. (c) Comparison of effects on the fluorescent intensity of cell bodies in D-type motor neurons between CdTe QDs and CdTe@ZnS QDs. (d) Comparison of effects on the neuronal loss of D-type motor neurons between CdTe QDs and CdTe@ZnS QDs. (e) Comparison of effects on the formation of gaps on ventral and dorsal cords between CdTe QDs and CdTe@ZnS QDs. (f) Effects of CdTe@ZnS QDs on the expression levels of *unc-30*, *unc-25*, and *unc-47* genes. The qRT-PCR results were expressed as the relative expression ratio between targeted gene and reference *tba-1* gene. (g) CdTe@ZnS QDs (1 μg/L) could not be colocalized with D-type motor neurons. (h) Effects of CdTe@ZnS QDs on the ROS production. Control, without CdTe QDs exposure. CdTe QDs was exposed from L1-larvae to young adults. Bars represent means ± SEM. **$P < 0.01$

10.3.4   Cellular Mechanism for ZnS Surface Coating in Reducing the Neurotoxicity of CdTe QDs on D-Type Motor Neurons

To determine the underlying cellular mechanism for ZnS surface coating in reducing the neurotoxicity of CdTe QDs, the distribution and the translocation of CdTe@ ZnS QDs in nematodes were examined. In nematodes, the fluorescence signals of CdTe QDs could be colocalized with the fluorescence signals labeling the D-type motor neurons [46]. It has further been shown that ZnS surface coating could effectively block the translocation of CdTe QDs into the D-type motor neurons (Fig. 10.6) [46]. Besides this, ZnS surface coating also inhibited the induction of ROS production caused by the CdTe QDs exposure [46]. Therefore, ZnS surface coating may reduce the neurotoxicity of CdTe QDs on the development and the function of D-type motor neurons by preventing the translocation of CdTe QDs into D-type motor neurons and the induction of ROS production.

10.4   FBS Surface Coating

In organisms, it is normally considered that the FBS surface coating can provide a protein corona on the surface of certain ENMs, which in turn is helpful for reducing the toxicity of ENMs [50].

10.4.1   Beneficial Effect of FBS Surface Coating in Reducing GO Toxicity in Nematodes

In nematodes, prolonged exposure (from L1-larvae to young adults) could cause the toxicity in reducing lifespan, in decreasing locomotion behavior, and in reducing brood size [29, 30]. The evidence has been raised that FBS surface coating could markedly inhibit the GO (100 mg/L) toxicity in reducing lifespan, in decreasing locomotion behavior, and in reducing brood size [30].

10.4.2   Cellular Basis for FBS Surface Coating in Reducing GO Toxicity in Nematodes

One cellular mechanism raised to explain the beneficial effect of FBS surface coating in reducing GO toxicity is that FBS surface coating could inhibit the induction of ROS production induced by prolonged exposure to GO (100 mg/L) [30]. Another important cellular mechanism to explain the beneficial effect of FBS surface coating in reducing GO toxicity is that FBS surface coating effectively prevented the translocation of GO into the secondary targeted organs, such as the reproductive organs, and the accumulation of GO in the intestine in nematodes [30].

10.4.3 Molecular Basis for FBS Surface Coating in Reducing GO Toxicity in Nematodes

The role of lncRNAs in the regulation of the function of FBS surface coating in reducing GO toxicity has been determined. FBS surface coating could suppress the increase in the expressions of *linc-37*, *linc-24*, *linc-14*, *XLOC_004416*, *XLOC_013698*, and *XLOC_012820* and the decrease in the expression of *XLOC_007959* induced by GO exposure [30]. More importantly, it was found that both the FBS surface coating and the PEG surface modification could decrease the expressions of *linc-37*, *linc-24*, *linc-14*, and *XLOC_004416* and increase the expression of *XLOC_007959* in GO-exposed nematodes [30], which implies that FBS surface coating or PEG surface modification may potentially reduce the GO toxicity by influencing the functions of some certain conserved molecular basis mediated by same lncRNAs in nematodes.

Nevertheless, after the comparison of the mRNA profiles in GO-PEG- or GO-FBS-exposed treated nematodes with that in GO-treated nematodes, it was found that FBS surface coating caused the downregulation of 298 genes and the upregulation of 184 genes compared with GO exposure [30]. Different from this, PEG surface modification resulted in the downregulation of 178 genes and the upregulation of 156 genes [30]. Most of the dysregulated genes in GO-FBS-treated nematodes were different from those in GO-PEG-treated nematodes, which implies that the mechanisms for the beneficial effect of FBS surface coating in reducing GO toxicity may be somewhat different from those of PEG surface modification.

10.5 COOH Modification

Carboxyl (–COOH) surface modification is also a commonly used chemical strategy for ENMs with the aims to modulate their application in targeted delivery of anticancer drugs and imaging [51–55].

10.5.1   Beneficial Effect of Carboxyl Surface Modification in Reducing Toxicity of MWCNTs at Environmentally Relevant Concentrations

The predicted environmentally relevant concentrations for CNTs were 6.6–31.5 ng/L for the sewage treatment plant effluent and 55 ng/m^3 for the urban air samples [56, 57]. After prolonged exposure from L1-larvae to adult day-1, carboxyl surface modification could significantly inhibit the toxicity of MWCNTs at concentrations of 0.01–10 µg/L in decreasing locomotion behavior and the toxicity of MWCNTs at concentrations of 0.1–10 µg/L in reducing brood size and in inducing intestinal ROS production [58]. Thus, carboxyl surface modification can effectively reduce the toxicity of MWCNTs at environmentally relevant concentrations or concentrations slightly higher than the environmentally relevant concentrations in nematodes.

10.5.2   Cellular Basis for the Beneficial Effect of Carboxyl Surface Modification in Reducing MWCNT Toxicity

The first cellular mechanism to explain the beneficial effect of carboxyl surface modification in reducing MWCNT (10 µg/L) toxicity is that carboxyl surface modification could block the translocation of MWCNTs into the secondary targeted organs, such as the spermatheca, and the embryos in the body through the intestinal barrier, and inhibit the severe accumulation of MWCNTs in the intestine in nematodes (Fig. 10.7) [58]. The second cellular mechanism to explain the beneficial effect of carboxyl surface modification in reducing MWCNTs (10 µg/L) toxicity is that carboxyl surface modification significantly suppressed the increase in intestinal permeability induced by prolonged exposure to MWCNTs (10 µg/L).

10.5.3   Molecular Basis for the Beneficial Effect of Carboxyl Surface Modification in Reducing MWCNT Toxicity

Prolonged exposure to MWCNTs (1 mg/L) could significantly dysregulate the transcriptional expressions of *sod-2, sod-3, mev-1, isp-1, gas-1,* and *clk-1*, which are required for the control of oxidative stress [58]. In contrast, carboxyl surface modification completely rescued the dysregulated expression patterns of these genes in MWCNT (10 µg/L)-exposed nematodes [58]. *sod-2* and *sod-3* encode Mn-SODs in mitochondria, *mev-1* encodes a subunit of the enzyme succinate dehydrogenase cytochrome b, *isp-1* encodes a "Rieske" iron-sulfur protein, *gas-1* encodes a subunit of mitochondrial complex I, and *clk-1* encodes a ubiquinone biosynthetic enzyme.

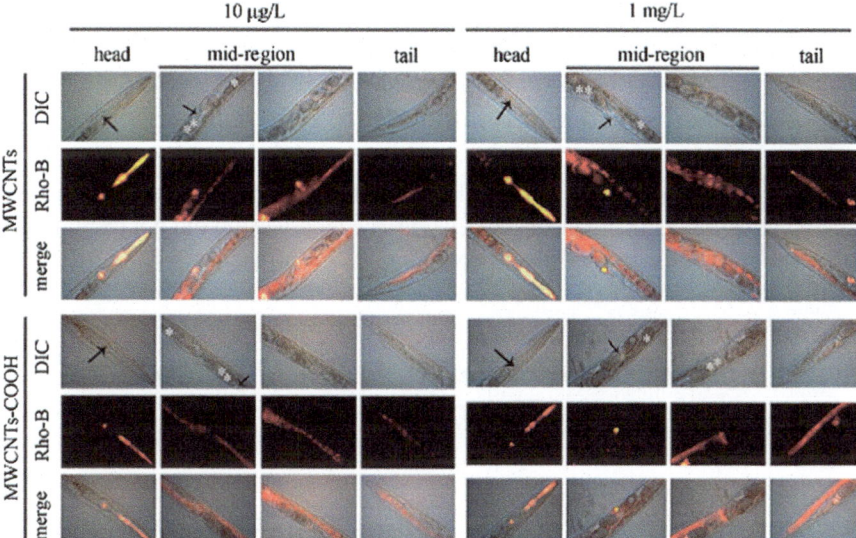

Fig. 10.7 Comparison of translocation and distribution of MWCNTs and MWCNTs-COOH after exposure at different concentrations from L1-larvae to adult day-1 [58]

Arrowheads indicate the pharynx and the spermatheca, respectively, at the head region or the mid-region of nematodes. The intestine (*) and the embryos (**) in the mid-region are also indicated. Exposure concentrations for MWCNTs and MWCNTs-COOH were 10 mg/L and 1 mg/L, respectively

10.5.4   Limitation of Carboxyl Surface Modification in Reducing MWCNT Toxicity

Although carboxyl surface modification significantly inhibited the toxicity of MWCNTs at concentrations ≤10 μg/L in decreasing locomotion behavior, in reducing brood size, and in inducing intestinal ROS production, carboxyl surface modification could not prevent the toxicity of MWCNTs at concentrations ≥100 μg/L [58]. Carboxyl surface modification could only partially rescue the dysregulated expression patterns of *sod-2, sod-3, mev-1, isp-1, gas-1,* and *clk-1* in MWCNT (1 mg/L)-exposed nematodes [58]. Moreover, in MWCNT-COOH (1 mg/L)-exposed nematodes, MWCNTs-COOH could still be detected in the secondary targeted organs, such as the spermatheca, and a large amount of MWCNTs-COOH was still accumulated in the intestine (Fig. 10.7) [58]. Additionally, carboxyl surface modification could only moderately inhibit the increase in intestinal permeability induced by prolonged exposure to MWCNTs (1 mg/L) [58].

10.6   Beneficial Effects of Other Surface Modifications

10.6.1   Effects of Aminated or Hydroxylated MWCNTs on Survival and Reproduction in Nematodes

After acute exposure (24 h), besides the carboxyl surface modification, aminated ($-NH_2$) or hydroxylated/oxygenated (O^+) surface modification did not display any specificity toward the MWCNT functionalization effects in nematodes [59].

10.6.2   Effect of Amine, Carboxyl, or Polyaniline Surface Modification on Toxicity of FeO Nanoparticles (FeO-NPs)

Based on the analysis on endpoints of reproduction and lifespan, amine, carboxyl, and polyaniline surface modification may be able to reduce the toxicity of FeO-NPs in nematodes [60]. Different from this, using the defecation rate as the endpoint, only amine surface modification reduced the FeO toxicity, whereas carboxyl or polyaniline surface modification did not obviously reduce the FeO-NP toxicity [60].

10.6.3   Effect of Albumin Surface Coating on Toxicity of Citrate-Coated SPIONs (C-SPIONs)

Based on the observation on cell internalization of nanoparticles, the more biocompatible property for BSA-C-SPIONs than that of uncoated C-SPIONs in particular at high dosage (500 µg/mL) was observed in nematodes [61].

10.7   Perspectives

We here focus on three important surface chemical modifications, polyethylene glycol (PEG) modification, ZnS or fetal bovine serum (FBS) surface coating, and carboxyl (–COOH) modification, to introduce their possible beneficial effects in reducing the toxicity of nanomaterials in nematodes. Nevertheless, the effects of many other types of surface chemical modifications or coating on the toxicity formation of ENMs are not clear in organisms. Especially, now most of used ENMs in the industrial and medical fields are nanocomposites. The effects of certain surface chemical modifications on the toxicity formation of different types of nanocomposites and the underlying mechanisms are largely unknown. More importantly, the long-term stability for the different surface chemical modifications or coatings in reducing toxicity of ENMs under different conditions needs to be carefully examined.

Additionally, so far, the data on the beneficial effects of certain surface chemical modification in reducing the toxicity of ENMs is usually performed in wild-type animals or under the normal conditions. And then, what's the possible effect of certain surface chemical modification on toxicity formation of ENMs in animals with the deficits with biological barrier, such as the intestinal barrier or the epidermal barrier? Also, what's the possible effect of certain surface chemical modification on toxicity formation of ENMs in organisms under the condition of different stresses?

Besides these, the elucidated information on the underlying cellular and molecular mechanisms for different surface chemical modifications in reducing the toxicity of ENMs is still very limited. Especially, the underlying chemical mechanisms for different surface chemical modifications in reducing the toxicity of ENMs are also needed to be further deeply determined.

References

1. Kawamoto M, He P, Ito Y (2017) Green processing of carbon nanomaterials. Adv Mater. https://doi.org/10.1002/adma.201602423
2. Zhang Q, Wu Z, Li N, Pu Y, Wang B, Zhang T, Tao J (2017) Advanced review of graphene-based nanomaterials in drug delivery systems: synthesis, modification, toxicity and application. Mater Sci Eng C Mater Biol Appl 77:1363–1375
3. Shi Y, Li X (2012) Biomedical applications and adverse health effects of nanomaterials. J Nanosci Nanotechnol 12:8231–8240
4. Fubini B, Ghiazza M, Fenoglio I (2010) Physico-chemical features of engineered nanoparticles relevant to their toxicity. Nanotoxicology 4:347–363
5. Biju V (2014) Chemical modifications and bioconjugate reactions of nanomaterials for sensing, imaging, drug delivery and therapy. Chem Soc Rev 43:744–764
6. Myung JH, Hsu HJ, Bugno J, Tam KA, Hong S (2017) Chemical structure and surface modification of dendritic nanomaterials tailored for therapeutic and diagnostic applications. Curr Top Med Chem 17:1542–1554
7. Duque Sánchez L, Brack N, Postma A, Pigram PJ, Meagher L (2016) Surface modification of electrospun fibres for biomedical applications: a focus on radical polymerization methods. Biomaterials 106:24–45
8. Hernández-Rivera M, Zaibaq NG, Wilson LJ (2016) Toward carbon nanotube-based imaging agents for the clinic. Biomaterials 101:229–240
9. Chen N, Kim do H, Kovacik P, Sojoudi H, Wang M, Gleason KK (2016) Polymer thin films and surface modification by chemical vapor deposition: recent progress. Annu Rev Chem Biomol Eng 7:373–393
10. Mao X, Xu J, Cui H (2016) Functional nanoparticles for magnetic resonance imaging. Wiley Interdiscip Rev Nanomed Nanobiotechnol 8:814–841
11. Ku SH, Jo SD, Lee YK, Kim K, Kim SH (2016) Chemical and structural modifications of RNAi therapeutics. Adv Drug Deliv Rev 104:16–28
12. Habibi Y (2014) Key advances in the chemical modification of nanocelluloses. Chem Soc Rev 43:1519–1542

13. De Volder MF, Tawfick SH, Baughman RH, Hart AJ (2013) Carbon nanotubes: present and future commercial applications. Science 339:535–539
14. Gao MR, Xu YF, Jiang J, Yu SH (2013) Nanostructured metal chalcogenides: synthesis, modification, and applications in energy conversion and storage devices. Chem Soc Rev 42:2986–3017
15. Saka C (2018) Overview on the surface functionalization mechanism and determination of surface functional groups of plasma treated carbon nanotubes. Crit Rev Anal Chem 48:1–14
16. Karousis N, Tagmatarchis N, Tasis D (2010) Current progress on the chemical modification of carbon nanotubes. Chem Rev 110:5366–5397
17. Guo Y, Xu K, Wu C, Zhao J, Xie Y (2015) Surface chemical-modification for engineering the intrinsic physical properties of inorganic two-dimensional nanomaterials. Chem Soc Rev 44:637–646
18. Lee YK, Choi EJ, Webster TJ, Kim SH, Khang D (2014) Effect of the protein corona on nanoparticles for modulating cytotoxicity and immunotoxicity. Int J Nanomed 10:97–113
19. van Vlerken LE, Vyas TK, Amiji MM (2007) Poly(ethylene glycol)-modified nanocarriers for tumor-targeted and intracellular delivery. Pharm Res 24:1405–1414
20. Wang R, Xiao R, Zeng Z, Xu L, Wang J (2012) Application of poly(ethylene glycol)-distearo-ylphosphatidylethanolamine (PEG-DSPE) block copolymers and their derivatives as nanomaterials in drug delivery. Int J Nanomed 7:4185–4198
21. Khung YL, Narducci D (2015) Surface modification strategies on mesoporous silica nanoparticles for anti-biofouling zwitterionic film grafting. Adv Colloid Interf Sci 226:166–186
22. Liu Y, Tee JK, Chiu GN (2015) Dendrimers in oral drug delivery application: current explorations, toxicity issues and strategies for improvement. Curr Pharm Des 21:2629–2642
23. Karakoti AS, Das S, Thevuthasan S, Seal S (2011) PEGylated inorganic nanoparticles. Angew Chem Int Ed Engl 50:1980–1994
24. Amoozgar Z, Yeo Y (2012) Recent advances in stealth coating of nanoparticle drug delivery systems. Wiley Interdiscip Rev Nanomed Nanobiotechnol 4:219–233
25. Xu M, Zhu J, Wang F, Xiong Y, Wu Y, Wang Q, Weng J, Zhang Z, Chen W, Liu S (2016) Improved *in vitro* and *in vivo* biocompatibility of graphene oxide through surface modification: poly(Acrylic Acid)-functionalization is superior to PEGylation. ACS Nano 10:3267–3281
26. Tan X, Feng L, Zhang J, Yang K, Zhang S, Liu Z, Peng R (2013) Functionalization of graphene oxide generates a unique interface for selective serum protein interactions. ACS Appl Mater Interf 5:1370–1377
27. Wu Q-L, Zhao Y-L, Fang J-P, Wang D-Y (2014) Immune response is required for the control of *in vivo* translocation and chronic toxicity of graphene oxide. Nanoscale 6:5894–5906
28. Zhao Y-L, Wu Q-L, Tang M, Wang D-Y (2014) The *in vivo* underlying mechanism for recovery response formation in nano-titanium dioxide exposed *Caenorhabditis elegans* after transfer to the normal condition. Nanomedicine 10:89–98
29. Wu Q-L, Yin L, Li X, Tang M, Zhang T, Wang D-Y (2013) Contributions of altered permeability of intestinal barrier and defecation behavior to toxicity formation from graphene oxide in nematode *Caenorhabditis elegans*. Nanoscale 5:9934–9943
30. Wu Q-L, Zhou X-F, Han X-X, Zhuo Y-Z, Zhu S-T, Zhao Y-L, Wang D-Y (2016) Genome-wide identification and functional analysis of long noncoding RNAs involved in the response to graphene oxide. Biomaterials 102:277–291
31. Wu Q-L, Li Y-X, Li Y-P, Zhao Y-L, Ge L, Wang H-F, Wang D-Y (2013) Crucial role of the biological barrier at the primary targeted organs in controlling the translocation and toxicity of multi-walled carbon nanotubes in the nematode *Caenorhabditis elegans*. Nanoscale 5:11166–11178
32. McGhee JD (2007) The *C. elegans* intestine. WormBook. https://doi.org/10.1895/wormbook.1.133.1
33. Yang W, Guo W, Gong X, Zhang B, Wang S, Chen N, Yang W, Tu Y, Fang X, Chang J (2015) Facile synthesis of Gd-Cu-In-S/ZnS bimodal quantum dots with optimized properties for tumor targeted fluorescence/MR *in vivo* imaging. ACS Appl Mater Interfaces 7:18759–18768

34. Wang Y, Geng Z, Guo M, Chen Y, Guo X, Wang X (2014) Electroaddressing of ZnS quantum dots by codeposition with chitosan to construct fluorescent and patterned device surface. ACS Appl Mater Interfaces 6:15510–15515

35. Senellart P, Solomon G, White A (2017) High-performance semiconductor quantum-dot single-photon sources. Nat Nanotechnol 12:1026–1039

36. Yao J, Li L, Li P, Yang M (2017) Quantum dots: from fluorescence to chemiluminescence, bio-luminescence, electrochemiluminescence, and electrochemistry. Nanoscale 9:13364–13383

37. Matea CT, Mocan T, Tabaran F, Pop T, Mosteanu O, Puia C, Iancu C, Mocan L (2017) Quantum dots in imaging, drug delivery and sensor applications. Int J Nanomedicine 12:5421–5431

38. Lee JJ, Saiful Yazan L, Che Abdullah CA (2017) A review on current nanomaterials and their drug conjugate for targeted breast cancer treatment. Int J Nanomedicine 12:2373–2384

39. Pietryga JM, Park YS, Lim J, Fidler AF, Bae WK, Brovelli S, Klimov VI (2016) Spectroscopic and device aspects of nanocrystal quantum dots. Chem Rev 116:10513–10622

40. Xu G, Zeng S, Zhang B, Swihart MT, Yong KT, Prasad PN (2016) New generation cadmium-free quantum dots for biophotonics and nanomedicine. Chem Rev 116:12234–12327

41. Kirchner C, Liedl T, Kudera S, Pellegrino T, Muñoz Javier A, Gaub HE, Stölzle S, Fertig N, Parak WJ (2005) Cytotoxicity of colloidal CdSe and CdSe/ZnS nanoparticles. Nano Lett 5:331–338

42. Galeone A, Vecchio G, Malvindi MA, Brunetti V, Cingolani R, Pompa PP (2012) In vivo assessment of CdSe-ZnS quantum dots: coating dependent bioaccumulation and genotoxicity. Nanoscale 4:6401–6407

43. Su Y, He Y, Lu H, Sai L, Li Q, Li W, Wang L, Shen P, Huang Q, Fan C (2009) The cytotoxicity of cadmium based, aqueous phase - synthesized, quantum dots and its modulation by surface coating. Biomaterials 30:19–25

44. McIntire SL, Jorgensen E, Kaplan J, Horvitz HR (1993) The GABAergic nervous sys-tem of *Caenorhabditis elegans*. Nature 364:337–341

45. Zhao Y-L, Wang X, Wu Q-L, Li Y-P, Wang D-Y (2015) Translocation and neurotoxicity of CdTe quantum dots in RMEs motor neurons in nematode *Caenorhabditis elegans*. J Hazard Mater 283:480–489

46. Zhao Y-L, Wang X, Wu Q-L, Li Y-P, Tang M, Wang D-Y (2015) Quantum dots exposure alters both development and function of D-type GABAergic motor neurons in nematode *Caenorhabditis elegans*. Toxicol Res 4:399–408

47. Jin Y, Jorgensen E, Hartwieg E, Horvitz HR (1999) The *Caenorhabditis elegans* gene *unc-25* encodes glutamic acid decarboxylase and is required for synaptic transmission but not synaptic development. J Neurosci 19:539–548

48. Eastman C, Horvitz HR, Jin Y (1999) Coordinated transcriptional regulation of the *unc-25* glu-tamic acid decarboxylase and the *unc-47* GABA vesicular transporter by the *Caenorhabditis elegans* UNC-30 homeodomain protein. J Neurosci 19:6225–6234

49. McIntire SL, Jorgensen E, Horvitz HR (1993) Genes required for GABA function in *Caenorhabditis elegans*. Nature 364:334–337

50. Hu W, Peng C, Lv M, Li X, Zhang Y, Chen N, Fan C, Huang Q (2011) Protein corona-mediated mitigation of cytotoxicity of graphene oxide. ACS Nano 5:3693–3700

51. Cheng C, Muller KH, Koziol KK, Skepper JN, Midgley PA, Welland ME, Porter AE (2009) Multi-walled carbon nanotubes inhibit regenerative axon growth of dorsal root ganglia neu-rons of mice. Biomaterials 30:4152–4160

52. Khandare JJ, Jalota-Badhwar A, Satavalekar SD, Bhansali SG, Aher ND, Kharas F, Banerjee SS (2012) PEG-conjugated highly dispersive multifunctional magnetic multi-walled carbon nanotubes for cellular imaging. Nanoscale 4:837–844

53. Zhang T, Tang M, Kong L, Li H, Zhang T, Zhang S-S, Xue Y-Y, Pu Y-P (2012) Systemic and immunotoxicity of pristine and PEGylated multi-walled carbon nanotubes in an intravenous 28 days repeated dose toxicity study. J Hazard Mater 219–220:203–212

54. Ren J, Shen S, Wang D, Xi Z, Guo L, Pang Z, Qian Y, Sun X, Jiang X (2012) The targeted delivery of anticancer drugs to brain glioma by PEGylated oxidized multi-walled carbon nanotubes modified with angiopep-2. Biomaterials 33:3324–3333
55. Bottini M, Rosato N, Bottini N (2011) Biomacromolecules PEG-modified carbon nanotubes in biomedicine: current status and challenges ahead. Biomacromolecules 12:3381–3393
56. Doudrick K, Herckes P, Westerhoff P (2012) Detection of carbon nanotubes in environmental matrices using programmed thermal analysis. Environ Sci Technol 46:12246–12253
57. Gottschalk F, Sonderer T, Scholz RW, Nowack E (2009) Modeled environmental concentrations of engineered nanomaterials (TiO$_2$, ZnO, Ag, CNT, fullerenes) for different regions. Environ Sci Technol 43:9216–9222
58. Nouara A, Wu Q-L, Li Y-X, Tang M, Wang H-F, Zhao Y-L, Wang D-Y (2013) Carboxylic acid functionalization prevents the translocation of multi-walled carbon nanotubes at predicted environmental relevant concentrations into targeted organs of nematode *Caenorhabditis elegans*. Nanoscale 5:6088–6096
59. Chatterjee N, Yang J, Kim H, Jo E, Kim P, Choi K, Choi J (2014) Potential toxicity of differential functionalized multiwalled carbon nanotubes (MWCNT) in human cell line (BEAS2B) and *Caenorhabditis elegans*. J Toxicol Environ Health A 77:1399–1408
60. Callaway MK, Ochoa JM, Perez EE, Ulrich PE, Alocilja EC, Vetrone SA (2013) Investigation of the toxicity of amine-coated, carboxyl-coated and polyaniline-coated FeO magnetic nanoparticles in *Caenorhabditis elegans*. J Biosens Bioelectron 4:145
61. Yu S-M, Gonzalez-Moragas L, Milla M, Kolovou A, Santarella-Mellwig R, Schwab Y, Laromaine A, Roig A (2016) Bio-identity and fate of albumin-coated SPIONs evaluated in cells and by the *C. elegans* model. Acta Biomater 43:348–357

Chapter 11
Pharmacological Prevention of the Toxicity Induced by Environmental Nanomaterials

Abstract An increasing evidence has suggested that surface chemical modifications may not be able to effectively or completely reduce the toxicity of nanomaterials under certain physiological conditions or genetic backgrounds. Along with the progress on different surface chemical modifications of nanomaterials, the design of effective pharmacological prevention strategies against the toxicity of nanomaterials has received the attention gradually. In this chapter, we will introduce and discuss the value and progress on using *C. elegans* as an in vivo assay system to design different effective pharmacological prevention strategies against the toxicity of nanomaterials.

Keywords Pharmacological strategy · Nanotoxicity · Prevention · *Caenorhabditis elegans*

11.1 Introduction

The engineered nanomaterials (ENMs), especially some metal oxide nanoparticles (NPs) such as TiO_2-NPs and carbon-based ENMs such as carbon nanotubes (CNTs), have been produced in a large amount and applied widely in different aspects already [1–5]. The ENMs can not only be applied in the industrial and the medical fields but also be applied as additives of food [6–9]. These facts imply the increasing availability of ENMs to both human beings and environmental organisms [10–13].

So far, a great effort to reduce the toxicity of ENMs is to design the suitable surface chemical modifications on ENMs [14–18]. One of such efforts is to design chemical binding of certain group(s) on the surface of candidate ENMs. With multiwalled carbon nanotubes (MWCNTs) and graphene oxide (GO) as examples, surface PEG modification could effectively reduce their toxicity in nematodes [19, 20]. Another important effort is to design the chemical coating on the surface of candidate ENMs. With GO and quantum dots (QDs) as examples, surface FBS coating could significantly reduce the GO toxicity, and surface ZnS coating could significantly reduce the CdTe QDs toxicity in nematodes [21–23]. However, most of these

© Springer Nature Singapore Pte Ltd. 2018
D. Wang, *Nanotoxicology in* Caenorhabditis elegans,
https://doi.org/10.1007/978-981-13-0233-6_11

data and observations have been obtained in wild-type organisms or under normal conditions. Recently, it has been found that PEG-modified GO still exhibited the toxicity in nematodes under the condition of oxidative stress induced by juglone exposure [24], implying that PEG surface modification may not be able to effectively reduce the GO toxicity in organisms under the condition of oxidative stress.

Additionally, a series of evidence has been raised that some chemical modifications may only have limited beneficial effects against the toxicity of ENMs. For example, long-term exposure to thiolated GO in the range of μg/L could result in the severe toxicity in nematodes [25]. Although carboxylic acid functionalization could reduce the toxicity of MWCNTs at environmentally relevant concentrations, carboxylic acid functionalization did not effectively decrease the MWCNTs toxicity at those more than environmentally relevant concentrations [26]. Moreover, it has been observed that long-term exposure (2 months) to PEGylation could only partially improve the biocompatibility of carbon-based ENMs in mice [27]. These results have suggested that, in addition to the chemical modifications, development of some pharmacological prevention strategies against the toxicity of ENMs is still required.

So far, it has been shown that *Caenorhabditis elegans* has the great potential in the drug screen and the pharmacological study of certain candidate drugs [28–30]. *C. elegans* is very useful for the screen of drugs from a large amount of candidate chemicals of extracts from different organisms such as plants [31–36]. For example, *C. elegans* has been employed to screen the candidate drugs against cell death, to suppress hypoxic injury, and to inhibit neurodevelopmental disorders [37–39]. *C. elegans* may be also helpful for preclinical drug screen and discovery, such as the anti-infective drugs [40]. Using *C. elegans* as an in vivo pharmacological study model, it is also useful to identify or confirm candidate drugs with the certain research aim(s) and to further elucidate the underlying mechanisms [41–46]. For example, it has been shown that folate could act in *E. coli* to accelerate aging independently of the bacterial biosynthesis in nematodes [47]. Colistin has been identified as a new immunomodulator that targets conserved pathways, such as p38/PMK-1 MAPK signaling pathway, FOXO transcription factor DAF-16, and transcription factor SKN-1 in nematodes [48]. Additionally, the potential drug targets could be identified in a *C. elegans* model with α1-antitrypsin deficiency based on the genome-wide RNAi screen [49].

In this chapter, we summarize the progress on the identification and the elucidation of pharmacological mechanisms for candidate drugs or prevention strategies against the toxicity of ENMs using *C. elegans* as an in vivo assay system. Using *C. elegans* assay system, the identified drugs or prevention strategies mainly include (1) antioxidants, (2) natural compounds, and (3) lactic acid.

11.2   Value of Antioxidants in Being Against the Toxicity of ENMs

Oxidation is normally considered as a chemical reaction producing free radicals, which will in turn lead to the chain reactions that may potentially cause the damage on cells or organisms [50–53]. The antioxidant is defined as the molecules with the function to inhibit the oxidation process of other molecules by terminating the chain reactions induced by oxidation-produced free radicals [54–56]. Therefore, the antioxidants are mainly from two sources: (1) industrial chemicals with the aim to prevent the excess oxidation and (2) natural chemicals found in foods and body tissues with the beneficial health effects [57–59]. Plants and animals have the complex systems of overlapping antioxidants, including glutathione and enzymes such as superoxide dismutases and catalases produced internally or derived from the dietary uptake, to balance the oxidative state in the body [60–63].

11.2.1   Beneficial Function of Ascorbic Acid Against the Toxicity of Aluminum Oxide Nanoparticles (Al_2O_3-NPs)

Ascorbic acid (vitamin C) is a monosaccharide oxidation-reduction (redox) catalyst. However, one of the enzymes needed to synthesize the ascorbic acid may be lost by mutation during the evolution; human beings need to obtain the ascorbic acid from the diet daily [64]. Nevertheless, most of other animals have the ability to synthesize the ascorbic acid by themselves and do not need to obtain this compound from the diets [65]. In the cells, the ascorbic acid can be maintained in the reduced form by reaction with glutathione, and this process is catalyzed by protein disulfide isomerase and glutaredoxins [66, 67]. Ascorbic acid, a redox catalyst, can directly reduce and neutralize the reactive oxygen species (ROS), such as the hydrogen peroxide [68, 69]. Besides this, ascorbic acid can also act as a substrate for the redox enzyme of ascorbate peroxidase [70], which is important for stress resistance in organisms [71].

11.2.1.1   Ascorbic Acid Administration Protects Against the Toxicity of Al_2O_3-NPs

Al_2O_3-NPs, one of the most abundantly produced ENMs, have already been widely used in both the industrial and the medical fields [72–75]. In *C. elegans*, chronic exposure (10 days) to Al_2O_3-NPs (60 nm) at concentrations of 0.01–23.1 mg/L induced a significant decrease in locomotion behavior, a severe stress response, and a severe oxidative stress [76]. The formation of oxidative stress in Al_2O_3-NP-exposed nematodes has been proved to be due to both an increase in ROS production and a suppression of ROS defense mechanisms [76]. Based on this, the day-9

Al_2O_3-NP-exposed (23.1 mg/L) adult nematodes were treated with 10 mm ascorbic acid for 24 h to retrieve the mitochondrial dysfunction [76].

Treatment with ascorbic acid (10 mm) could effectively suppress the induction of stress response, induction of oxidative damage, increase in ROS generation, and decrease in SOD activities in Al_2O_3-NP-exposed (23.1 mg/L) nematodes (Fig. 11.1) [76]. Additionally, treatment with ascorbic acid (10 mm) could also significantly inhibit the neurotoxic effects on locomotion behaviors induced by Al_2O_3-NPs (23.1 mg/L) exposure in nematodes (Fig. 11.1) [76]. Therefore, treatment with ascorbic acid has the beneficial effect for nematodes against the toxicity of Al_2O_3-NPs in inducing the oxidative stress and in decreasing the locomotion behaviors in nematodes.

11.2.1.2   Safety Assessment of Ascorbic Acid in Nematodes

In nematodes, treatment with 10 mm ascorbate did not influence the survival and the locomotion behavior [76]. Meanwhile, treatment with 10 mm ascorbate would not induce the oxidative stress [76].

11.2.2   Vitamin E Ameliorates Neurodegeneration-Related Phenotypes Caused by Neurotoxicity of Al_2O_3-NPs

Vitamin E is a group of compounds with eight lipid soluble substances consisting of a chromanol ring and a saturated (tocopherols) or an unsaturated (tocotrienols) carbon side chain [77, 78]. Vitamin E, especially the α-tocopherol, has been widely studied for their fat-soluble, antioxidant, and body preferentially absorbing and metabolizing properties [79–81]. The α-tocopherol can protect cellular membranes from the oxidation by reacting with lipid radicals generated in the lipid peroxidation chain reaction [79–81]. Administration with α-tocopherol will be helpful to remove the free radical intermediates, and the produced oxidized α-tocopheroxyl radicals can be further recycled back to active reduced form after reduction by other antioxidants [79–81]. Thus, α-tocopherol may serve as a free radical scavenger and protects membranes from oxidative damages in organisms.

11.2.2.1   Vitamin E Ameliorates the Neurodegeneration Caused by Al_2O_3-NP Exposure

At least three lines of evidence has been raised to prove the beneficial function of α-tocopherol in ameliorating the neurodegeneration caused by Al_2O_3-NP exposure. In nematodes, exposure to Al_2O_3-NPs (60 nm) at concentrations more than 1 mg/L resulted in the significant decrease in fluorescence intensity of D-type GABAergic neurons, the formation of abnormal axon guidance, the induction of severe neuronal

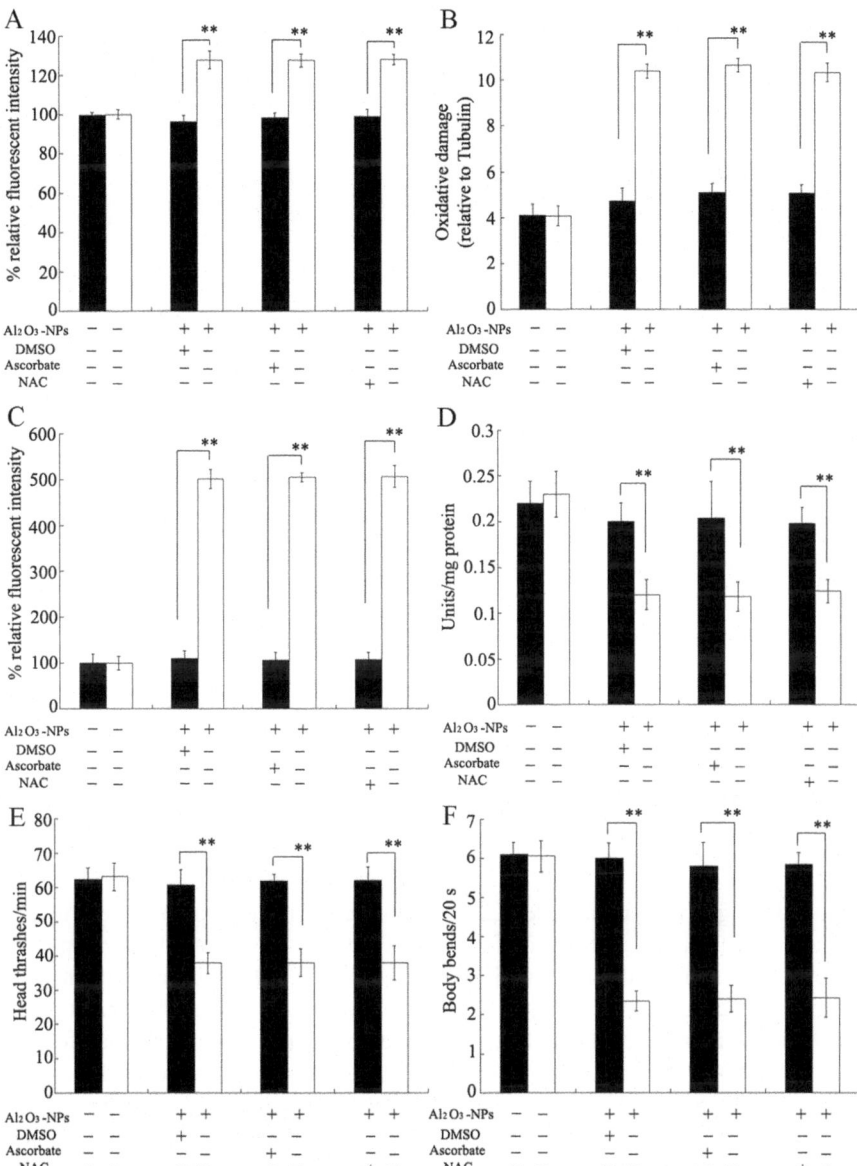

Fig. 11.1 Effects of antioxidant treatment on stress response (**a**), oxidative damage (**b**), ROS production (**c**), SOD activity (**d**), head thrash (**e**), and body bend (**f**) in nematodes chronically exposed to Al_2O_3-NPs (23.1 mg/L) [76]

Increased relative fluorescent intensities for P*hsp-16.2::gfp* expression were observed to reflect stress response. Examination of carbonylated proteins reveals increased oxidative damage. Molecular probe of CM-H$_2$DCFDA was used to detect ROS production, and semiquantified ROS was expressed as relative fluorescent units (RFU). Day-9 Al_2O_3-NP-exposed nematodes were treated with the antioxidant (10 mm ascorbate or 5 mm N-acetyl-l-cysteine (NAC)) for 24 h. Bars represent mean ± SEM. **$P < 0.01$

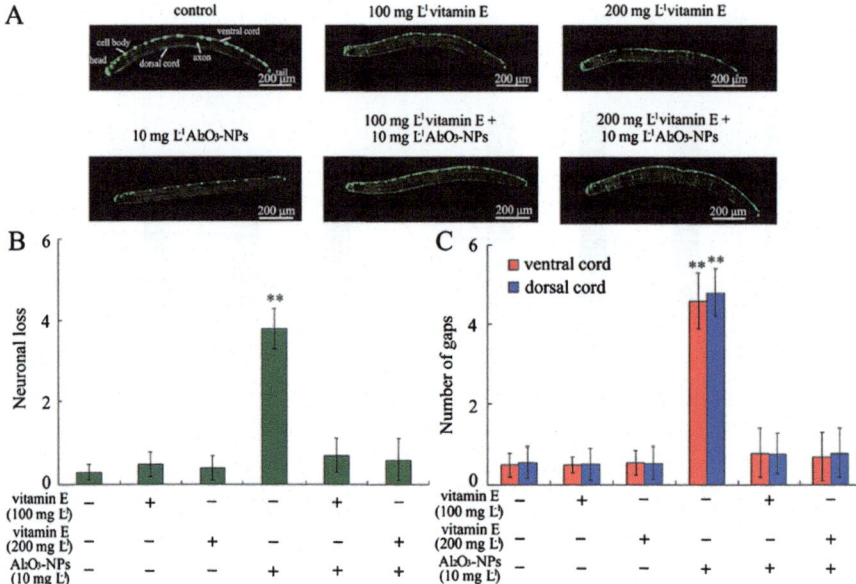

Fig. 11.2 Effects of pretreatment with α-tocopherol on the neurotoxicity of Al₂O₃-NPs in inducing neurodegeneration in *oxIs12* transgenic nematodes [82]
(**a**) Effects of α-tocopherol pretreatment on the development of D-type GABAergic neurons in Al₂O₃-NP-exposed *oxIs12* transgenic nematodes. (**b**) Effects of α-tocopherol pretreatment on neuronal loss of D-type GABAergic neurons in Al₂O₃-NP-exposed *oxIs12* transgenic nematodes. (**c**) Effects of α-tocopherol pretreatment on gap formation on nerve cords for D-type GABAergic nervous system in Al₂O₃-NP-exposed *oxIs12* transgenic nematodes. α-tocopherol pretreatment was performed at the L2-larvae stage for 24 h. Al₂O₃-NPs exposure was performed from L4-larvae for 24 h. Bars represent means ± SEM. **$P < 0.01$

loss, and the gap formation or decrease in fluorescent intensity on both ventral and dorsal cords [82]. In contrast, pretreatment with 100 or 200 mg/L α-tocopherol effectively prevented the neurotoxicity of Al₂O₃-NPs (10 mg/L) in inducing the decrease in fluorescence intensity of D-type GABAergic neurons, the formation of abnormal axon guidance, the induction of severe neuronal loss, and the gap formation or decrease in fluorescent intensity on both ventral and dorsal cords (Fig. 11.2) [82]. Additionally, Al₂O₃-NPs (10 mg/L) exposure resulted in the reduction of fluorescence size of the cell body, the decrease in fluorescence size of the cilia, and the formation of a wrinkled dendrite in AFD sensory neurons [82]. Al₂O₃-NPs (10 mg/L) exposure also caused the reduction of fluorescence intensity of the cell body and the decrease in fluorescence size of the cell body in AIY interneurons [82]. In contrast, pretreatment with α-tocopherol (200 mg/L) could prevent the neurotoxicity of Al₂O₃-NPs on the development of the cell body, the cilia, and the dendrite in AFD sensory neurons or AIY interneurons in nematodes [82]. Al₂O₃-NPs (10 mg/L) exposure significantly decreased the expressions of *unc-30*, *ttx-1* and *ttx-3* genes [82]. The *unc-30* is involved in the regulation of cell identity of GABAergic motor neurons that control the locomotion behavior, the *ttx-1* is responsible for the regulation of cell identity of AFD sensory neurons, and the *ttx-3* is required for the control of cell identity of AIY interneurons [83–85]. Moreover, pretreatment with α-tocopherol (200 mg/L) could prevent the decrease in expressions of *unc-30*, *ttx-1*, and *ttx-3* in Al$_2$O$_3$-NP-exposed nematodes [82].

11.2.2.2   Vitamin E Ameliorates the Deficits in Behavioral Performance Caused by Neurotoxicity of Al$_2$O$_3$-NPs

In nematodes, 1 or 10 mg/L Al$_2$O$_3$-NPs decreased the thermotaxis learning, 10 mg/L Al$_2$O$_3$-NPs decreased the percentage of animals performing isothermal tracking (IT) behavior and increased the percentage of animals performing abnormal thermotaxis behavior, and 10 mg/L Al$_2$O$_3$-NPs decreased both the head thrash and the body bend [82]. In contrast, pretreatment with α-tocopherol (200 mg/L) could prevent the neurotoxicity of Al$_2$O$_3$-NPs (10 mg/L) on thermotaxis learning, thermotaxis perception, and locomotion behavior in nematodes (Fig. 11.3) [82].

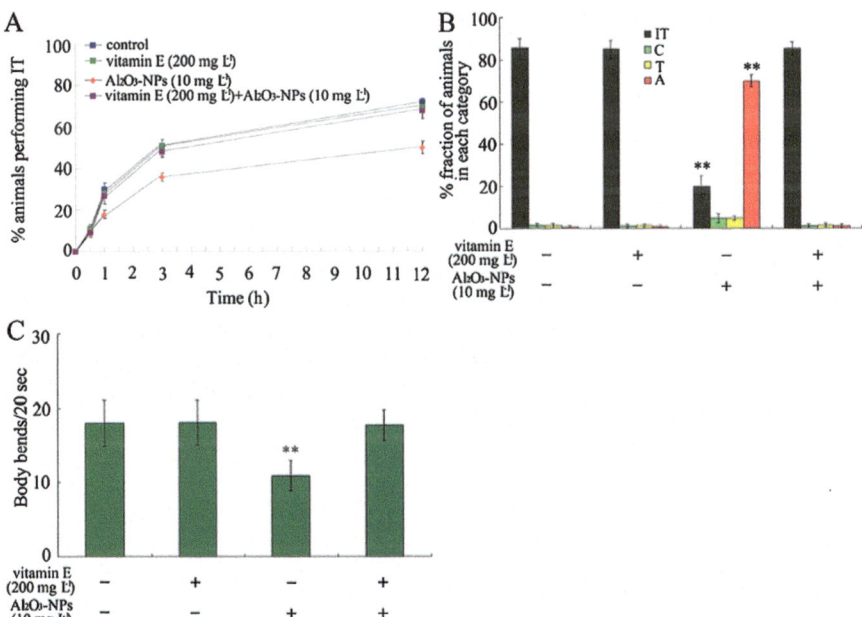

Fig. 11.3   Effects of pretreatment with α-tocopherol on the behavioral performance in Al$_2$O$_3$-NPs-exposed nematodes [82]
(**a**) Effects of α-tocopherol pretreatment on thermotaxis learning in Al$_2$O$_3$-NPs-exposed nematodes. (**b**) Effects of α-tocopherol pretreatment on thermotaxis perception in Al$_2$O$_3$-NPs-exposed nematodes. T, thermophilic; C, movement to 17 °C; A, movement across the thermal gradient (17 °C/25 °C); IT, movement at 20 °C. (**c**) Effects of α-tocopherol pretreatment on body bend in Al$_2$O$_3$-NPs-exposed nematodes. α-Tocopherol pretreatment was performed at the L2-larvae stage for 24 h. Al$_2$O$_3$-NPs exposure was performed from L4-larvae for 24 h. Bars represent means ± SEM. $^{**}P < 0.01$

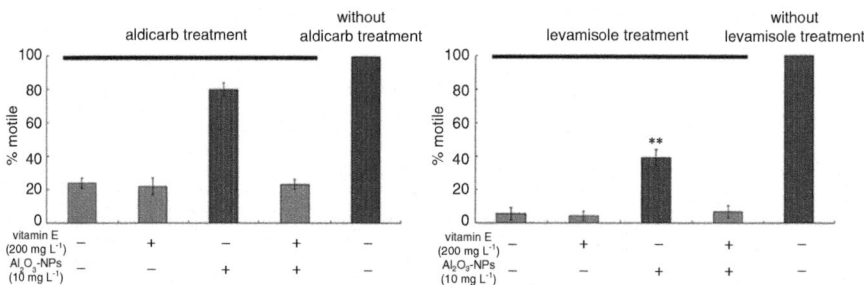

Fig. 11.4 Effects of pretreatment with α-tocopherol on neurotransmission in Al₂O₃-NPs-exposed nematodes [82]

Presynaptic function was evaluated by aldicarb resistance, and postsynaptic function was evaluated by levamisole resistance. α-tocopherol pretreatment was performed at L2-larvae stage for 24 h. Al₂O₃-NPs exposure was performed from L4-larvae for 24 h. Bars represent means ± SEM. **$P < 0.01$**

11.2.2.3 Vitamin E Ameliorates the Deficits in Neurotransmission Caused by Neurotoxicity of Al₂O₃-NPs

In *C. elegans*, aldicarb, an acetylcholinesterase (AChE) inhibitor, or levamisole, a nicotinic acetylcholine receptor (AChR) agonist, induces the formation of hyperactive cholinergic synapses, the muscle hypercontraction, and the paralysis [86, 87]. The synaptic transmission in nematodes can be determined based on aldicarb and levamisole resistance assays, since nematodes with the deficit in presynaptic Ca^{2+}-dependent vesicle release are resistant to aldicarb and nematodes lacking a functional AChR are resistant to levamisole [86, 87]. Exposure to Al₂O₃-NPs (10 mg/L) caused the deficits in both the presynaptic and the postsynaptic functions as indicated by their resistance to both the aldicarb and the levamisole (Fig. 11.4) [82]. In contrast, pretreatment with α-tocopherol (200 mg/L) recovered these deficits in presynaptic and postsynaptic functions in Al₂O₃-NPs (10 mg/L)-exposed nematodes, because pretreatment with α-tocopherol (200 mg/L) prevented the resistance in Al₂O₃-NPs-exposed nematodes to aldicarb and levamisole (Fig. 11.4) [82].

11.2.2.4 Vitamin E Ameliorates the Neurotoxicity of Al₂O₃-NPs in Inducing Oxidative Stress and in Enhancing Intestinal Permeability

Al₂O₃-NPs (10 mg/L) exposure could cause the significant induction of intestinal ROS production and the enhancement in intestinal permeability [76, 82]. Moreover, it was reported that α-tocopherol (200 mg/L) pretreatment suppressed the induction of intestinal ROS production and the enhancement in intestinal permeability in Al₂O₃-NPs (10 mg/L) exposed nematodes [82].

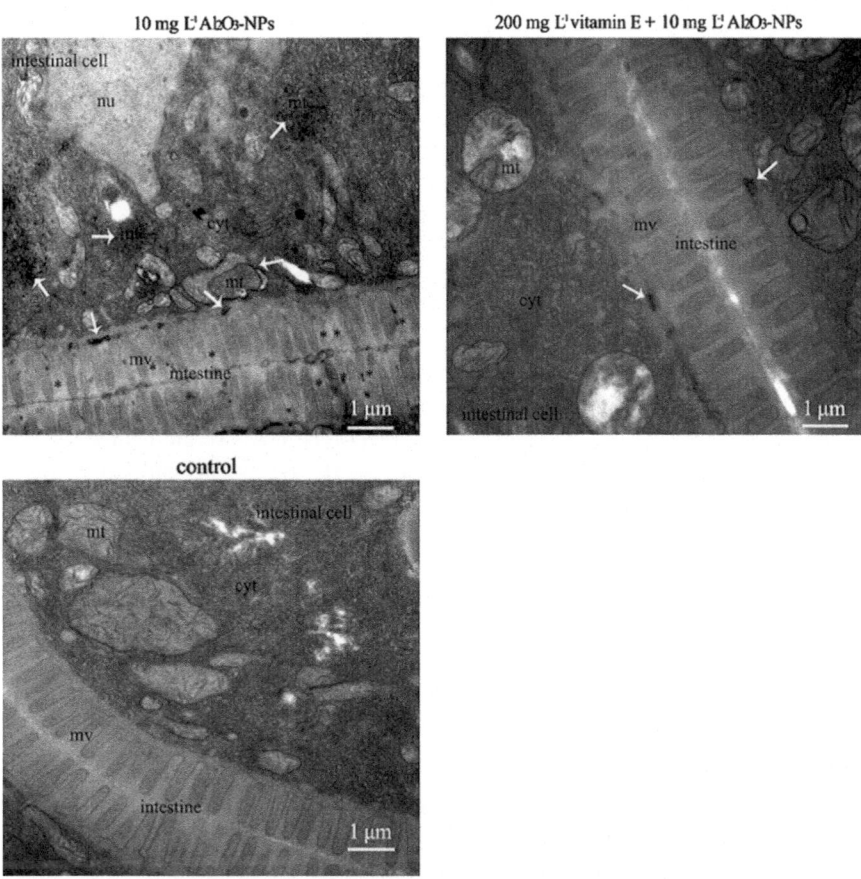

Fig. 11.5 Subcellular distribution of Al_2O_3-NPs in intestinal cells of nematodes [82] *mt* mitochondria, *cyt* cytosol, *nu* nucleus, *mv* microvilli. Asterisks indicate the position with loss of microvilli. α-tocopherol pretreatment was performed at L2-larvae stage for 24 h. Al_2O_3-NP exposure was performed from L4-larvae for 24 h. Arrowheads indicate the accumulation positions of Al_2O_3-NPs. Control, without addition of α-tocopherol or Al_2O_3-NPs

11.2.2.5 Vitamin E Pretreatment Suppresses the Translocation of Al_2O_3-NPs into the Body of Nematodes

Based on the transmission electron microscopy (TEM) analysis, Al_2O_3-NPs could be translocated into the intestinal cells and be deposited in the mitochondria and the cytosol (Fig. 11.5) [82]. Al_2O_3-NPs could also be observed around the mitochondria or adjacent to the intestinal microvilli in intestinal cells (Fig. 11.5) [82]. Meanwhile, exposure to Al_2O_3-NPs led to the developmental deficits in intestinal microvilli, such as the loss of the microvilli (Fig. 11.5) [82]. In contrast, pretreatment with α-tocopherol (200 mg/L) could effectively block the translocation and the accumulation of Al_2O_3-NPs into the cytosol and the mitochondria of intestinal cells

(Fig. 11.5) [82]. Only a trace amount of Al$_2$O$_3$-NPs was occasionally observed in the intestinal cells adjacent to the intestinal microvilli after α-tocopherol pretreatment [82]. Additionally, the development of intestinal microvilli became relatively normal after α-tocopherol pretreatment (Fig. 11.5) [82].

11.2.2.6   Safety Assessment of α-Tocopherol in Nematodes

In nematodes, treatment with 100 or 200 mg/L α-tocopherol did not noticeably influence the development or induce the neurodegeneration in D-type GABAergic neurons, the development of both AFD sensory neurons and AIY interneurons, and the synaptic neurotransmission [82]. Meanwhile, treatment with 200 mg/L α-tocopherol could not alter the expression levels of *unc-30*, *ttx-1*, and *ttx-3* [82], which are required for the fate and identity of D-type GABAergic neurons, AFD sensory neurons, and AIY interneurons [83–85]. Treatment with 200 mg/L α-tocopherol also did not affect thermotaxis learning, thermotaxis perception, and locomotion behavior [82]. Treatment with 200 mg/L α-tocopherol could not induce the intestinal ROS production and alter the intestinal permeability [82].

Nevertheless, it has been found that treatment with 400 mg/L α-tocopherol obviously affected both the thermosensation and the thermotaxis learning, and the observed decrease in thermotaxis learning in 400 mg/L α-tocopherol-treated nematodes was partially due to the moderate but significant deficits in thermosensation [88]. Moreover, treatment with 400 mg/L α-tocopherol decreased fluorescent intensities of cell bodies in AFD sensory neurons and AIY interneurons and altered presynaptic function of neurons [88].

11.2.3   Beneficial Function of N-Acetyl-L-Cysteine (NAC) Against the Toxicity of Al$_2$O$_3$-NPs

NAC, a derivative of the amino acid L-cysteine, is another important antioxidant and free radical scavenger by acting as the precursor in the formation of glutathione in the body of animals and humans [89, 90]. Clinically, NAC administration will be helpful for increasing glutathione levels and binding with toxic breakdown products of paracetamol [91, 92]. In nematodes, treatment with 5 mm NAC could effectively suppress the induction of stress response and oxidative damage, the increase in ROS generation, and the decrease in SOD activities in Al$_2$O$_3$-NPs (23.1 mg/L)-exposed nematodes [76]. Moreover, treatment with 5 mm NAC could inhibit the neurotoxicity of Al$_2$O$_3$-NPs (23.1 mg/L) on locomotion behaviors in nematodes [76]. In nematodes, no adverse side effects on NAC have been reported so far.

11.3 Value of Natural Compounds in Being Against the Toxicity of ENMs

Besides some of the industrial chemicals having the function to prevent the oxidation, some natural compounds found in the foods, the organisms, or the body tissues have also been detected to obtain the beneficial health effects with the antioxidant property [93–95]. Next, we introduce some of the natural compounds, such as glycyrrhizic acid (GA) and paeonol, in being against the toxicity of ENMs [96, 97].

11.3.1 Beneficial Function of Glycyrrhizae Radix (GR) Extract Against the GO Toxicity

GR extract, a medicinal herb used in traditional Chinese medical formulas, has been widely known to have the protective effects for organisms [98, 99]. The GR extract in the Chinese medical formulas contain various important components, such as GA, liquiritin, isoliquiritin, etc. [100, 101].

11.3.1.1 Beneficial Function of GR Extract Against the GO Toxicity

After pretreatment with GR at the concentration of 62.5, 125, or 250 mg/mL from L1-larvae to L4-larvae in the presence of food, GR could significantly inhibit the GO (100 mg/L) toxicity on lifespan, brood size, and locomotion behavior (Fig. 11.6) [96]. In contrast, pretreatment with 31.25 mg/mL of GR from L1-larvae to L4-larvae did not obviously suppress the GO (100 mg/L) toxicity on lifespan, brood size, and locomotion behavior (Fig. 11.6) [96].

11.3.1.2 Administration with GR Suppressed the Accumulation of GO in the Body of Nematodes

The molecular probe Rhodamine B (Rho B) was used to label GO in order to determine the accumulation and the translocation of GO in the body of nematodes [96]. Under the condition without the GR administration, GO/Rho B was found to be distributed in the primary targeted organs such as the pharynx and intestine and the secondary targeted organs such as the gonad and spermatheca (Fig. 11.7) [96]. Different from this, after GR (250 mg/mL) pretreatment, GO/Rho B was observed to be mainly distributed in the pharynx and the intestine (Fig. 11.7) [96], which implies that GR administration can suppress the translocation of GO into the secondary targeted organs through the intestinal barrier. Under the condition without the GR administration, a large amount of GO/Rho B was also accumulated in the tail of nematodes (Fig. 11.7) [96]. After GR (250 mg/mL) pretreatment, both GO/

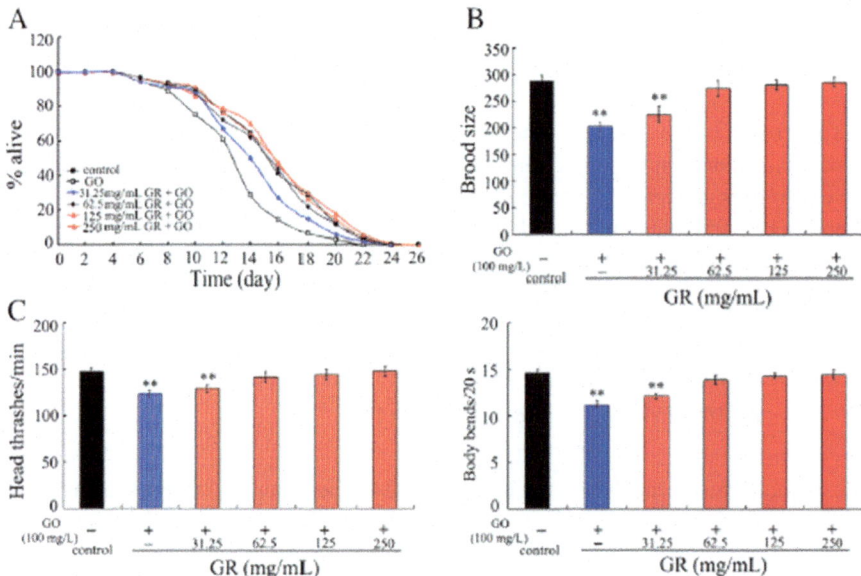

Fig. 11.6 Effects of pretreatment with GR on GO toxicity in nematodes [96]
(**a**) Effects of pretreatment with GR on lifespan in GO (100 mg/L)-exposed nematodes. (**b**) Effects of pretreatment with GR on brood size in GO-exposed nematodes. (**c**) Effects of pretreatment with GR on head thrash and body bend in GO-exposed nematodes. GR, *Glycyrrhizae radix*. Nematodes were pretreated with GR from L1-larvae to L4-larvae in the presence of food and then exposed to GO from L4-laevae for 24 h. Bars represent means ± SEM. **$P < 0.01$ vs control

Rho B in the intestine and GO/Rho B in the tail were significantly decreased compared with those without GR pretreatment (Fig. 11.7) [96], which further implies that GR administration may be helpful for excreting the GO out of the body of nematodes.

11.3.1.3   Beneficial Effects of GR Administration Against the Induction of Oxidative Stress Induced by GO Exposure

In nematodes, oxidative stress is a crucial cellular mechanism underlying the toxicity formation of ENMs [13]. GO (100 mg/L) exposure from L4-larvae for 24 h could cause the significant intestinal ROS production [96]. In contrast, GR (250 mg/mL) pretreatment prevented this induction of intestinal ROS production induced by GO exposure [96].

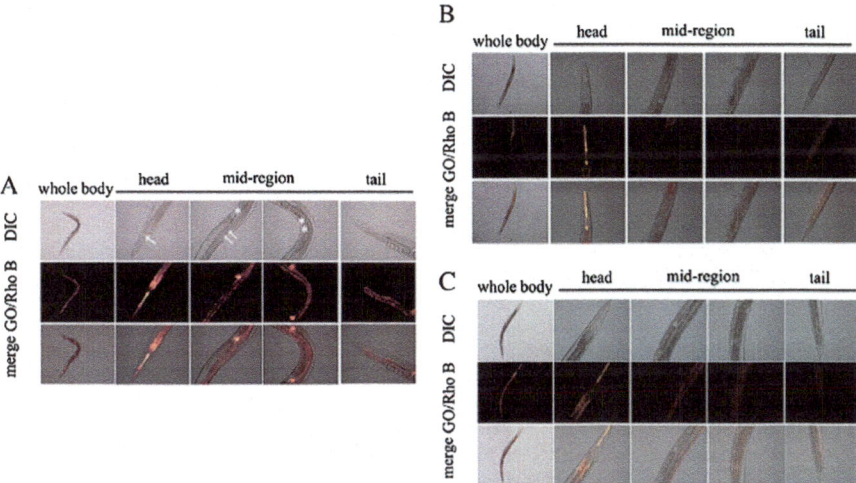

Fig. 11.7 Effects of pretreatment with GR (250 mg/mL) or GA (9.7 mg/mL) on translocation of GO/Rho B (100 mg/L) in nematodes [96]

(**a**) Nematodes were exposed to GO/Rho B from L4-laevae for 24 h. (**b**) Nematodes were pretreated with GR from L1-larvae to L4-larvae in the presence of food and then exposed to GO/Rho B from L4-laevae for 24 h. (**c**) Nematodes were pretreated with GA from L1-larvae to L4-larvae in the presence of food and then exposed to GO/Rho B from L4-laevae for 24 h. Arrowheads indicate the pharynx (single arrowhead) and intestine (double arrowhead), and asterisks indicate the spermatheca (*) and gonad (**). *GR Glycyrrhizae radix*

11.3.1.4 Safety Assessment of GR Extract

In nematodes, at least at these aspects, GR extract is safe to animals. Firstly, GR (250 mg/mL) treatment could not significantly alter the brood size and the locomotion behavior [96]. Secondly, GR (250 mg/mL) treatment did not induce the significant intestinal ROS production in nematodes [96]. Moreover, some evidence further implies the beneficial effect of GR extract in extending longevity. GR (250 mg/mL) treatment could increase the lifespan of nematodes [96].

11.3.2 Beneficial Function of GA Against the GO Toxicity

Based on the chromatographic fingerprint analysis, three main active components were identified in the GR extract, and they are liquiritin, isoliquiritin, and GA [96]. In a 250 mg/mL GR solution, 9.7 mg/mL GA, 1.5 mg/mL liquiritin, and 0.5 mg/mL isoliquiritin were detected [96]. Among these three active components, pretreatment

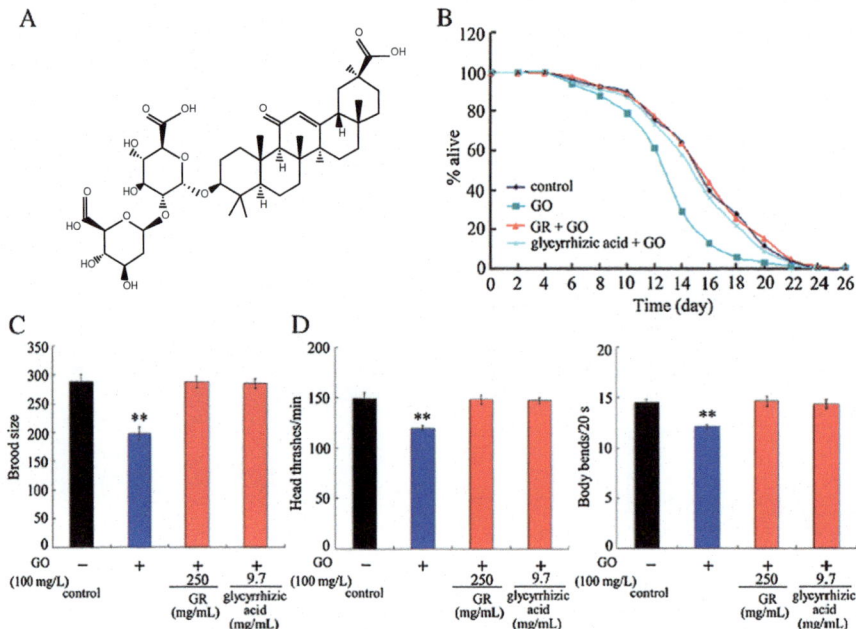

Fig. 11.8 Effects of pretreatment with glycyrrhizic acid on GO toxicity in exposed nematodes [96]
(**a**) Chemical structure of glycyrrhizic acid. (**b**) Effects of pretreatment with glycyrrhizic acid (9.7 mg/mL) on lifespan in GO-exposed nematodes. (**c**) Effects of pretreatment with glycyrrhizic acid on brood size in GO-exposed nematodes. (**d**) Effects of pretreatment with glycyrrhizic acid on head thrash and body bend in GO-exposed nematodes. *GR Glycyrrhizae radix*. Nematodes were pretreated with glycyrrhizic acid from L1-larvae to L4-larvae in the presence of food and then exposed to GO (100 mg/L) from L4-laevae for 24-h. Bars represent means ± SEM **$P < 0.01$ vs control

with liquiritin (1.5 mg/mL) or isoliquiritin (0.5 mg/mL) exhibited the minimal impact on the reduction of GO (100 mg/L) toxicity with respect to lifespan, brood size, and locomotion behavior [96], implying that both the liquiritin and the isoliquiritin may contribute minimally to the beneficial effects of GR in reducing GO toxicity.

11.3.2.1 Beneficial Effect of GA Administration Against the GO Toxicity

Pretreatment with GA (9.7 mg/mL) from L1-larvae to L4-larvae effectively prevented the GO (100 mg/L) toxicity on lifespan, brood size, and locomotion behavior, and the beneficial effects of GA (9.7 mg/mL) in reducing the GO toxicity were similar to that of GR solution (250 mg/mL) (Fig. 11.8) [96]. Therefore, among the three main active compounds in GR, GA has the beneficial effect in reducing the GO toxicity in nematodes.

11.3.2.2 Beneficial Effect of GA Administration in Reducing the Accumulation of GO in the Body of Nematodes

Like the observed function of GR extract, pretreatment with 9.7 mg/mL GA also caused the limited distribution of GO/Rho B in the pharynx and the intestine (Fig. 11.7) [96]. Moreover, after pretreatment with 9.7 mg/mL GA, both the amount of GO/Rho B in the intestine and the amount of GO/Rho B in the tail region were significantly reduced (Fig. 11.7) [96].

11.3.2.3 Beneficial Effect of GA Administration in Suppressing the Induction of Oxidative Stress in GO-Exposed Nematodes

Similar to the beneficial effect of GR administration, pretreatment with 9.7 mg/mL GA obviously prevented the induction of intestinal ROS production induced by GO (100 mg/L) [96]. In contrast, pretreatment with 1.5 mg/mL liquiritin or 0.5 mg/mL isoliquiritin had no such beneficial effect [96].

In nematodes, GO exposure could alter the expression patterns of some important genes required for the control of oxidative stress [96]. GO exposure (from L4-larvae for 24 h) significantly increased the expression levels of *sod-1*, *sod-2*, *sod-3*, *sod-4*, *isp-1*, and *clk-1* and decreased the expression level of *gas-1* [96]. In contrast, pretreatment with 9.7 mg/mL GA could suppress the increase in expression levels of *sod-1*, *sod-2*, *sod-3*, *sod-4*, *isp-1*, and *clk-1* and inhibit the decrease in expression level of *gas-1* in GO (100 mg/L)-exposed nematodes [96]. An additional evidence was also raised that pretreatment with GA (9.7 mg/mL) inhibited the increase in SOD-2::GFP expression in GO (100 mg/L)-exposed nematodes [96]. *sod-1* and *sod-4* encode copper/zinc superoxide dismutases, *sod-2* and *sod-3* encode manganese superoxide dismutases, *gas-1* encodes a subunit of mitochondrial complex I, *isp-1* encodes a subunit of mitochondrial complex III, and *clk-1* encodes a demethoxyubiquinone hydroxylase.

11.3.2.4 GA Administration Altered the Expressions of Some miRNAs

In nematodes, GO exposure could upregulate the miRNAs including *mir-259*, *mir-1820*, *mir-36*, *mir-82*, *mir-239*, *mir-246*, *mir-247*, *mir-392*, *mir-4806*, *mir-2217*, *mir-360*, *mir-4810*, *mir-4807*, *mir-1822*, *mir-4805*, *mir-800*, *mir-1830*, *mir-236*, *mir-244*, *mir-235*, *mir-4937*, *mir-4812*, and *mir-43*, and downregulate the miRNAs including *mir-1834*, *mir-800*, *mir-231*, *mir-5546*, *mir-42*, *mir-2214*, *mir-2210*, and *mir-73* [102]. Among these miRNAs, under the GO exposure condition, *mir-4805* and *mir-1820* may target SOD-1, *mir-360* and *mir-246* may target SOD-2, *mir-392* may target CLK-1, and *mir-4810* may target GAS-1 [96]. Meanwhile it was found that pretreatment with GA (9.7 mg/mL) could inhibit the increase in expression levels of *mir-1820*, *mir-4805*, *mir-246*, *mir-360*, *mir-392*, and *mir-4810* induced by GO (100 mg/L) exposure [96]. Also, pretreatment with GA (9.7 mg/mL) could suppress the increase in *mir-360p::GFP* expression induced by GO (100 mg/L) exposure [96].

So far, a molecular mechanism has been raised to explain the formation of beneficial function of GA in reducing GO toxicity. First of all, *mir-360* plays an important role in the formation of beneficial function of GA in reducing GO toxicity. Mutation of *mir-360* enhanced the beneficial effect of GR or GA in reducing GO toxicity on lifespan and in inducing intestinal ROS production [96]. Secondly, *mir-360* may function with SOD-2 as its molecular target to negatively regulate the beneficial effect of GA in reducing GO toxicity. Mutation of *sod-2* suppressed the beneficial effect of GA in reducing GO toxicity on locomotion behavior and in inducing intestinal ROS production observed in *mir-360(n4635)* mutant [96].

11.3.2.5   Safety Assessment of GA in Nematodes

At least at these aspects, GA is safe to nematodes. Firstly, GA (9.7 mg/mL) treatment did not affect the brood size and the locomotion behavior [96]. Secondly, GA (9.7 mg/mL) treatment will not induce the significant intestinal ROS production in nematodes [96]. Moreover, GA (9.7 mg/mL) treatment could increase the lifespan of nematodes [96]. Different from this, treatment with liquiritin (1.5 mg/mL) or isoliquiritin (0.5 mg/mL) did not show such a beneficial function [96].

11.3.2.6   Beneficial Effects of GA Against Toxicity of TiO₂ Nanoparticles (TiO₂-NPs) or MWCNTs

TiO₂-NPs and MWCNTs are two widely applied ENMs so far. In nematodes, exposure to TiO₂-NPs (25 mg/L) or MWCNTs (10 mg/L) from L4-laevae for 24 h could reduce the lifespan and brood size and decrease the locomotion behavior [96]. It has been further found that pretreatment with GA (9.7 mg/mL) could also prevent the toxicity of TiO₂-NPs (25 mg/L) or MWCNTs (10 mg/L) on lifespan, brood size, and locomotion behavior in nematodes [96].

11.3.3   Beneficial Function of Paeonol Against the GO Toxicity

Paeonol is a natural compound from the root bark of Moutan. Some evidence has suggested various pharmacological activities of paeonol, such as antioxidation and neuroprotection [103, 104].

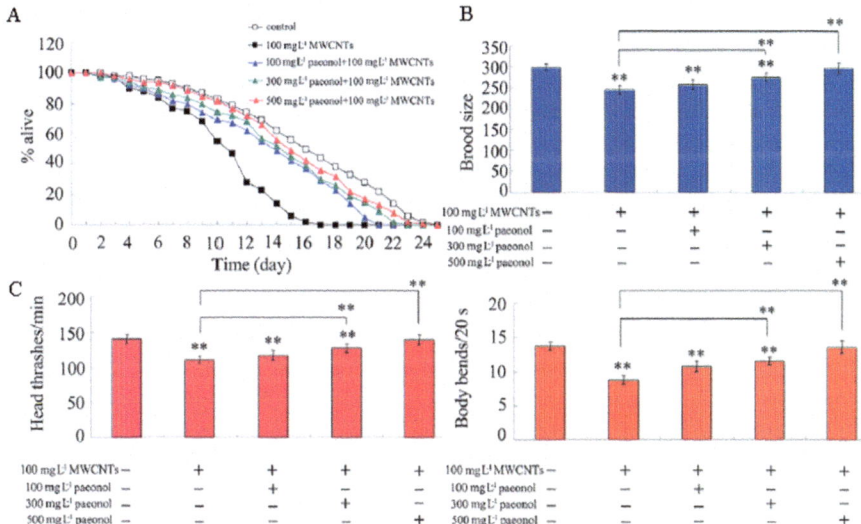

Fig. 11.9 Effect of paeonol administration on MWCNTs toxicity in nematodes [97] (**a**) Effects of pretreatment with paeonol on lifespan in MWCNTs-exposed nematodes. (**b**) Effects of pretreatment with paeonol on brood size in MWCNTs-exposed nematodes. (**c**) Effects of pretreatment with paeonol on head thrash or body bend in MWCNTs-exposed nematodes. Nematodes were pretreated with paeonol from L1-larvae to L4-larvae stage in the presence of food and then were exposed to MWCNTs from L4-larvae stage for 24 h. Bars represent means ± SEM. $^{**}P < 0.01$ vs control (if not specially indicated)

11.3.3.1   Effect of Paeonol Administration on MWCNTs Toxicity

In nematodes, acute exposure to MWCNTs (from larvae for 24 h) at concentrations of 1–100 mg/L caused the toxicity on nematodes as indicated by the endpoints of lifespan, brood size, locomotion behavior, and induction of ROS production [97]. Pretreatment with 300 or 500 mg/L paeonol from L1-larvae to L4-larvae could inhibit the toxicity of MWCNTs (100 mg/L) in reducing lifespan, in decreasing brood size, in suppressing locomotion behavior, and in inducing intestinal ROS production (Fig. 11.9) [97].

11.3.3.2   Effect of Paeonol Administration on MWCNTs Translocation

Under the condition without paeonol administration, MWCNTs/Rho B were distributed in the pharynx, the intestine, the tail, and the reproductive organs, such as the spermatheca and the gonad [97]. In contrast, pretreatment with paeonol (500 mg/L) caused the accumulation and distribution of MWCNTs only in the pharynx and the intestine [97]. That is, pretreatment with paeonol may be helpful for suppressing or blocking the translocation of MWCNTs into the secondary targeted organs through the intestinal barrier.

11.3.3.3   Effect of Paeonol Administration on Intestinal Permeability in MWCNTs-Exposed Nematodes

Under the condition without paeonol administration, MWCNT exposure at concentrations more than 100 mg/L induced an enhanced intestinal permeability as reflected by the signal of lipophilic fluorescent dye, Nile red [97]. Under the condition with paeonol administration, the enhancement in intestinal permeability in MWCNTs-exposed nematodes was obviously suppressed [97].

In nematodes, exposure to MWCNTs (100 mg/L) also dysregulated the expression patterns of some genes required for the control of intestinal development. Exposure to MWCNTs (100 mg/L) significantly increased the expression level of *lin-7* and decreased the expression levels of *pgp-3*, *gem-4*, *par-3*, *pkc-3*, *ajm-1*, *inx-3*, and *abts-4* [97]. In contrast, pretreatment with paeonol (500 mg/L) could increase the expression levels of *pgp-3*, *gem-4*, *par-3*, *pkc-3*, *ajm-1*, *inx-3*, and *abts-4* and decreased the expression level of *lin-7* in MWCNT (100 mg/L)-exposed nematodes [97]. *gem-4*, *par-6*, *pkc-3*, and *pgp-3* are required for the development of microvilli on intestinal cells, *inx-3* and *abts-4* are required for the development of basolateral domain of intestine, and *ajm-1* and *lin-7* are associated with the development of apical junction of intestine.

11.3.3.4   Effect of Paeonol Administration on Defecation Behavior in MWCNTs-Exposed Nematodes

In nematodes, both intestinal permeability and defecation behavior are crucial for the accumulation and the translocation of ENMs in the body [13]. MWCNTs (100 mg/L) exposure could further prolong the mean defecation cycle length [97], which implies that the MWCNTs may need a longer time to be excreted out of the body. Moreover, exposure to MWCNTs (100 mg/L) reduced the relative size of fluorescent puncta for the cell body of AVL or DVB neurons, which are required for the defecation behavior [97]. Very different from this, pretreatment with paeonol (500 mg/L) could significantly suppress the increase in mean defecation cycle length induced by MWCNT (100 mg/L) exposure [97]. Additionally, pretreatment with paeonol (500 mg/L) also suppressed the decrease in relative size of fluorescent puncta for the cell body of AVL or DVB neurons induced by MWCNT (100 mg/L) exposure [97].

11.3.3.5   Safety Assessment of Paeonol in Nematodes

So far, several evidence has been raised to support the possible safety of paeonol in nematodes. Firstly, paeonol (300–500 mg/L) treatment from L1-larvae to young adult did not result in any alterations in both the brood size and the locomotion behavior [97]. Secondly, paeonol (300–500 mg/L) treatment could not induce the significant ROS production in the intestine [97]. More importantly, paeonol (100–500 mg/L) treatment could significantly increase the lifespan of nematodes [97].

11.4 Beneficial Function of Lactic Acid (LAB) Against the GO Toxicity by Maintaining Normal Intestinal Permeability

LAB is a compound with the formula $CH_3CH(OH)COOH$. It can be produced both naturally and synthetically. In a form of its conjugate base called lactate, it obtains several biochemical processes. So far, LAB has been widely considered as a potential probiotic bacteria, and is safe and useful for food and feed fermentation [105–107]. In *C. elegans*, some reports indicate that feeding with specific LAB strains could be resistant to pathogenic infection [108, 109].

11.4.1 LAB Administration Prevents the GO Toxicity

Several aspects of evidence have been raised to support the beneficial function of LAB (*L. bulgaricus*) administration against GO toxicity. Firstly, pretreatment with LAB significantly inhibited the induction of intestinal ROS production induced by GO (100 mg/L) exposure (Fig. 11.10) [110]. Secondly, pretreatment with LAB suppresses the decrease in locomotion behavior induced by GO (100 mg/L) exposure (Fig. 11.10) [110]. Therefore, LAB pretreatment may be beneficial for being against the GO toxicity on the functions of both primary and secondary targeted organs in nematodes.

Moreover, it has been observed that the GO translocation pattern was altered by the LAB pretreatment. After LAB pretreatment, GO was mainly distributed in the pharynx and intestine, and no signals were detected in the secondary targeted organs (Fig. 11.10) [110]. Under the condition without LAB pretreatment, GO could be translocated into the secondary targeted organs, such as the reproductive organs (Fig. 11.10) [110].

11.4.2 LAB Administration Is Helpful for Maintaining the Normal State of Intestinal Permeability or Defecation Behavior in GO-Exposed Nematodes

Based on the Nile red labeling data, pretreatment with LAB could effectively block the increase in intestinal permeability induced by GO (100 mg/L) exposure [110]. In other words, LAB pretreatment potentially suppresses the formation of

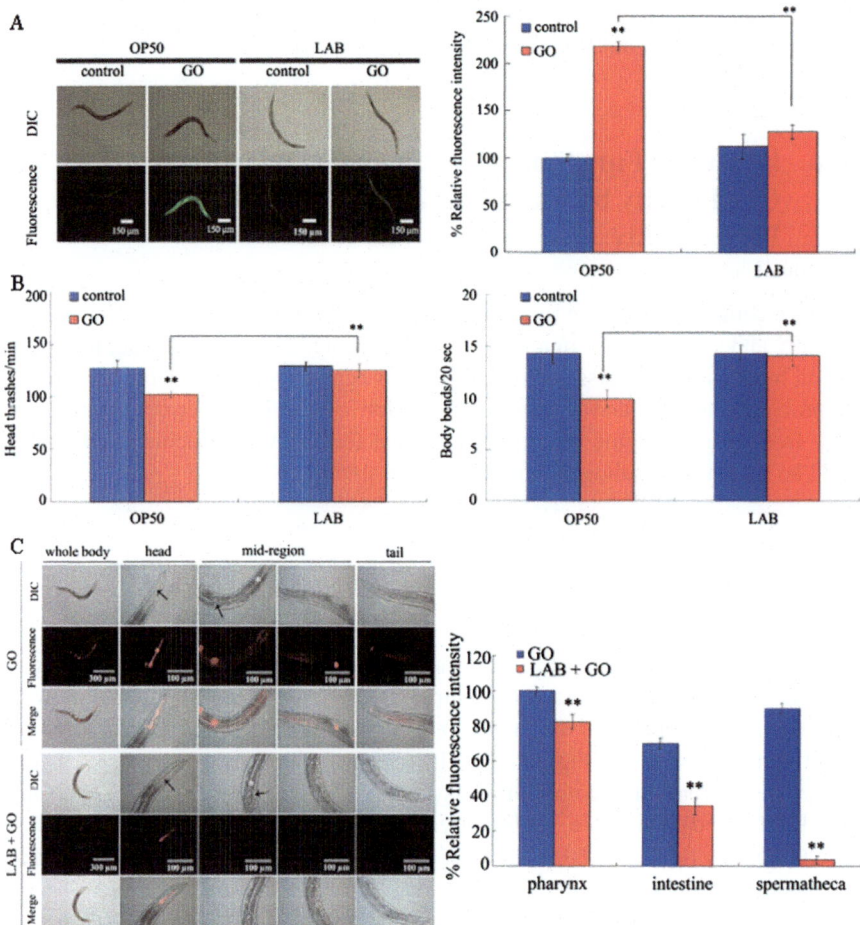

Fig. 11.10 LAB administration prevented the GO toxicity in nematodes [110]
(**a**) LAB administration prevented the induction of intestinal ROS production induced by GO exposure. (**b**) LAB administration prevented the toxicity of GO on locomotion behavior. GO exposure concentration was 100 mg/L. The used LAB strain was *L. bulgaricus*. L4-larvae were pretreated with LAB for 12 h and then exposed to GO for 24 h at 20 °C. Bars represent means ± SEM. $^{**}P < 0.01$ vs control. (**c**) GO distribution in wild-type nematodes. GO/Rho B was used to visualize the distribution of GO in nematodes. The arrowheads indicate the pharynx and spermatheca, respectively, at the head region or mid-region. The intestine (*) in the mid-region was also indicated. GO/Rho B exposure concentration was 100 mg/L

hyper-permeable intestinal barrier in GO-exposed nematodes. The molecular basis for altered intestinal permeability in GO-exposed nematodes by LAB pretreatment was also raised. In nematodes, GO exposure could dysregulate the expressions of *pkc-3*, *nhx-2*, and *par-6*, which are required for the control of intestinal development [111]. Pretreatment with LAB inhibited the dysregulation of these genes required for the control of intestinal development in GO (100 mg/L-exposed nematodes

[110]. In nematodes, *pkc-3* encodes an atypical protein kinase, *nhx-2* encodes a sodium/proton exchanger, and *par-6* encodes a PDZ-domain-containing protein.

Besides the maintenance of normal state of intestinal permeability, it was further found that LAB administration is helpful for the maintenance of normal state of defecation behavior in GO-exposed nematodes. In nematodes, exposure to GO (100 mg/L) could significantly increase the mean defecation cycle length and reduce the relative fluorescence size of cell body for AVL or DVB neurons, which are essential for the control of defecation behavior [110]. In contrast, LAB pretreatment obviously inhibited the GO (100 mg/L) toxicity on defecation behavior and the development of AVL and DVB neurons in nematodes [110].

Together, these observations will help us understanding the beneficial function of LAB administration against the GO toxicity and in suppressing the accumulation and the translocation of GO in the body of nematodes. That is, LAB administration can be helpful to maintain the normal function of intestinal barrier and defecation behavior, which blocks the translocation of GO particles into the secondary targeted organs and the accumulation of GO in the body of nematodes.

11.4.3 *LAB Administration Prevented the GO Toxicity in Nematodes with Mutations of Susceptible Gene*

In nematodes, mutations of some genes required for the control of oxidative stress, such as *sod-2*, *sod-3*, *gas-1*, or *aak-2*, could result in the susceptibility of nematodes to GO toxicity [112]. Mutation of *sod-2*, *sod-3*, *gas-1*, or *aak-2* caused the more severe induction of intestinal ROS production and decrease in locomotion behavior in GO (100 mg/L)-exposed nematodes compared with those in wild-type nematodes [112]. In contrast, LAB pretreatment effectively inhibited both the induction of intestinal ROS production and the decrease in locomotion behavior in GO (100 mg/L)-exposed *sod-2*, *sod-3*, *gas-1*, or *aak-2* mutant nematodes [110]. *sod-2* and *sod-3* genes encode mitochondrial manganese-superoxide dismutases, *gas-1* encodes a subunit of mitochondrial complex I, and *aak-2* encodes a catalytic alpha subunit of AMP-activated protein kinase. Therefore, LAB administration can not only prevent the GO toxicity in wild-type nematodes but also suppress the GO toxicity in nematodes with mutations of susceptible genes.

The cellular basis for the beneficial function of LAB administration in preventing GO toxicity has been determined. One of the possibilities is that LAB pretreatment can inhibit the increase in intestinal permeability induced by GO (100 mg/L) exposure in *sod-2*, *sod-3*, *gas-1*, or *aak-2* mutants [110]. The *sod-2*, *sod-3*, *gas-1*, and *aak-2* mutants had the similar intestinal permeability to that in wild-type nematodes, and GO exposure enhanced the intestinal permeability in these mutant nematodes based on Nile red signal detection [110]. Another possibility is that LAB pretreatment can also inhibit the extension of mean defecation cycle length induced by GO (100 mg/L) exposure in *sod-2*, *sod-3*, *gas-1*, or *aak-2* mutants [110]. The *sod-2*, *sod-3*, *gas-1*, or *aak-2* mutant had the similar mean defecation cycle length to that in wild-type nematodes, and GO (100 mg/L) exposure resulted in the formation of more prolonged mean defecation cycle length in *sod-2*, *sod-3*, *gas-1*, or *aak-2* mutant than GO (100 mg/L)-exposed wild-type nematodes [110].

11.4.4 Effects of acs-22 Mutation on Beneficial Effects of LAB Administration Against GO Toxicity

In *C. elegans*, *acs-22* encodes a protein homologous to mammalian FATP4 (fatty acid transport protein 4) and is primarily expressed in the intestine [113]. Mutation of *acs-22* can induce an enhanced intestinal permeability based on Nile red signal detection [110]. Meanwhile, mutation of *acs-22* could not induce significant intestinal ROS production and alteration in locomotion behavior in nematodes [110]. These results imply that *acs-22* mutant can be employed as a genetic tool with the deficit in function of intestinal barrier.

In nematodes, GO exposure significantly decreased the expression level of *acs-22*, whereas LAB pretreatment maintained the normal expression of *acs-22* in GO (100 mg/L)-exposed nematodes [110]. These results provide the further molecular basis for LAB administration in being against GO toxicity and in suppressing the translocation of GO into the secondary targeted organs.

It has been further observed that LAB administration could not effectively suppress the enhancement in intestinal permeability and prevent the toxicity in inducing intestinal ROS production and in decreasing locomotion behavior in GO (100 mg/L)-exposed *acs-22* mutant nematodes [110]. This observation implies that the beneficial function of LAB administration against GO toxicity is dependent on the ACS-22. More importantly, the role of LAB administration in reducing toxicity of ENMs in organisms with severe deficits in biological barrier may be somewhat limited.

11.4.5 Safety Assessment of LAB Administration in Nematodes

So far, no data were raised to reflect the potential adverse effects of LAB administration on nematodes. In nematodes, LAB administration could not induce significant intestinal ROS production and alter the locomotion behaviors [110]. LAB administration could not affect the defecation behavior and the development of neurons (AVL and DVB) required for the defecation behavior [110].

11.5 Perspectives

In this chapter, the progress on pharmacological strategies against the toxicity of ENMs has been summarized. The raise of pharmacological strategy provides an important and essential remediation for the limitation of chemical modification in reducing the toxicity of ENMs. With the concern on the human health, we suggest the combinational use of chemical modification and pharmacological prevention to prevent or inhibit the possible toxicity of environmental ENMs. An additional attempt was raised to modify the ENMs with certain beneficial compounds such as antioxidant to prepare nanocomposites. However, it is still largely unclear whether the prepared nanocomposites are stable enough and safe in the complex environment of the organisms including the humans.

So far, only limited compounds have been examined using *C. elegans* assay system to identify the compounds with the potential use in the future clinical against the toxicity of environmental ENMs. Based on this fact, the relatively specific underlying mechanisms for environmental ENMs in inducing toxicity should be further clearly and deeply defined. Moreover, after thoroughly considering the properties of nanotoxicity, it is necessary to perform the large-scale screen to identify more candidate compounds or pharmacological prevention strategies, which will then provide us more choice and judgment among the candidate compounds or pharmacological prevention strategies.

References

1. Samanta A, Medintz IL (2016) Nanoparticles and DNA – a powerful and growing functional combination in bionanotechnology. Nanoscale 8(17):9037–9095
2. Prateek TVK, Gupta RK (2016) Recent progress on ferroelectric polymer-based nanocomposites for high energy density capacitors: synthesis, dielectric properties, and future aspects. Chem Rev 116(7):4260–4317
3. Du X, Li X, Xiong L, Zhang X, Kleitz F, Qiao S (2016) Mesoporous silica nanoparticles with organo-bridged silsesquioxane framework as innovative platforms for bioimaging and therapeutic agent delivery. Biomaterials 91:90–127
4. Liu J, Wang H, Antonietti M (2016) Graphitic carbon nitride "reloaded": emerging applications beyond photocatalysis. Chem Soc Rev 45(8):2308–2326
5. Lofrano G, Carotenuto M, Libralato G, Domingos RF, Markus A, Dini L, Gautam RK, Baldantoni D, Rossi M, Sharma SK, Chattopadhyaya MC, Giugni M, Meric S (2016) Polymer functionalized nanocomposites for metals removal from water and wastewater: an overview. Water Res 92:22–37
6. Shin SR, Li YC, Jang HL, Khoshakhlagh P, Akbari M, Nasajpour A, Zhang YS, Tamayol A, Khademhosseini A (2016) Graphene-based materials for tissue engineering. Adv Drug Deliv Rev 105:255–274
7. Mukherjee S, Patra CR (2016) Therapeutic application of anti-angiogenic nanomaterials in cancers. Nanoscale 8(25):12444–12470
8. Ji H, Sun H, Qu X (2016) Antibacterial applications of graphene-based nanomaterials: recent achievements and challenges. Adv Drug Deliv Rev 105:176–189

9. Piperigkou Z, Karamanou K, Engin AB, Gialeli C, Docea AO, Vynios DH, Pavão MS, Golokhvast KS, Shtilman MI, Argiris A, Shishatskaya E, Tsatsakis AM (2016) Emerging aspects of nanotoxicology in health and disease: from agriculture and food sector to cancer therapeutics. Food Chem Toxicol 91:42–57
10. Hu X, Li D, Gao Y, Mu L, Zhou Q (2016) Knowledge gaps between nanotoxicological research and nanomaterial safety. Environ Int 94:8–23
11. Wang Z, Zhu W, Qiu Y, Yi X, von dem Bussche A, Kane A, Gao H, Koski K, Hurt R (2016) Biological and environmental interactions of emerging two-dimensional nanomaterials. Chem Soc Rev 45(6):1750–1780
12. Wang D-Y (2016) Biological effects, translocation, and metabolism of quantum dots in nematode *Caenorhabditis elegans*. Toxicol Res 5:1003–1011
13. Zhao Y-L, Wu Q-L, Li Y-P, Wang D-Y (2013) Translocation, transfer, and in vivo safety evaluation of engineered nanomaterials in the non-mammalian alternative toxicity assay model of nematode *Caenorhabditis elegans*. RSC Adv 3:5741–5757
14. Bai X, Liu F, Liu Y, Li C, Wang S, Zhou H, Wang W, Zhu H, Winkler DA, Yan B (2017) Toward a systematic exploration of nano-bio interactions. Toxicol Appl Pharmacol 323:66–73
15. Zhang Q, Wu Z, Li N, Pu Y, Wang B, Zhang T, Tao J (2017) Advanced review of graphene-based nanomaterials in drug delivery systems: synthesis, modification, toxicity and application. Mater Sci Eng C Mater Biol Appl 77:1363–1375
16. Fubini B, Ghiazza M, Fenoglio I (2010) Physico-chemical features of engineered nanoparticles relevant to their toxicity. Nanotoxicology 4:347–363
17. Amin ML, Joo JY, Yi DK, An SS (2015) Surface modification and local orientations of surface molecules in nanotherapeutics. J Control Release 207:131–142
18. Cheng LC, Jiang X, Wang J, Chen C, Liu RS (2013) Nano-bio effects: interaction of nanomaterials with cells. Nanoscale 5(9):3547–2569
19. Wu Q-L, Li Y-X, Li Y-P, Zhao Y-L, He L, Wang H-F, Wang D-Y (2013) Crucial role of biological barrier at the primary targeted organs in controlling translocation and toxicity of multi-walled carbon nanotubes in nematode *Caenorhabditis elegans*. Nanoscale 5:11166–11178
20. Wu Q-L, Zhao Y-L, Fang J-P, Wang D-Y (2014) Immune response is required for the control of *in vivo* translocation and chronic toxicity of graphene oxide. Nanoscale 6:5894–5906
21. Zhao Y-L, Wang X, Wu Q-L, Li Y-P, Wang D-Y (2015) Translocation and neurotoxicity of CdTe quantum dots in RMEs motor neurons in nematode *Caenorhabditis elegans*. J Hazard Mater 283:480–489
22. Zhao Y-L, Wang X, Wu Q-L, Li Y-P, Tang M, Wang D-Y (2015) Quantum dots exposure alters both development and function of D-type GABAergic motor neurons in nematode *Caenorhabditis elegans*. Toxicol Res 4:399–408
23. Wu Q-L, Zhou X-F, Han X-X, Zhuo Y-Z, Zhu S-T, Zhao Y-L, Wang D-Y (2016) Genome-wide identification and functional analysis of long noncoding RNAs involved in the response to graphene oxide. Biomaterials 102:277–291
24. Zhang W, Wang C, Li Z, Lu Z, Li Y, Yin JJ, Zhou YT, Gao X, Fang Y, Nie G, Zhao Y (2012) Unraveling stress-induced toxicity properties of graphene oxide and the underlying mechanism. Adv Mater 24:5391–5397
25. Ding X-C, Wang J, Rui Q, Wang D-Y (2018) Long-term exposure to thiolated graphene oxide in the range of μg/L induces toxicity in nematode *Caenorhabditis elegans*. Sci Total Environ 616-617:29–37
26. Nouara A, Wu Q-L, Li Y-X, Tang M, Wang H-F, Zhao Y-L, Wang D-Y (2013) Carboxylic acid functionalization prevents the translocation of multi-walled carbon nanotubes at predicted environmental relevant concentrations into targeted organs of nematode *Caenorhabditis elegans*. Nanoscale 5:6088–6096
27. Zhang D, Deng X, Ji Z, Shen X, Dong L, Wu M, Gu T, Liu Y (2010) Long-term hepatotoxicity of polyethylene-glycol functionalized multi-walled carbon nanotubes in mice. Nanotechnology 21:175101
28. Anastassopoulou CG, Fuchs BB, Mylonakis E (2011) *Caenorhabditis elegans*-based model systems for antifungal drug discovery. Curr Pharm Des 17(13):1225–1233

29. Muhammed M, Arvanitis M, Mylonakis E (2016) Whole animal HTS of small molecules for antifungal compounds. Expert Opin Drug Discov 11(2):177–1784

30. Pukkila-Worley R, Holson E, Wagner F, Mylonakis E (2009) Antifungal drug discovery through the study of invertebrate model hosts. Curr Med Chem 16(13):1588–1595

31. Mathew MD, Mathew ND, Miller A, Simpson M, Au V, Garland S, Gestin M, Edgley ML, Flibotte S, Balgi A, Chiang J, Giaever G, Dean P, Tung A, Roberge M, Roskelley C, Forge T, Nislow C, Moerman D (2016) Using *C. elegans* forward and reverse genetics to identify new compounds with anthelmintic activity. PLoS Negl Trop Dis 10(10):e0005058

32. Burns AR, Luciani GM, Musso G, Bagg R, Yeo M, Zhang Y, Rajendran L, Glavin J, Hunter R, Redman E, Stasiuk S, Schertzberg M, Angus McQuibban G, Caffrey CR, Cutler SR, Tyers M, Giaever G, Nislow C, Fraser AG, MacRae CA, Gilleard J, Roy PJ (2015) *Caenorhabditis elegans* is a useful model for anthelmintic discovery. Nat Commun 6:7485

33. Rajamuthiah R, Fuchs BB, Jayamani E, Kim Y, Larkins-Ford J, Conery A, Ausubel FM, Mylonakis E (2014) Whole animal automated platform for drug discovery against multi-drug resistant *Staphylococcus aureus*. PLoS One 9(2):e89189

34. Bae YK, Sung JY, Kim YN, Kim S, Hong KM, Kim HT, Choi MS, Kwon JY, Shim J (2012) An *in vivo C. elegans* model system for screening EGFR-inhibiting anti-cancer drugs. PLoS One 7(9):e42441

35. Gosai SJ, Kwak JH, Luke CJ, Long OS, King DE, Kovatch KJ, Johnston PA, Shun TY, Lazo JS, Perlmutter DH, Silverman GA, Pak SC (2010) Automated high-content live animal drug screening using *C. elegans* expressing the aggregation prone serpin α1-antitrypsin Z. PLoS One 5(11):e15460

36. Sleigh JN, Buckingham SD, Esmaeili B, Viswanathan M, Cuppen E, Westlund BM, Sattelle DB (2011) A novel *Caenorhabditis elegans* allele, *smn-1(cb131)*, mimicking a mild form of spinal muscular atrophy, provides a convenient drug screening platform highlighting new and pre-approved compounds. Hum Mol Genet 20(2):245–260

37. Schwendeman AR, Shaham S (2016) A high-throughput small molecule screen for *C. elegans* linker cell death inhibitors. PLoS One 11(10):e0164595

38. Schmeisser K, Fardghassemi Y, Parker JA (2017) A rapid chemical-genetic screen utilizing impaired movement phenotypes in *C. elegans*: input into genetics of neurodevelopmental disorders. Exp Neurol 293:101–114

39. Sun CL, Zhang H, Liu M, Wang W, Crowder CM (2017) A screen for protective drugs against delayed hypoxic injury. PLoS One 12(4):e0176061

40. Kim W, Hendricks GL, Lee K, Mylonakis E (2017) An update on the use of *C. elegans* for preclinical drug discovery: screening and identifying anti-infective drugs. Expert Opin Drug Discov 12(6):625–633

41. Grotewiel M, Bettinger JC (2015) Drosophila and *Caenorhabditis elegans* as discovery platforms for genes involved in human alcohol use disorder. Alcohol Clin Exp Res 39(8):1292–1311

42. Wolozin B, Gabel C, Ferree A, Guillily M, Ebata A (2011) Watching worms whither: modeling neurodegeneration in *C elegans*. Prog Mol Biol Transl Sci 100:499–514

43. Link EM, Hardiman G, Sluder AE, Johnson CD, Liu LX (2000) Therapeutic target discovery using *Caenorhabditis elegans*. Pharmacogenomics 1(2):203–217

44. Ruan Q-L, Qiao Y, Zhao Y-L, Xu Y, Wang M, Duan J-A, Wang D-Y (2016) Beneficial effects of *Glycyrrhizae radix* extract in preventing oxidative damage and extending the lifespan of *Caenorhabditis elegans*. J Ethnopharmacol 177:101–110

45. Zhang W-M, Lv T, Li M, Wu Q-L, Yang L-S, Liu H, Sun D-F, Sun L-M, Zhuang Z-H, Wang D-Y (2013) Beneficial effects of wheat gluten hydrolysate to extend lifespan and induce stress resistance in nematode *Caenorhabditis elegans*. PLoS One 8(9):e74553

46. Rui Q, Lu Q, Wang D-Y (2009) Administration of *Bushenkangshuai Tang* alleviates the UV irradiation- and oxidative stress-induced lifespan defects in nematode *Caenorhabditis elegans*. Front Med China 3(1):76–90

47. Virk B, Jia J, Maynard CA, Raimundo A, Lefebvre J, Richards SA, Chetina N, Liang Y, Helliwell N, Cipinska M, Weinkove D (2016) Folate acts in *E. coli* to accelerate *C. elegans* aging independently of bacterial biosynthesis. Cell Rep 14(7):1611–1620

48. Cai Y, Cao X, Aballay A (2014) Whole-animal chemical screen identifies colistin as a new immunomodulator that targets conserved pathways. MBio 5(4):e01235–e01214

49. O'Reilly LP, Long OS, Cobanoglu MC, Benson JA, Luke CJ, Miedel MT, Hale P, Perlmutter DH, Bahar I, Silverman GA, Pak SC (2014) A genome-wide RNAi screen identifies potential drug targets in a *C. elegans* model of α1-antitrypsin deficiency. Hum Mol Genet 23(19):5123–5132

50. Dormandy TL (1980) Free-radical reaction in biological systems. Ann R Coll Surg Engl 62(3):188–194

51. Mason RP, Chignell CF (1981) Free radicals in pharmacology and toxicology-selected topics. Pharmacol Rev 33(4):189–211

52. Fridovich I (1983) Superoxide radical: an endogenous toxicant. Annu Rev Pharmacol Toxicol 23:239–257

53. DiGuiseppi J, Fridovich I (1984) The toxicology of molecular oxygen. Crit Rev Toxicol 12(4):315–342

54. Tappel AL (1970) Biological antioxidant protection against lipid peroxidation damage. Am J Clin Nutr 23(8):1137–1139

55. Logani MK, Davies RE (1980) Lipid oxidation: biologic effects and antioxidants-a review. Lipids 15(6):485–495

56. Sen CK, Packer L (1996) Antioxidant and redox regulation of gene transcription. FASEB J 10(7):709–720

57. Poot M (1991) Oxidants and antioxidants in proliferative senescence. Mutat Res 256(2–6):177–189

58. Gerster H (1991) Review: antioxidant protection of the ageing macula. Age Ageing 20(1):60–69

59. Diplock AT (1990) The role of antioxidants in clinical practice. Br J Clin Pract 44(7):257–258

60. Halliwell B (1978) Biochemical mechanisms accounting for the toxic action of oxygen on living organisms: the key role of superoxide dismutase. Cell Biol Int Rep 2(2):113–128

61. Sies H, Berndt C, Jones DP (2017) Oxidative stress. Annu Rev Biochem 86:715–748

62. Deisseroth A, Dounce AL (1970) Catalase: physical and chemical properties, mechanism of catalysis, and physiological role. Physiol Rev 50(3):319–375

63. Sies H (1974) Biochemistry of the peroxisome in the liver cell. Angew Chem Int Ed Engl 13(11):706–718

64. Smirnoff N (2001) L-ascorbic acid biosynthesis. Vitam Horm 61:241–266

65. Linster CL, Van Schaftingen E (2007) Vitamin C. Biosynthesis, recycling and degradation in mammals. FEBS J 274(1):1–22

66. Meister A (1994) Glutathione-ascorbic acid antioxidant system in animals. J Biol Chem 269(13):9397–9400

67. Englard S, Seifter S (1986) The biochemical functions of ascorbic acid. Annu Rev Nutr 6:365–406

68. Smirnoff N, Pallanca JE (1996) Ascorbate metabolism in relation to oxidative stress. Biochem Soc Trans 24(2):472–478

69. Wells WW, Xu DP, Yang YF, Rocque PA (1990) Mammalian thioltransferase (glutaredoxin) and protein disulfide isomerase have dehydroascorbate reductase activity. J Biol Chem 265(26):15361–15364

70. Padayatty SJ, Katz A, Wang Y, Eck P, Kwon O, Lee JH, Chen S, Corpe C, Dutta A, Dutta SK, Levine M (2003) Vitamin C as an antioxidant: evaluation of its role in disease prevention. J Am College Nut 22:18–35

71. Shigeoka S, Ishikawa T, Tamoi M, Miyagawa Y, Takeda T, Yabuta Y, Yoshimura K (2002) Regulation and function of ascorbate peroxidase isoenzymes. J Exp Bot 53(372):1305–1319

72. Hua M, Zhang S, Pan B, Zhang W, Lv L, Zhang Q (2012) Heavy metal removal from water/wastewater by nanosized metal oxides: a review. J Hazard Mater 211-212:317–331

73. Loomba L, Scarabelli T (2013) Metallic nanoparticles and their medicinal potential. Part II: aluminosilicates, nanobiomagnets, quantum dots and cochleates. Ther Deliv 4(9):1179–1196

74. Pelgrift RY, Friedman AJ (2013) Nanotechnology as a therapeutic tool to combat microbial resistance. Adv Drug Deliv Rev 65(13–14):1803–1815

75. Karimi Z, Karimi L, Shokrollahi H (2013) Nano-magnetic particles used in biomedicine: core and coating materials. Mater Sci Eng C Mater Biol Appl 33(5):2465–2475

76. Li Y-X, Yu S-H, Wu Q-L, Tang M, Pu Y-P, Wang D-Y (2012) Chronic Al_2O_3-nanoparticle exposure causes neurotoxic effects on locomotion behaviors by inducing severe ROS production and disruption of ROS defense mechanisms in nematode *Caenorhabditis elegans*. J Hazard Mater 219-220:221–230

77. Brigelius-Flohé R, Traber MG (1999) Vitamin E: function and metabolism. FASEB J 13:1145–1155

78. Bieri JG, Evarts RP (1974) γ-tocopherol: metabolism, biological activity and significance in human vitamin E nutrition. Am J Clin Nutr 27:980–986

79. Green J (1972) Vitamin E and the biological antioxidant theory. Ann NY Acad Sci 203:29–44

80. Tappel AL (1972) Vitamin E and free radical peroxidation of lipids. Ann NY Acad Sci 203:12–28

81. McCay PB (1985) Vitamin E: interactions with free radicals and ascorbate. Annu Rev Nutr 5:323–340

82. Yu X-M, Guan X-M, Wu Q-L, Zhao Y-L, Wang D-Y (2015) Vitamin E ameliorates the neurodegeneration related phenotypes caused by neurotoxicity of Al_2O_3-nanoparticles in *C. elegans*. Toxicol Res 4:1269–1281

83. Satterlee JS, Sasakura H, Kuhara A, Berkeley M, Mori I, Sengupta P (2001) Specification of thermosensory neuron fate in *C. elegans* requires *ttx-1*, a homolog of otd/Otx. Neuron 31(6):943–956

84. Hobert O, Mori I, Yamashita Y, Honda H, Ohshima Y, Liu Y, Ruvkun G (1997) Regulation of interneuron function in the *C. elegans* thermoregulatory pathway by the *ttx-3* LIM homeobox gene. Neuron 19(2):345–357

85. Eastman C, Horvitz HR, Jin Y (1999) Coordinated transcriptional regulation of the *unc-25* glutamic acid decarboxylase and the *unc-47* GABA vesicular transporter by the *Caenorhabditis elegans* UNC-30 homeodomain protein. J Neurosci 19(15):6225–6234

86. Miller KG, Alfonso A, Nguyen M, Crowell JA, Johnson CD, Rand JB (1996) A genetic selection for *Caenorhabditis elegans* synaptic transmission mutants. Proc Natl Acad Sci USA 93(22):12593–12598

87. Waggoner LE, Dickinson KA, Poole DS, Tabuse Y, Miwa J, Schafer WR (2000) Long-term nicotine adaptation in *Caenorhabditis elegans* involves PKC-dependent changes in nicotinic receptor abundance. J Neurosci 20(23):8802–8811

88. Li Y-P, Li Y-X, Wu Q-L, Ye H-Y, Sun L-M, Ye B-P, Wang D-Y (2013) High concentration of vitamin E decreases thermosensation and thermotaxis learning and the underlying mechanisms in nematode *Caenorhabditis elegans*. PLoS One 8(8):e71180

89. Dludla PV, Nkambule BB, Dias SC, Johnson R (2017) Cardioprotective potential of N-acetyl cysteine against hyperglycaemia-induced oxidative damage: a protocol for a systematic review. Syst Rev 6(1):96

90. Mokhtari V, Afsharian P, Shahhoseini M, Kalantar SM, Moini A (2017) A review on various uses of N-acetyl cysteine. Cell J 19(1):11–17

91. Elbini Dhouib I, Jallouli M, Annabi A, Gharbi N, Elfazaa S, Lasram MM (2016) A minireview on N-acetylcysteine: an old drug with new approaches. Life Sci 151:359–363

92. Neuwelt AJ, Nguyen T, Wu YJ, Donson AM, Vibhakar R, Venkatamaran S, Amani V, Neuwelt EA, Rapkin LB, Foreman NK (2014) Preclinical high-dose acetaminophen with N-acetylcysteine rescue enhances the efficacy of cisplatin chemotherapy in atypical teratoid rhabdoid tumors. Pediatr Blood Cancer 61(1):120–127

93. Hasima N, Ozpolat B (2014) Regulation of autophagy by polyphenolic compounds as a potential therapeutic strategy for cancer. Cell Death Dis 5:e1509
94. Singh AA, Singh S, Agrawal M, Agrawal SB (2015) Assessment of ethylene diurea-induced protection in plants against ozone phytotoxicity. Rev Environ Contam Toxicol 233:129–184
95. Sarkar D, Shetty K (2014) Metabolic stimulation of plant phenolics for food preservation and health. Annu Rev Food Sci Technol 5:395–413
96. Zhao Y-L, Jia R-H, Qiao Y, Wang D-Y (2016) Glycyrrhizic acid, active component from *Glycyrrhizae radix*, prevents toxicity of graphene oxide by influencing functions of microRNAs in nematode *Caenorhabditis elegans*. Nanomedicine 12:735–744
97. Shu C-J, Yu X-M, Wu Q-L, Zhuang Z-H, Zhang W-M, Wang D-Y (2015) Pretreatment with paeonol prevents the adverse effects and alters the translocation of multi-walled carbon nanotubes in nematode *Caenorhabditis elegans*. RSC Adv 5:8942–8951
98. Amagaya S, Sugishita E, Ogihara Y, Ogawa S, Okada K, Aizawa T (1984) Comparative studies of the stereoisomers of glycyrrhetinic acid on anti-inflammatory activities. J Pharmacobiodyn 7(12):923–928
99. Hasegawa A, Kawaguchi Y, Nakasa H, Nakamura H, Ohmori S, Ishii I, Kitada M (2002) Effects of Kampo extracts on drug metabolism in rat liver microsomes: Rhei Rhizoma extract and *Glycyrrhizae Radix* extract inhibit drug oxidation. Jpn J Pharmacol 89(2):164–170
100. Yokozawa T, Liu ZW, Chen CP (2000) Protective effects of *Glycyrrhizae radix* extract and its compounds in a renal hypoxia (ischemia)-reoxygenation (reperfusion) model. Phytomedicine 6(6):439–445
101. Sugishita E, Amagaya S, Ogihara Y (1984) Studies on the combination of *Glycyrrhizae Radix* in Shakuyakukanzo-To. J Pharmacobiodyn 7(7):427–435
102. Wu Q-L, Zhao Y-L, Zhao G, Wang D-Y (2014) microRNAs control of *in vivo* toxicity from graphene oxide in *Caenorhabditis elegans*. Nanomedicine 10:1401–1410
103. Gong X, Yang Y, Huang L, Zhang Q, Wan RZ, Zhang P, Zhang B (2017) Antioxidation, anti-inflammation and anti-apoptosis by paeonol in LPS/d-GalN-induced acute liver failure in mice. Int Immunopharmacol 46:124–132
104. Lin B (2011) Polyphenols and neuroprotection against ischemia and neurodegeneration. Mini Rev Med Chem 11(14):1222–1238
105. Boor KJ, Wiedmann M, Murphy S, Alcaine S (2017) A 100-year review: microbiology and safety of milk handling. J Dairy Sci 100(12):9933–9951
106. Guerin J, Burgain J, Gomand F, Scher J, Gaiani C (2017) Milk fat globule membrane glycoproteins: valuable ingredients for lactic acid bacteria encapsulation? Crit Rev Food Sci Nutr 4:1–13
107. Arqués JL, Rodríguez E, Langa S, Landete JM, Medina M (2015) Antimicrobial activity of lactic acid bacteria in dairy products and gut: effect on pathogens. Biomed Res Int 2015:584183
108. Ikeda T, Yasui C, Hoshino K, Airkawa K, Nishikawa Y (2007) Influence of lactic acid bacteria on longevity of *Caenorhabditis elegans* and host defense against *Salmonella enterica* serovar enteritidis. Appl Environ Microbiol 73:6404–6409
109. Lee J, Yun HS, Cho KW, Oh S, Kim SH, Chun T, Kim B, Whang KY (2011) Evaluation of probiotic characteristics of newly isolated Lactobacillus spp.: immune modulation and longevity. Int J Food Microbiol 148:80–86
110. Zhao Y-L, Yu X-M, Jia R-H, Yang R-L, Rui Q, Wang D-Y (2015) Wang, lactic acid bacteria protects *Caenorhabditis elegans* from toxicity of graphene oxide by maintaining normal intestinal permeability under different genetic backgrounds. Sci Rep 5:17233
111. Wu Q-L, Yin L, Li X, Tang M, Zhang T, Wang D-Y (2013) Contributions of altered permeability of intestinal barrier and defecation behavior to toxicity formation from graphene oxide in nematode *Caenorhabditis elegans*. Nanoscale 5(20):9934–9943
112. Wu Q-L, Zhao Y-L, Li Y-P, Wang D-Y (2014) Molecular signals regulating translocation and toxicity of graphene oxide in nematode *Caenorhabditis elegans*. Nanoscale 6:11204–11212
113. Kage-Nakadai E, Kobuna H, Kimura M, Gengyo-Ando K, Inoue T, Arai H, Mitani S (2010) Two very long chain fatty acid acyl-CoA synthetase genes, *acs-20* and *acs-22*, have roles in the cuticle surface barrier in *Caenorhabditis elegans*. PLoS One 5:e8857

Printed by Printforce, the Netherlands